MARGINAL MAN:
THE DARK VISION OF HAROLD INNIS

ALEXANDER JOHN WATSON

Marginal Man: The Dark Vision of Harold Innis

UNIVERSITY OF TORONTO PRESS
Toronto Buffalo London

Toronto Buffalo London
Printed in Canada

ISBN 0-8020-3916-2

Printed on acid-free paper

Library and Archives Canada Cataloguing in Publication

Watson, A. John (Alexander John), 1948–
Marginal man : the dark vision of Harold Innis / Alexander John Watson.

Includes bibliographical references and index.
ISBN 0-8020-3916-2

1. Innis, Harold A., 1894–1952. 2. Mass media specialists – Canada – Biography. 3. Economists – Canada – Biography. I. Title.

FC151.I55W38 2006 302.23′092 C2005-903120-4

University of Toronto Press acknowledges the financial assistance to its publishing program of the Canada Council for the Arts and the Ontario Arts Council.

University of Toronto Press acknowledges the financial support for its publishing activities of the Government of Canada through the Book Publishing Industry Development Program (BPIDP).

This book has been published with the help of a grant from the Canadian Federation for the Humanities and Social Sciences, through the Aid to Scholarly Publications Programme, using funds provided by the Social Sciences and Humanities Research Council of Canada.

For my parents,
Alexander Watson
(1894–1982)
and
Helenora McIntosh Slater
(1903–1997),
who moved from one margin to another

Contents

Acknowledgments

Looking back over the lifetime of scholarship of Harold Innis, one is struck by his commitment to exhaustive 'dirt' research and his bias towards the oral tradition. In writing this intellectual biography of Innis, I have attempted to follow the same path: to apply Innis to Innis if you will.[1] This has involved great effort to track down individuals and obscure archival sources that may cast light on some aspects of the man. It has also meant that, in writing the biography, I have tried as far as possible to let the characters, including Innis, speak for themselves. For this reason, there are far more quotations and less paraphrasing than one might normally find in a work of this nature. It also means that I am indebted to a wide range of people who have assisted me along the way.

This book began many years ago as a doctoral thesis and I would not have completed it without the patience, encouragement, and positive criticism of my supervising committee: Peter Russell, Abraham Rotstein, and Mel Watkins. In this regard, Mel deserves a particular vote of thanks, for, without his continuing interest in my work and his steady contention that it should be published, I doubt that this book would have seen the light of day. No one person has done more over the years to keep interest in Innis alive.

I must also thank all the friends, family, and colleagues of Innis who have taken the time to talk to me and other researchers about the man. Particular thanks go out to Anne Dagg and other members of the Innis family for their interest and help, their kind provision of family photographs, and their generous consent to the extensive use of quotations and photographs from the Innis Personal Records, held in the University of Toronto Archives. I thank as well Lib Spry for her perspective on the whole life that her mother, Irene, chose to live, and for the use of her photograph.

Archives are of no use to any researcher without the work of archivists and here I am heavily indebted to the many professionals whose quiet day-to-day work makes access to the precious records of the past possible. I am particularly indebted to Garron Wells, Harold Averill, Marnee Gamble, and other staff at the University of Toronto, who, over the twenty-five-year period of my dealings with them, have been unfailingly helpful in guiding me on my long travels through the main depository of Innis-related records. I would also like to thank the many staff who helped me as I waded through Innis material at the National Archives of Canada, the Canadian Baptist Archives, and the Norwich Township Archives. I am especially indebted to the South Norwich Historical Society and Gail Lewis, who was kind enough to give me an insider's tour of the Otterville Museum and of Innis's boyhood haunts as they survive in the present day. Thanks also to the women of the Otterville Fellowship Baptist Church for their efforts in tracking down Innis records. I am indebted to Barry Eastwood of Lamont, Alberta, who made me aware of the fate of the North Star School (where Innis was the first teacher in the summer of 1915) as well as its proximity to several old fur trade forts. Finally, I would be remiss if I did not also thank the many faceless staff in archives I could not visit first hand, who responded professionally to my requests for obscure material related to Innis that might be found among their holdings.

I am grateful also for the cooperation of Innis College at the University of Toronto, and for the college's hearty endorsement of this project.

Thanks also to Laurenda Daniells for agreeing to the inclusion of the poem 'To Harold Innis' by her husband, Roy Daniells, 'a great admirer of Professor Innis.'

The excerpt from Dennis Lee's 'The Death of Harold Ladoo' from *Nightwatch: New and Selected Poems 1968–1996* appears courtesy of the author and with the permission of McClelland & Stewart Ltd.

Quotations from Wyndham Lewis, *Time and Western Man,* © Wyndham Lewis and the estate of the late Mrs G.A. Wyndham Lewis, appear by kind permission of the Windham Lewis Memorial Trust (a registered charity).

No writer (and especially not one who has dealt with Innis's crabbed script) should ignore the painstaking work of translating messy handwritten hieroglyphics to neat print on a page (and keeping it there during an era of rapidly changing word-processing software), and in this regard I wish to thank Julie Weedmark, Deirdre McKennirey, John Redfern, and Hélène Mousseau. Thanks are also owed to Ellie Prepas, who encouraged me to make use of computer technology in the form of the university's mainframe – for text editing in the days before personal computers and word-processing software existed.

Martin Weaver, who has passed on too soon, deserves credit for encouraging the resurrection of this project over an unforgettable Indian meal on a rainy night in Manhattan. Rosemary Tassie and the staff of CARE Canada have my thanks for carrying on business as usual while the 'boss' was on part-time status. In this regard, I am grateful to Nancy Gordon both for ably deputizing for me at CARE and for her often-stated conviction that this project should be done. Nancy also was kind enough to read and comment on the manuscript at the final draft stage. The board of CARE Canada has my gratitude for permitting the sabbatical which resulted in this book.

Don Bastian provided invaluable editorial assistance in the rewriting and restructuring of the text; I could not have made the changes and additions necessary without his encouragement. It should be noted that Don, as a lifelong Methodist, provided valuable personal insight that helped this lapsed Presbyterian come to grips with the story of a rural Baptist turned agnostic. Thanks also to Omar and Famida Majeed who gave me insight into the persistence of the oral tradition and the contingency of religion as empires displace populations and vernaculars disappear. As well, I thank Omar and Usha Majeed for providing me with a base in Toronto during the final stage of my research.

Finally, and most important, this book would not have been possible without the ongoing support and enthusiasm of the love of my life, Rosanna Majeed. Her professional editorial competence has led to a vastly improved final product, and her moral support when this project seemed to have become impossibly complex allowed me to persevere through to publication.

A world that denies
the gods, the gods
make mad. And they choose their
instruments with care ...
The artist they favour
becomes a priest indeed, he meditates
the sacraments of limbo ...
... it is his gorgeous vocation
to bludgeon the corpse for signs of life, achieving
impossible feats of resuscitation, returning, pronouncing it
dead again. Opening new
fever paths in the death heaps of civilization.
And he names the disease, again and again he makes great
art of it, squandering
what little heritage of health and meaning remains,
although his diagnoses are true, they are
truly part of the disease
and they worsen it, leaving
less of life than they found; yet in our time
an art that does not go that route
is deaf and blind, a coward's pastorale,
unless there be grace in words ...

Harold, how shall I exorcize you?
This is not for blame.
I know that
it lived *you*, there was no
choice; some men do carry this century
malignant in their cells from birth
like the tick of genetic stigmata,
and it is no longer
whether it brings them down, but only
when.

– Dennis Lee, *The Death of Harold Ladoo*

Introduction: The Innisian Puzzle

I once had to choose between going into university work or into politics and I decided to go into politics.

H.A. Innis[1]

'Who the hell is Harold Innis?' So goes the school song of the Toronto college named after Canada's first internationally recognized post-colonial intellectual.[2] Innis would have chuckled at the irony of the students' cheekiness. This scholarly pioneer of Canadian economic history and communications theory had a dry wit which carried him through an extraordinary life touched by poverty and comfort, loneliness and public accolades, peace and war, religious zeal and the loss of faith, the passions of nationalism and internationalism, the joy and anguish of love gained and lost, the awe and puzzlement of colleagues and students, intellectual brilliance and mental collapse, and, above all, the gift and burden of vision. This book, then, is an attempt to answer the students' question, 'Who the hell is Harold Innis?'

The research for *Marginal Man* started many years ago in an effort to elucidate the cryptic communications works that Innis, working alone, produced during the last ten years of his life.[3] My plan was to gain new insights into these later works by thoroughly understanding the biographical context out of which they arose. Since Innis's death, communications studies has developed as a separate discipline and his later work has taken on a 'canonic' significance.[4] His work represents the old testament of communications theory, which, when paired with Marshall McLuhan's new testament, forms the 'Toronto School' of communications. Despite the fact that Innis would have deplored this type of talk, it seems clear that what he wrote late in life is still speaking to social scientists grappling with understanding our world so many years later.

As Innis well knew, a research path pursued honestly and diligently often has the odd trait of taking one to an unexpected destination. I had hoped to clarify the communications works by exploring Innis's biography. It now seems to me that those works will never be made clear in the sense I had hoped at the start. Instead, I trust that, by using the communications works as a window into his intellectual biography, I have produced a book that leads to a more profound understanding of Innis himself.

I should have known that elucidating Innis's communications works through understanding the biographical context of that work was bound to create a high degree of frustration, given the cryptic character of those studies and the enigmatic personality of their author. More than two generations of scholars have already tried to piece together the Innis material since his death in 1952, and still one has the impression of being left with a series of badly assembled puzzles. Each attempt seems to expose a different picture of the man and his work. Yet these pictures can be maintained only by ignoring a number of embarrassingly extraneous jigsaw pieces. Innis himself looked on his life very much as a puzzle to be solved: instead of offering prescriptive endeavours, or what he would have called 'preaching,' he was concerned with finding a hidden pattern and manipulating already existing bits and pieces of his world so that this pattern would become evident to others. In terms of Innis's life, I call this hidden pattern his 'project'; in the less secular terminology of his childhood, it would have been termed a 'mission' or a 'calling.' The project went far beyond the career plans of even a brilliant intellectual. This is why, in tracing its extent, one should term it a 'political' project. Without understanding the passionate intensity with which Innis pursued this project, we will never be in a position to make the communications works clear. Put briefly, we cannot assess their content without a prior understanding of the context.

This is so because Innis failed to complete his political project. Time and the times were against him. Furthermore, because the project was not prescriptive, it became almost invisible in its failure. He did not create a Utopia. Hence, to reconstruct the pattern towards which he was working, we must undertake a painstaking re-examination of the way in which he manipulated events and ideas, especially during his later years. The communications work is to be seen as a final but incomplete stage in this overall project. In this sense, the Innisian puzzle will never be definitively solved because Innis himself failed to complete the final assembly of the great intellectual synthesis he was attempting. Before embarking on this effort, Innis had published enough brilliant material to justify the reputation of several gifted scholar combined. He was nationally recognized for this work and was one of Canada's most renowned intellectuals at the international level. Eulogies generally tend to emphasize 'success,' and, in the case of Innis, this tendency has been carried on to the present day by friends, colleagues, institutions, and, indeed, a country eager to celebrate the memory of a great scholar.

Innis was fond of saying that 'to the founder of a school, everything may be forgiven except his school.' Speakers on Innis are equally fond of saying that he would have been pleased that no Innisian 'school' has developed. But they overlook the fact that they are usually speaking in the context of some occasion that reinforces the 'cult' of Innis as the quintessential scholar: the Innis of plaques and Innis College; of memorabilia and anecdotes; of CBC radio documentaries and 'Innisfree' Farm; and of the Creighton biography. In short, the cult of Innis has become a significant obstacle to critical follow-up of his work. Funds have been spent mainly on enshrining his memory rather than sponsoring new scholarship in the areas in which he had been active. And the effects on those who knew him best have been even more various. They were superficially interviewed so many times that a form of reverberating eulogy was set in motion, with the result that their critical perspective on the man was virtually nil. In fact, Innis's stature among Canadian intellectuals is so great that one feels the tendency simply to assume that his work is fundamentally sound. Saying that he was 'ahead of his time' expresses the belief that, with patience and diligence, we will eventually get at what was so clear in his mind. I do not now believe this to be so.

The early and continued recognition of Innis's academic success has tended to obscure the basic fact that Innis looked on his work as a failure. Innis was not a happy man. Reviewing the archival material, we find evidence both of overwork and of a recurring and profound sense of depression and anxiety, a sense that goes far beyond what we would expect to find in a scholar with a healthy desire for perfection in his own research. There is a dark side to Innis's personality that is not an insignificant psychological quirk. It is a fundamental characteristic of Innis stemming from the size and intensity of his project; paradoxically, one of the greatest of several impediments to the recognition of the overall project is its daunting size.

What he had hoped to accomplish in life was simply too much for one man. When his colleagues and others judged him to be tremendously successful, they were comparing his output and his influence to the accomplishments of fellow academics. Innis, however, judged himself to be a failure using as his yardstick the crucial areas of his project that always remained yet to be completed.

The Promethean extent of his project has been effectively hidden by his extremely reserved nature. By all accounts, he retained a shy personality even at the height of his career. Furthermore, this shyness has carried over into the guarded style of his writing and to a 'bias' imposed on his archival material through the editing out of many documents of a controversial or personal nature. A reserved individual will attempt to keep private that to which he holds most passionately. This Innis did with his overall project. On those few occasions when he did speak out with intense feeling on the difficulties he faced in undertaking his work, his comments often were interpreted by his audience as humorous irony rather than deadly serious sarcasm. These events in themselves tended to reinforce his secretive nature. Thus, Innis's character is of a sort that easily eludes the intellectual biographer. Yet, without the prerequisite of an adequate intellectual biography, it is unlikely that the logic of the communications period can be fully appreciated.

Careful examination of existing material shows that, although Innis judged himself to be a failure, he was also well aware of his rising reputation among fellow scholars. Yet what struck Innis most forcefully as far as his own communications work was concerned was the limitation of this great reputation. After all, embarking on the communications studies meant leaving behind colleagues in whose field his reputation had been established: Canadian economic history. In his new area of research, he was obliged to justify his efforts in two quite different ways. First, in a positive sense, within the field of scholarship, he attempted to link his new research with his old reputation and with the concerns of his colleagues in economic history by describing it as an extension of the earlier work on staples he had done as a means of understanding Canadian economic development. Second, in a negative, more defensive sense, Innis had to justify to himself his expanded role at this new stage of his project. He was extremely sensitive to the criticism that he was attempting to accomplish too much. It was generally thought during the last ten years of his life that he was 'getting beyond his depth,' partly because of the demands on his time caused by his administrative duties but also because of the wide range of specialist disciplines on which the communications work encroached. His obsession with this criticism was so marked that his works on communication theory devote a great deal of attention to the theme of the important role played by peripheral intellectuals with a generalist bent in advancing Western civilization.

The first aim was accomplished with some success. He was interpreted as a 'media determinist,' just as he had been typed a 'staples determinist' in his earlier works. However, as a result of this limited understanding, colleagues did not follow Innis into the area of communications. Rather, they tended to consider the 'media as staple' approach to imperial history to be an unwarranted over-extension of the theme Innis had devised to explain the recent economic history of 'new' lands.

The second aim went virtually unrecognized. Innis's conception of scholarly dedication could grasped only be as a personal compulsion. To raise it to the level of the intellectual everyman representing the keystone of Western civilization would, at best, have been considered farfetched. Yet this is what Innis was doing.

In his treatment of this theme, particularly in his collapsing of general theory into the case of the individual scholar, Innis is justifying the final stage of his life's work. Central to his project is the idea that Western civilization can be renewed only by intellectual developments on a periphery that, in turn, becomes a new centre for cultural florescence. In other words, Innis is asserting that, although these revitalized efforts in the field of knowledge may originate in the decadence of the imperial centre, they can be pursued effectively only if they become allied to a new and independent politico-economic structure or force at the margin of contemporary civilization. In the past, this force may have been styled as a tribe, court, or city, but in our day it is the nation-state. Innis traced the general movement of this centre of Western civilization from the East to classical Greece to Western Europe and eventually, in modern times, to the New World. His theory moved marginal man, particularly the 'colonial' intellectual, to the centre stage of Western culture.

Innis presents a picture in which significant work in either area – knowledge or force, scholarship or politics – can be effective in the long run only if it is coupled with work in the other. Yet, paradoxically, he sees both as distinctly different types of endeavours that cannot be pursued in unison. Hence, specific institutions are needed to provide the structural basis for this division of labour. The institutionalization of force is manifest in the army, police force, courts, the bureaucracy, the mechanized workplace, and the political structures of society. These have as their *raison d'être* the transmission of orders, obedience, and an absence of potentially contestatory thinking. This is why Innis viewed power as always anathema to intelligence. The institutionalization of intellectual work, therefore, is of a totally different order and seeks to guarantee the uninhibited dialogue of the 'best minds' of a society with their peers. For Innis, the university as ivory tower was the key institution that made this possible; other related institutions are concerned with the seeking out, initial training, and guiding of intelligent youth towards this sheltered environment.

However, in Innis's view, even a well-developed pre-university and university system did not guarantee the maintenance of this type of dialogue. The cultural borrowing of the new milieu from the old must be of a limited or selective nature. Without a sense of incompleteness in the cultural milieu, rote learning replaces the oral tradition. Thus, the potential for development of a new and distinctive perspective is aborted. Innis traced the contemporary crisis of the West to the fact that this sense of incompleteness is less and less possible given the increasingly perfect technologies for the transmission of cultural 'baggage': the various modern media of communications.

Valid or not, this is how Innis saw his world. The vision is extraordinary, for, by placing intellectual endeavour linked to protection by force as the cornerstone of Western civilization, he was justifying a personal project that had ramifications ranging from his most profound thoughts to the national politics of the day. The consequent sense of responsibility that he felt was enormous.

Looking at Innis from this perspective helps us to avoid some of the quandaries faced by previous commentators. The most obvious deficiency in existing materials on Innis is their contradictory nature. This phenomenon goes far beyond the discrepancies that one would expect to find as a result of the different subjectivities of the various commentators. Furthermore, it encompasses everything from biographical details to the most subtle of Innis's ideas. Many of these apparent contradictions arise when only one level of his project is being considered. For instance, Innis, from the beginning, sought to establish a scholarly reputation, and his increasing long list of honorary degrees indicated institutional or professional success in this area. It does not follow from this, however, that Innis considered himself a success. In his mind, such a judgment depended not so much on the ends that had already been met as on the goals of the overall project that continued to elude his efforts.

Confusion has also arisen through the failure of commentators to understand how the priority assigned to any particular level of the project changed over time as external circumstances and personal developments unfolded. The tremendous difference in Innis's attitude towards cultural issues between the Depression and the early 1950s would be an example. The final essays do indeed seem like jeremiads, containing something alien to Innis's spirit from a perspective that stresses the research inclinations of the individual scholar. Yet, in *Marginal Man*, consideration of changes in institutional structures and the climate of national and international politics place the essays squarely within the logic of his overall project.

There are other conflicting views of Innis that find their source not so much in the misunderstanding of Innis's project as in the retrojection of other paradigms onto his work. The internal consistency of the paradigm involved allows for the imposition of a comprehensive and orthodox version of Innis. But, while scattered and minor contradictions are thereby avoided, different paradigms construct different versions of Innis that are fundamentally at odds with one another. This problem developed largely as the result of the interregnum of more than a decade following his death. During this period, concern with his work was at an extremely low ebb. The curious paradox is that, during the same period, the myth of Innis as the complete scholar was spreading among the wide group of acquaintances, colleagues, students, and friends who so fondly remembered the man.

When Innis's thought began to be rediscovered in the mid-1960s, it was not out of intellectual curiosity but as a corollary of increasingly strident discussions concerning foreign ownership on the one hand and the psychological effects of media on the other. Specialization in the social sciences had proceeded to the point at which these discussions were initiated in two distinct fields: economic history and communications theory. Out of each of these politically charged debates has come an 'orthodox' interpretation of Innis's contribution to knowledge. These two separate debates, disciplines, and orthodoxies each tend to stress that stage of Innis's work during which he was dealing with themes familiar to the respective disciplines.

In the process, Innis becomes the victim of the very specialization in the social sciences about which he warned. An intellectual tyranny is set up in which the 'late' Innis dominates the 'early' or vice versa. This presentation of Innis's life work as a kind of retroactive intellectual schizophrenia is accepted all too readily for, ironically enough, it provides the built-in coherence that goes with a specialized branch of the social sciences, be that economic history or communications theory. The problem is that the best of the commentaries have been too effective. They have resurrected an interest in Innis by placing his work in the present-day political context. This offers us a scenario that is far more easily appreciated than the original, outmoded project. Who today, for instance, would seriously put forward the university as the key institution for the maintenance of Western civilization?

The most powerful of these orthodoxies is the staples theory, best described in the seminal article by Mel Watkins, 'A Staple Theory of Economic Growth.'[5] While Innis never based his communications work on a prior recognition of this theory, there are, not surprisingly, many elements in his 'staples' phase that are of central importance to his project. In a like manner, Marshall McLuhan's 'the medium is the message' selected and developed many of the seminal ideas of Innis's communications work. Largely independently of the intentions of Watkins and McLuhan, their own work has been simplified and popularized to the extent that it has led to the development of two 'orthodoxies' with regard to Innisian theory.

The staples-theory approach has a tendency to emphasize those aspects of Innis that have an air of rigid determinism. It picks up on Innis's habit of exaggerating causality to provide a clearer picture of the logic of economic development. The McLuhan approach also emphasizes a determinist element – the nature of the media. Beyond this, the key weakness in McLuhanist commentary is the extent to which it makes a virtue of an obscure style of writing. The defence of this style, on the grounds that it coincides with the non-linear mentality of modern media, is more often than not a weak rationalization for muddled thinking. The ambiguity in Innis's later work arose during a period in which the project he pursued encountered a distinctly hostile environment. As such, his abstruseness was a weakness imposed, at least partly, by his strategy of avoiding attack while realizing his intellectual goals. Thus, subjecting the orthodox versions of Innis to biographical verification emphasizes their lack of adequacy in accounting for the complexity of the man and his theories.

Over the past twenty-five years, scholarship on Innis has expanded considerably and, to some extent, the dichotomy among Innis commentators between those who approach him mainly from a staples/economic-history perspective and those who do so from a communications theory perspective has broken down. More accurately, it has become increasingly fragmented, reflecting the general trend towards specialization in the social sciences. In particular, the portrayal of Innis as a technological determinist has waned somewhat as it 'has been challenged by other perspectives – such as feminism, cultural studies, cultural geography, social constructionism and post-modernism.'[6] Reflecting the trend to specialization, scholars have been intent on analysing smaller and smaller elements in Innis's life and opus. They have produced studies on Innis's thinking about a wide variety of subjects including the north, the Maritimes, Quebec, gender, Mary Quayle, politics, publishing, the role of the intellectual, religion, orality, the Chicago School, post-modernism, the Frankfurt School, ecological theory, evolutionary theory, dependency theory, Marxism, the communication 'canon,' the Canadian Social Science Research Council (CSSRC), and Canadian cultural policy.[7] At their best, these scholars have returned to a careful review of primary resources and have made significant incremental contributions to our understanding of Innis.[8]

Some of these scholars have responded to the absence of a clearly articulated philosophical approach or methodology of research in Innis's works by importing the framework of other paradigms to 'explain' better what Innis was getting at. This strategy has had mixed results. On the one hand, it has rejuvenated interest in Innis by 'translating' his terminology into the vocabulary of currently popular 'schools.' On the other hand, it has added little to our understanding of how Innis's mind worked because it uses intellectual scaffolding which, if the lack of evidence in the primary-source material is any quide, had no direct influence on Innis during his lifetime.[9] This has led to an impasse in the progress of Innis scholarship, directly related to the increased specialization in the social sciences about which he expressed so much scepticism. I hope to go beyond this impasse by applying Innis to Innis. By returning to a detailed examination of the massive primary-source material, I attempt to present what has been noted by many commentators as a missing element in Innisian studies: a full-scale intellectual biography.

With the First World War, Innis's personal project became intensely and irrevocably political, this term defined in a broader sense than commonly understood. The Greek city-states had no state apparatus as we understand it, and 'political' gained its meaning in the activities of a community's citizens. In these societies, 'political' designated the collective search, conducted through free dialogue of individual members, for 'good' in a community. It was, by definition, a personal activity. This meaning is still in general evidence today, but in a hidden form. Today the word 'politician' carries almost universally negative connotations, a fact that, in itself, indicates our continued understanding and appreciation of the older sense of the term in an age in which politics has been usurped by a professional group dealing in power and administrative secrecy rather than general principles.[10] The time-honoured definition is a high standard to meet.

The 'political' element in Innis's career was not something that arose automatically from his biographical background. His mother's determination to produce an upwardly mobile son was typical of many other farm families at the time. Such plans for personal betterment might have produced an individual who defined his problem as the permanent, inherent cultural poverty of a childhood milieu on the periphery of empire. This point of view, in turn, might have led to a career strategy that emphasized deracination as a positive response. Following this strategy, the individual would move away from his early roots both spiritually and physically, drawn to the centres of intellectual activity where his chief goal will be mastery of the manner and methodology of contemporary metropolitan paradigms. Success will be measured by the extent to which one is not recognized as coming from the periphery. The chief difficulty posed by this strategy is the effective suppression of a person's background. There is no priority put on influencing the society left behind in this strictly personal quest for recognition through cultural mimicry.

In contrast, Innis's project for personal betterment was political in two senses. First, it valorized the colonial perspective as the fundamentally critical one. Second, it included a commitment to remaining in the peripheral community. Accordingly, Innis's efforts were to be seen not just in terms of individual betterment. He represented an important character of the colonial milieu. If he succeeded, others from a similar background could and would follow. This, in turn, meant that his community was entering a new stage during which it was capable of taking on complete responsibility for its own direction. It was a context in which the personal and the political merged naturally.

Yet we must be careful in our understanding of Innis's background. Although the importance of his rural roots is mentioned by virtually all commentators, including Innis himself, this can be done in such a way as to obscure the political aspect of his work. Innis is not a metropolitan intellectual of the order of Robert Burns or Thomas Hardy, engaged in translating the memories of a countryside still close by in time and space to the increasingly industrialized and urbanized populace. Nor does he belong to a rural polity whose indigenous cultural frame continues to exist while being overlaid with a ruling group whose values derive from Western civilization (as in the Third World). Innis's project does not involve the politics of a nostalgic nationalism or a revolutionary anti-colonialism. Rather, he comes from the peculiar milieu of the European-settler colony. It is obvious that the mentality of individuals in this type of polity is an extension of Western civilization. Nevertheless, it is an anaemic variety of Western culture, for it has involved for generations the purchase, from the centre, of cultural artefacts in return for raw materials produced in the periphery. Furthermore, significant adjustments in the original metropolitan mentality of the settlers have been necessary in order for them to survive and to exploit these resources. These adjustments include borrowing from the cultural heritage of peripheral non-European peoples. Innis's recognition that his polity was, on the one hand, an extension of Western civilization, and, on the other, a unique and precarious variety of that civilization tempered his endeavours. To become more 'cultured' without abandoning the perspective afforded by a position on the periphery became the central aim, in his understanding of both the individual and his community.

Innis's family, his rural and colonial background, the Baptist church, his schooling, and the Great War had the combined effect of fixing for him a personal project of a totally encompassing nature before he reached his mid-twenties. The project pervaded every aspect of his life with an immense sense of responsibility and with consequent anxieties. It set for Innis tremendously difficult goals for intellectual training and linked the progress of his career with far more general developments at an institutional and national level. The boundaries of his project can be set out by identifying those areas to which it applied and those phases through which it would pass (see table 1).

It was the ironic combination of an anaemic culture and a unique perspective in the European-settler state that determined the outline of Innis's intellectual training. The first priority was a period of academic apprenticeship that would allow someone from a rural and colonial background to become acquainted and feel at ease with that measure

Table 1: The Scope of Innis's Project

	Personal betterment	Institution building	National politics
Phase 1	Basic education	Extension and localization of pre-university education	Colonial status
Phase 2	Canadian knowledge	Extension and localization of university education	Dominion status
Phase 3	Universal knowledge	'Thickening' of higher education structures	Independent status

of cultural baggage carried by the educated individual at a metropolitan institution. Since it is in essence a period of catching-up, the dominant anxiety is one of personal inferiority, a suspicion that dogged Innis throughout this period of his basic education. We can consider this phase, although not the emotions it engendered, complete with the attainment of his PhD in 1921.

The next phase of Innis's project involved intensive familiarization with the peripheral environment. There were extensive field trips and collective work with the new generation of Canadian academics on virgin archival documentation. The basic approaches learned during the first phase were now honed to produce explanations of developments on the periphery that, in their excellence, far surpassed previous work. However, for Innis, it was not the accumulation of important published works on various aspects of the peripheral economy that was the final goal of this stage. Rather, it was the opportunity this work gave to the intellectual community of the hinterland to develop a critical methodology of research. While this methodology would benefit from the unique perspective of the intellectual on the periphery of empire, it was not restricted to handling only subject matter from this region. In this regard, the wide-ranging conclusions of his book *The Cod Fisheries* can be conveniently taken as the final point of the second phase of Innis's intellectual progress. Since we are talking about a time in which young Canadian academics were establishing their reputations and taking over from (mainly) British staff members, the dominant anxiety was one of insecurity. This disquiet was further heightened by the Depression.

The political nature of Innis's personal project becomes obvious when he applies the above methodology to fields of knowledge of a more universal nature. It is one thing for 'colonials' to develop a familiarity with the political economy of their region. It is quite another for them to maintain that the methodology, however ingenious, that is the by-product of their efforts has some sort of special critical and universal significance. Given the radically new departure implicit in this final phase, it is not surprising that not one of Innis's economic-history colleagues followed him in his jump to the application of this methodology to altogether different domains of study. The dominant anxiety of this stage, therefore, is one of intellectual isolation. As long as the personal project retains a political cast, this loneliness is unmitigated, for to join in the discussions of other scholars already active in these new areas of his personal concern would have entailed a prior acceptance of paradigms forged in the metropole. This in turn would have implied abandonment of the belief in a novel hinterland perspective. Indeed, the political nature of the project became most manifest at this juncture, for the very extent to which the peripheral perspective was recognized determined the amount of hostility it engendered among the established intellectual community of the metropole. Innis' efforts suffered a dual fate: either they were completely ignored or they were denounced as colonial precocity.

Even at the individual level, therefore, Innis's project was inherently political. Because he was among the first indigenous academics in his field, his life's work represented a model for others to follow rather than a never-to-be-repeated touch of genius. If the colonial farm boy could complete these efforts for personal improvement, others from the same context could and would follow. His level of individual achievement would automatically be viewed as exemplary, the atomic component in the overall decolonization process.

In another sense, it was impossible to view such a project only at the individual level, for it assumed prior developments in the structure of the periphery before it could take place. It was very different from the relatively simple process of sending the children of a colonial elite back to the centre of empire to be educated at established institutions already serving metropolitan needs. In contrast, it involved the development of a far-flung school system to identify gifted children wherever they were and to provide them with a path to cultural enrichment without the consequent introduction of deracination. Institution building was inevitably a key element in the overall project. Innis's work at building up the educational institutions of the periphery was undertaken with a view to providing others with the opportunity to follow the same sort of individual academic career as his. As with his personal project, there are three different stages in the institution-building level.

The first stage was ensuring that there was a network of primary and secondary education that stretched to the very periphery of settlement. Only through the effective operation of these one-room rural schools could the best students be identified at an early age and encouraged to continue their efforts. The localization of both personnel and content in these scattered schools was important for two reasons. First, trained metropolitan personnel were simply not available in sufficient numbers. Second, the relevance of locally produced study material and the familiarity of the teacher with the context from which the students came ensured that appropriately gifted individuals were accurately identified. Even those who did not go on in their studies benefited from a civic education free from a colonial spirit. This could be assured only by the preparation of both material and personnel in the periphery.

In practical terms, Innis participated in this process throughout his career but in particular during its initial period. Examples of activities of this type would be his teaching at one-room rural schools in Norwich, Ontario, and Landonville, Alberta; his participation in the Workers' Educational Association (WEA), Khaki College, and Frontier College; his work on popular publications such as *Peter Pond* for the Canadian Men of Action Series; and his participation in the Manitoba Royal Commission on Adult Education. However, his most important contribution came indirectly, as his work on economic history percolated down to be included in standard school texts and as teachers taught by Innis (or influenced by his approach) took up positions in the system.

The second stage of the institution-building process was the localization of personnel in Canadian universities and the production of new material on the Canadian political economy. This stage accounted for the bulk of Innis's efforts in institution building and encompassed a wide range of activities. He orchestrated the training of Canadian personnel at metropolitan institutions through the control of funds from such foundations as the Rockefeller, Carnegie, and Ford. He influenced research and publications policy in the same way, facilitating the establishment of various professional bodies, conferences, and journals. He fostered the reintegration of young Canadian academics following their training abroad and the phasing-in of new institutes, departments, and universities. No one, before or since, has had such singular control over the placement of individuals in the social sciences in Canada.

The third stage was an extension of the second. It involved the 'thickening' of the recently established educational structures through more staff with a greater variety of perspectives on similar topics. The key distinction, in Innis's mind, was that, at this point, the institutions involved would become accepted in their own right at the international level rather than being colonial institutions strong only in those areas of knowledge pertaining to the colonies.

In this regard, increased enrolment at Canadian graduate schools and the continued good record of Canadian graduates studying abroad were important. Also important to this stage were the holding of international conferences in Canada and the participation of Canadian academics in the intellectual (Beit, Cust, and Stamp Foundation Lectures) and administrative (presidency of the American Economic Association) functions of international institutions. Only when this level of institutional development had been achieved could one be certain that an individual from the periphery could sustain intense cultural training leading to an indigenous critical perspective without consequent deracination.

Here we have the crux of Innis's project. On the one hand, only institutional support on the periphery could ensure that the training of an intellectual from these marginal regions would result in an indigenous perspective that would not cut him off from his background. Yet, for the first generation of local scholars, this institutional support was not in place. They found themselves in an exposed position. They had to build up a new critical perspective to understand their country and to resist the temptation of importing conveniently available, fully developed paradigms from the centre. At the same time, they had to build the institutions that would allow others to follow in their path without being subject to the same pressures of alienation from their home context.

Just as institutional strengthening was dependent on the ability of many individuals to undergo a high level of cultural training without experiencing deracination, so the influence of this new wave of intellectuals on national policy was dependent on the growing importance of their institutions, in particular the university. In Innis's project it was always the institution that mediated the effect of the individual on national policy and vice versa. Since effective influence on national policy depended on the success of the personal and institutional project, this was something that could not be rushed. The formation of an understanding of the domestic political economy was a collective effort that would involve the intensive 'dirt' research of many years. In the meantime, premature influence on national policy by peripheral intellectuals could lead to the destruction of their personal research projects as well as their attempts to set up first-rate educational institutions. Repeated requests from those engaged in practical politics posed a continuing danger, in that peripheral intellectuals might eventually be convinced that they should have the 'solutions' to social problems. Innis's fear was that this would lead to the appropriation of schematics available through Keynesianism, Marxism, or Fabian socialism and their application in a cardboard fashion to the Canadian scene. He recognized the temptation of influence-for-the-sake-of-influence as one of the most serious problems to be faced by intellectuals. His only justification for academic influence in society at large was on the basis of a hard-won critical understanding of the peripheral polity.

Again, this part of the project can be broken down into stages: a 'colonial' stage in which final decision-making powers in crucial areas are located outside the peripheral polity, either at the centre of empire or physically in the polity but in the person of a foreign officer; a 'dominion' stage in which decision making is undertaken essentially within the peripheral society but by personnel still viewing the world through metropolitan paradigms; and, finally, a fully independent stage in which responsible government has been achieved both in structure and in spirit. At this stage, not only are the decisions concerning the polity made within the periphery but they are made on the basis of a critical mentality that is the final outcome of the work of the indigenous intellectuals.

The key to this sense of participation in national politics by the intellectuals is the prerequisite development of a critical consciousness. This quest and the demands of effectively administering the polity are fundamentally at odds, for the one is concerned with the continued pursuit of truth while the other demands a 'truth' of the moment to act on. This is why Innis maintained that the only legitimate type of influence of the intellectual and his institutions on national politics was a fundamentally conservative or reactive one that operated within severe limitations. First, the free pursuit of knowledge within the polity must be seen to continue, adequately buffered by the organization of the institutions of higher education. Second, influence was best achieved in the long term by the inculcation, in the future leaders of all social sectors passing through the university, of a non-fanatical and indigenous perspective that stressed the complexity of the domestic political economy. This approach emphasized the limits of decision making as opposed to preaching fixed solutions. Finally, influence was achieved by responding directly to the requests from the political level with the best advice possible, while modestly emphasizing the limitations implied by the academic point of view.

The activities of Innis in this field are far more significant than is usually recognized. This is so because his criticism of the contributions of certain other scholars who chose to serve in the bureaucracy or political parties per se have been taken to imply a general condemnation of intellectuals working outside their institutions. The fallacy of this point of view is evident from Innis's participation in a variety of royal commissions, his contributions to political party summer schools, the Workers' Educational Association, and the Bankers' Educational Association (BEA), his recommendation of various individuals for work in the public service, and his numerous public-speaking engagements on subjects such as the Rowell-Sirois Commission recommendations (of which he was a strong critic) and post–Second World War Russia. But most of all, it is demonstrated by the massive correspondence he kept up with non-university people ranging from prime ministers to Cyrus Eaton and the editor of the *Financial Post*. The destruction, after Innis's death, of the bulk of this correspondence with influential people has further hidden this aspect of his project. If we reproduce this comprehensive project schematically using the categories outlined above, we can better understand the paradox of Innis's final years: outward success coupled with an intense inward sense of failure (see table 2).

Innis's reputation and recognition rested on his success in pursuing the project into its second phase, a phase recognized as valid by the intellectual communities of both the hinterland and the metropolitan centres. It seemed appropriate that a division of labour in the task of writing imperial history should leave the detailed understanding of hinterland developments to former colonials. What these colonial intellectuals lacked in terms of sophisticated methodology, they made up for in the fervour with which they attacked subject matter so close to home. The validity of this stage of the project was widely recognized because it corresponded to the time in which Canada was clearly drifting away from the British sphere of influence. However, this dynamic was not to lead inevitably to an independent polity with its own self-confident perspective on the world. Instead, there was an intense, although unnoticed, dependence setting in on the United States as a metropolitan

Table 2: The 'Political' Project of Harold Innis

Phase	Personal betterment	Institution building	National politics
Phase 1	*Basic education* – Making up for colonial background without deracination – Norwich, Otterville, McMaster, Chicago – First World War & Khaki College – From the ministry to law to political economy – Ends with PhD, 1921	*Pre–higher education* – Establishing rural schools, teacher training, and Canadian curriculum – Teachers' college, South Norwich school, Londonville school – *Canadian Man of Action* Series – Influence of staples works on school texts	*Colonial status* – Major decision-making functions held outside country, and Canadian personnel subject to foreign supervision – First World War – Political Economy Department pre-1937
Phase 2	*Canadian knowledge* – Archives research and field trips – Staples works – From the CPR study to *The Cod Fisheries* (1940)	*Higher education* – Canadian personnel replace foreign personnel – Growth of the universities and the development of an indigenous perspective – Publication of the economic history of Canada	*Dominion status* – Decision making within the hinterland by indigenous personnel using metropolitan paradigms – Diatribes against preachers of Marx, Keynes, Fabianism – Speaking engagements – royal commissions

Table 2: (*Concluded*)

Phase	Personal betterment	Institution building	National politics
Phase 3	*Universal knowledge* – A contribution to general knowledge recognized by metropolitan institutions but coming from outside those institutions, from an intellectual who has remained in the hinterland and who has developed a perspective and methodology in research on hinterland topics – From the 'Perspective Powers' and the conclusion of *The Cod Fisheries* to the communications work – Leaving his original colleagues behind yet not accepted by the metropolitan-oriented groups in whose domain he has begun working; the dominant sentiment is one of loneliness, 1940–52	*'Thickening' of higher education system* – Setting up new departments, research councils, conferences – More Canadian graduate students in training both in Canada and at metropolitan institutions – More Canadians recognized with honorary degrees at metropolitan institutions and participating as officers in instutional learned societies – The success of this stage is best determined by the extent to which contributions to general knowledge are made by indigenous intellectuals working in Canadian institutions or the recognition at the metropolitan level of an authentically Canadian national culture	*Independent status* – The final attainment of resonsible government – Decision making undertaken by politicians, a public service, a business sector, etc. who have a thorough grasp of the complexities of their milleu and therefore the limits to their decision-making ability. The development of this mature national perspective depends on the advent of Phase II. – A pool of intellectual advisers who can respond to requests for advice from the political level without abandoning their priorities for the continued pursuit of knowledge and the training of cadres in perspective rather than administrative techniques.

power. The same forces that had worked in Innis's favour during the second phase of his project worked against him as he moved on to the third phase. Increased dependence on the United States set up inauspicious conditions for Innis in the Canadian cultural milieu in general and in the academic milieu in particular. These ways of thinking strongly militated against his practical concerns, such as the maintenance of a generalist bias in the structure of the university (that is, no special 'centres' or 'institutes') and the hiring of Canadian, in preference to foreign, faculty.

This type of policy came to be characterized as parochial to such an extent that the more universalistic rationale that lay behind it, in Innis's case, was forgotten. When Innis's colleagues speak of his actions in later life as being 'crankish,' they are really suggesting that these positions were motivated only by personal whim. In other words, Innis's project in its third phase had become invisible to most of his peers. People did not contest Innis's stands on issues by stating their disagreement with the world-view lying behind his actions because, by this period, that world-view was completely out of step with the times. The very idea of expecting the development in hinterland intellectual institutions of a critical paradigm by which the measure of Western civilization could be taken was unthinkable. Academic reputations were made by adapting derivative methodologies to Canadian 'content' or by becoming recognized internationally for intellectual contributions that made no claim to a distinctively Canadian perspective.

At the institutional level, all of the developments that Innis abhorred were rampant. Increased emphasis was placed on the role of universities as centres for the teaching of administrative technique and for the solutions of practical problems. Structurally, continued division into ever more specialized and compartmentalized departments, institutes, and research centres meant that scholarship was becoming more and more concerned with meeting the immediate research demands of the public and private sector. Finally, any hope of developing an intellectual class with a distinct perspective was aborted by an unprecedented influx of foreign faculty accompanied by the transmission of metropolitan paradigms to the large numbers of Canadian graduate students studying abroad. As Innis recognized at the time of his death, Canada had moved 'from colony to nation to colony.'

The failure of Innis's project – indeed, the existence of the project itself – has not been thoroughly recognized because we live in comfort with the biases of a failed national policy. During the last ten years of his life, Innis pursued his project with the relentless realization that his time was short, that he was increasingly alone in understanding this project and that trends towards dependency were waxing. His sense of isolation during these years meant that he had to pursue, personally, all three levels of the project. This crushing workload in a highly charged political atmosphere had extremely negative ramifications for his intellectual contribution. The strident nature of his final essays can be understood only in the context of the tremendous anxiety Innis felt at the prospect of his failing efforts. In tracing the life and thought of Innis, we are describing at the individual level the failed potential of Canada as a whole.

But it would be misleading to leave the matter at this level. Innis was never a nationalist in the parochial sense. He believed that the continued vitality of Western civilization depended on the efforts of individual thinkers whose marginal position in relation to the great centres of that civilization allowed them to develop new critical perspectives. The intellectual synthesis produced by these marginal men from generation to generation represented the lifestream of Western culture. When Innis, as a marginal intellectual, found himself unable to complete and disseminate his new critical synthesis – his communications work – he was led to consider a radical new possibility: the final failure of Western civilization itself.

This book's re-examination of Innis's life and thought is appropriate at a time in which the prospect of such a definitive return to barbarism seems less and less exotic.

Part One

From the Margin, 1894–1939

CHAPTER ONE

The 'Herald' of Otterville, 1894–1913

I came from a class of ministers, professors, school teachers, lawyers, and doctors who lived in that essentially agricultural and commercial community ... This was a class which had no future as Canada became integrated into American capitalism. Nothing so much can drive one to philosophy as being part of a class which is disappearing.

George Grant, 1978[1]

Harold Adams Innis remains very much an enigma despite a significant amount of biographical work done on him. While the figure of the man has been sketched out, its relationship to its 'ground' has either been overlooked or oversimplified.[2] His wife, Mary Quayle Innis, identified precisely this area of the figure/ground relationship as one of the upsetting aspects of the biographical treatments of her husband.[3] She protested, for instance, that those attempting to understand Innis made too much of his rural background. 'After all,' she would remark hyperbolically, 'didn't everyone come from the farm in those days?'[4] By this she meant that Innis's brilliance could not be explained by projecting what would be an exceptional background for a young scholar today (growing up on a poor farm) onto an era in which this was the norm. The tendency has been to treat Innis in an exaggerated fashion as a character *sui generis*. This singularity is then explained by looking for what was atypical in his background. Further confusion sets in when, in an unconscious manner, 'atypical' factors are judged on the basis of today's norms.

I believe that Innis was not atypical but archetypical in that he fits the character of the intellectually precocious child from a relatively impoverished background who succeeds beyond the confines of his community. Let us consider his farming and family background, in particular, looking at the elements in his milieu that help account for the basic foundation on which he would build his reputation as a social scientist.

Farm and Family

Innis was born on a farm just outside the town of Otterville in Oxford County, southwestern Ontario, on 5 November 1894. The Innises had been in Canada since his great-great-grandfather, James, a Scottish soldier in the American Revolutionary War, had been granted land in New Brunswick as a Loyalist refugee. Innis's great-grandfather, Isaac, had moved to Upper Canada and cleared the land to establish a pioneer farm. His grandfather, Sam, was also a pioneer farmer, taking up and clearing land close to Otterville. Innis's father, William, the eighth child of Sam's family of twelve, was raised on a farm adjacent to the one on which Harold was born. Mary Adams, Harold's mother, was of more recently arrived Scottish stock, her father, William, having immigrated to Canada directly from Scotland.[5] William and Mary Innis were married in 1893; he was the first of the Innises not to start out as a pioneer farmer. With a loan from his father, he paid cash for a rudimentary but functioning farm, immediately north of his father's, that had already been largely cleared of bush.

The Innis farm is located on a gentle rise of sandy land north of Lake Erie, about one mile east of the village of Otterville. A map of the region from 1879 reveals a number of striking features.[6] Although it depicts pioneer farms cleared out of the wilderness in the generations immediately proceeding Innis's birth, there is no sense of haphazard development. These farms were the result of a government titling program on crown land. Surveyors preceded the farmers, with the whole region divided neatly into counties, townships, concessions, and lots in a regular checkerboard pattern that has survived virtually unchanged to this day. A standard farm would run between concessions (a distance of a mile and a quarter), with a frontage of 440 yards. This represented 200 acres per average farm. Lots were often divided in two lengthwise to make smaller 100-acre farms.[7] A look at the 1879 map, and others of similar vintage, highlights two additional features of the landscape of Innis's early years. The first is the nostalgia of homesick British colonial administrators who applied the place names of the old country to the wilderness of Canada: Oxford County, South Norwich

Township, Woodstock, and Port Dover, to name a few. The second is the land pride of the small farmers. Literally every plot and subplot is identified with the owner's name. They may have been relatively poor pioneers to start, but they were propertied in a way that would have seemed impossible to most of their ancestors.

The area itself embodied the phases of investment and development, from water to rail, that Innis would later write about. Otterville was founded because its immediate geographic features made it the ideal site for a mill, around which related industries could grow. (The town continues to be based on industry, even though the mix has changed over the years.) In Innis's childhood, the level of local artisanal skill and innovation was greater than what our more integrated economy requires. The town itself boasted about ten blacksmiths and carriage makers, and virtually every farm had a small forge for simple repairs not needing the full skills of a professional blacksmith.[8] William Innis's neighbour developed and patented a stump-pulling contraption that can still be seen in pride of place at the Otterville Museum. In a period in which transportation was quite difficult, local entrepreneurship and innovation flourished. Over four hundred patents were issued to individuals from Oxford County over the years, mainly for improvements in agricultural implements.[9]

One of the key milestones in the history of the development of the community was the building of the Port Dover and Lake Huron Railway, which ran from Port Dover on Lake Erie north through Hawtrey (where Innis's uncle's impressive 'General Store' building still stands in the now post-railway ghost town), Otterville, Norwich, Woodstock, and Stratford, where it connected with another line running through small towns all the way north to Southampton on Lake Huron.[10] This line was typical of the small feeder lines that criss-crossed the farmland of southwestern Ontario and represented a key element in an economy whose roads were primitive and whose vehicles were powered by draught animals. Without it, Innis would not have been able to complete high school. The line also provided an outlet for locally manufactured consumer goods such as 'Otter Sweepers' for floors, hand pumps, toys, screens, and corn planters produced by the Otterville Manufacturing Company.[11] To this day, the historical society in Otterville receives requests for information on the provenance of products such as particularly well-made piano stools that had been produced in Otterville for export all over eastern North America.[12]

The community was a mix of Loyalists and late Loyalists, mainly of

Scots/Irish descent but with a smattering of Germans, Quakers, and ex-slaves who had followed the underground railway north. The key element of social interaction was religion, and the region was dotted with small congregations, particularly Baptists, Methodists, and Quakers, that specialized in meeting the needs of rural flocks. People took their religion seriously in a context where going against a moral code could lead to the locally applied excommunication of 'shunning,' which could effectively end one's already limited access to social life.

To modern Canadians, Innis's background may seem inauspicious and relatively impoverished. But we should be careful as to what we mean by 'relatively impoverished.' The question is not what we believe impoverishment to be but what Innis's own community recognized as such. It should also be kept in mind that his community recognized its own impoverishment vis-à-vis what it expected its future economic potential would be.

But, even within a general context of relative deprivation, the Innis family felt, at least to some extent, especially impoverished relative to the surrounding community. William's farm, a half-lot of one hundred acres, was plagued by persistent drainage problems and was not one of the more productive in the region. The farm had already changed ownership twice before Mary Adams Innis arranged for her and William to move onto it after her marriage, a move that left them financially indebted for many years to her father-in-law. This feeling of relative impoverishment is reflected in Harold's earlier memories, which come from the difficult years of the 1890s rather than the brighter times of the pre-war boom.

The farm on which Harold was reared was a farm in transition. The last quarter of the nineteenth century had been a period of depression, and, while economic conditions were improving, the farms around Otterville were changing from wheat production to dairy and livestock. It was not an area producing a clearly established cash crop; only well into the future would tobacco and ginseng, which now flourish there, become dominant. The family was casting around incessantly for the right mix of crops, livestock, and activities that would allow them to break out of a subsistence existence. In short, it was a rich practical background for a future economic historian.

The strategy of farming in Innis's childhood was basically twofold. First, meet as much of one's basic needs as possible from the farm itself. The less one had to depend on 'store-bought goods' from the Hawtrey General Store, the better the household economy. Second, look for any opportunity to use the farm and its manpower to generate a cash income. Wood and wood products such as cordwood for fuel, or butter boxes and barrels are examples. Cash crops were key but fickle given the vicissitudes of the weather and the agricultural economy; they were also of relatively low value. Services such as hauling milk from the local farms to the cheese factory in Bookton provided a more steady income but required a heavy investment in time and effort.[13]

The Innis farm was to furnish Innis with a focal point for understanding the economic history of the periphery. The farm occupied a position that we could term the 'middle of the margin.' Being between the colonial metropoles and the new agricultural frontiers, it was extremely sensitive to the changing relationship of the former to the latter. To cite only the most obvious example: the extension of the transcontinental railway and the settlement of the west brought cheap wheat onto the market and necessitated the changeover to other types of agriculture in southwestern Ontario – largely corn, mixed livestock, and dairy production. So Innis grew up in a region buffeted by the geographical extremes of the empire. The farm also effectively taught the broader lessons of economic history, lessons that were not only theoretical. Because the farm was fairly marginal agricultural land, no effort was spared in determining which crops or supplementary products could be expected to raise overall productivity. Misunderstanding or misjudging the economic situation or the details implied by crop or produce changes meant not only a reduction in Innis's chances for an education but more labour and a less enticing diet (especially during the winter months) for the whole family. The serious nature of these issues is evident in the precise way Innis recalls the changing products of the farm during his boyhood years: fur, wheat, oak, white pine, cattle, hens, pigs, turkeys, related forage crops, fruit, dairy products, eggs, nuts, maple syrup, and gold all figure in his 'Autobiography.' Innis's necessary boyhood familiarity with these products and their consequences provided a secure base for his work in economic history, in particular, for the staples research of his early scholarship.

The farm was also in the middle of the margin in the way it was influenced by both Great Britain and the United States. Commercially, some staples were destined for Europe, others for shipment south. Demographically, the community's origins were partly British (some directly, some via other regions of Canada) and partly American (United Empire Loyalists and other later immigrants). Culturally, the religious denominations in Canada, especially the Baptists, were subject to strong and often contradictory pressures from their maternal organizations in both countries.

There is a third way in which Otterville during Innis's childhood was at the middle of the margin: in terms of time. The pioneer days had passed, but those who had lived through them were still present in the community. The last large stumps of the original forest were still being cleared, and the stump fences of the original settlers were being replaced by those of a more orthodox variety. A spartan but effective school system was in place, and the farms were beginning to move beyond subsistence. The farm surplus, coupled with increased mechanization, made it feasible for certain families to allow some of their children to pursue non-agricultural careers. The bizarre juxtaposition of three items in the fields of the Innis farm effectively summarizes the outstanding perspective of the area on the fortunes of empire: under the nut trees he found Indian arrowheads; remaining massive white-pine stumps still dotted the land; and yet traces of the railway that had been built to carry these trees south to the United States had all but disappeared.[14]

While the members of the young Innis family may have felt economically deprived compared with the surrounding community, they did not feel that way culturally. The importance of cultural refinement to Innis's mother is evident in the fact that, despite her far-from-secure background, she chose to spend a thousand-dollar family inheritance on a year's attendance at De Mille Ladies College in Whitby to the east of Toronto. Although her academic accomplishments were minimal, the sketching ability acquired at this finishing school made her the 'Wonder of Otterville.' In fact, this skill did far more than high academic marks to publicize in her community the benefits of education. Certainly, the young Innis was impressed by her ability to reproduce farm animals on paper.[15] (In fact, the school she went to was a far more serious institution than this anecdote would indicate. Established by the Methodists, it was one of the first post-elementary institutions for women. Consequently, many of the remarkable 'first generation' of women who struggled for professional qualifications passed through it. It continues to exist as Trafalgar Castle School.[16])

As for his extended family, Innis descended from a large group of ancestors notable for their Scots background, their pioneering efforts, and their dispersion. The family history in a very direct sense was caught up with the history of the colonies. The sole non-Scot among his grandparents was a Pennsylvania Quaker who soon became a Baptist and was assimilated into the environment of the Innises, the Mac-Donalds (the maternal grandmother's family), and the Adamses (the maternal grandfather's family). The Innises were not new arrivals but had a settler history of several generations in the New World. Their family itinerary is a study in the expanding frontier. Innis mentions Newfoundland, New England (Boston), New Brunswick, southwestern Ontario, Michigan, Manitoba, and the gold fields of the west coast as stopping points along the way for various elements of the family.[17] By the time of Innis's generation, success in the New World could be measured by the family's rapidly increasing size rather than in the advancement of individual members into positions of prestige, power, or wealth. His two grandmothers raised twenty-nine children between them.

Among these individuals Innis found an extreme diversity of religious beliefs, types of employment, and levels of education. They provided a much wider latitude of practical examples for his life's plans than did his immediate family, which was devout and literate but farm-based and relatively unschooled. We should be on guard against assuming that Innis's references to individuals in his 'Autobiography,' written at the end of his life, mirror accurately his attitude towards them as a boy. However, it is evident that, when certain opportunities implying a change in plans or attitudes presented themselves to him, he could always assess these by finding an example among those relatives he respected. For instance, although his immediate family was fervently Baptist, Innis eventually lost his faith. In so doing he came to appreciate the agnostic eccentrics within the family such as his Uncle Charles and his Great-Grandfather Isaac. In a similar manner, Innis was obliged to come to an early appreciation of the oral tradition and non-literate skills in a large family, the majority of whom were illiterate or semi-literate agricultural workers.

The further Innis was removed from the family farm by his studies, the more he would have felt the deprived nature of his childhood background. On the one hand, there was the familiar tendency of the relatively deprived to focus on sections of the community who were even poorer. In this case, they were designated by the local community as being of 'low class,' which Innis suggests was defined not in terms of the absence of material possessions but by the presence of head lice. On the other hand, he notes his difficulty in participating in high school sports on account of a lack of equipment.[18] Anyone who has experienced the childhood shame of showing up for a game with inadequate equipment can appreciate the intensity of feeling experienced over such a seemingly insignificant matter as, in Innis's case, being obliged to use old-fashioned strap-on 'spring' skates to play hockey.[19]

Innis was not the first of his family to break out of this pioneer mould; he benefited greatly from his mother's brush with formal education and was guided by other members of his extended family. He was influenced in a general way by his Uncle Sam's transformation into a local merchant. Another uncle, Harvey, who had started out in partnership with Sam running the Hawtrey store, went on to homestead in Manitoba, ran cheese factories there and in Ontario, and ended up training as a medical doctor and practising in Grand Rapids, Michigan. He seems to have been the first professional in the family. Contrary to what one would expect from the metropolitan counterpart of this successful individual, he stayed in touch with the illiterate and semi-literate members of the family, who were still, by and large, farmers. He later influenced (and funded) another family member's training as a medical doctor.[20] Yet another uncle, Mary Adams's brother William, was brought in to persuade Innis to continue at McMaster University after his disastrous first term; at the time, William was himself studying medicine at the University of Western Ontario.[21]

Great Expectations: Innis as a Type

Innis had four siblings, Lillian, Hughena, Sam, and William (who died in infancy), but from the start his parents seemed to have something special in mind for him. Despite being the first-born son, he was given not an ancestral name but one unique in the history of both sides of the family.[22] Both his parents treated him from the start as a young man of great expectations. Not surprisingly, given the heavy Scots background of his family, this was couched in terms of advancement through education. It appears that his father wanted him to become a teacher rather than take up the hard life of a farmer. His mother, on the other hand, seems to have favoured a future for him in the Baptist clergy. (Both parents, however, were relatively circumspect in passing on their hopes for his future to Harold, for, true to Baptist belief, they knew that he would have to declare his own 'calling' in life when he reached adulthood.)

From the beginning, their inquisitive, shy, hard-working, and awkward son repaid their faith in his advancement through education. He attended S.S.#1 South Norwich, a typical rural one-room elementary school located just down the road from the farm. Harold was a fast learner and took advantage of the rapid advancement possible in a one-room school by following the lessons of the older children. At the age of eleven, he qualified in the provincial exam for entry to secondary school.

In the fall of 1906 he entered Otterville High School, a 'continuation school' that offered the first few years of post-elementary education but not up to junior matriculation level. (The birth in 1905 of Sam, a second son who could take over the family farm, may have confirmed his parents' belief that Harold should continue his education.) This was a government school and his chief memory of it seems to have been his exclusion from organized sports because of cost: 'Consequently, a great deal of time was spent in piling up leaves in the fall, in erecting forts and throwing snowballs in the winter and in any sort of activity in which numbers could participate but which did not involve expense.'[23]

Innis's educational path became increasingly challenging in terms of the distance he had to travel to get to school. The Otterville school was two miles from the farm, which meant a hike of four miles daily no matter what the weather, and on top of his farm chores. This, added to the fact that he had neither the time nor the family income to permit his participation in team sports and social activities, helped to mould his personality as a studious loner.

By the end of the 1907–8 school year, he had completed his courses at Otterville. The nearest institution offering high school senior matriculation was Woodstock Collegiate Institute (WCI) twenty miles away. This entailed significantly increased costs and time in pursuing his education. Nevertheless, his family decided to continue supporting his educational endeavours, which were now focused on a career in teaching. They could do so because the farm was generating more of a surplus than in earlier years; the fact that WCI was a Baptist institution probably also influenced their decision. A residential school and a department of McMaster University, WCI advertised its 'Large Library, two Literary Societies, Reading Rooms, Fine Grounds. Fees very Reasonable. Teachers all University Graduates.' Innis's parents may have been more taken by one other item in the local advertisements: 'Pupils Arriving by Train ... from the South by the GTR morning train from Port Dover and intervening stations arrive in ample time to begin work and can return home on the evening trains.' The return fare for students was ten cents daily.[24]

It was an exhausting life. Every school day from September 1908 to June 1912, Harold walked two miles to the railway station in Otterville, took the 7 A.M. train to Woodstock, and walked a mile to the institute. The itinerary was repeated in reverse in the late afternoon. He did well at Woodstock, taking the final two-year preparatory course for teacher's training in one. Still, the toll was heavy. Who can imagine a teenager in our time walking four thousand plus miles and taking well over one thousand train trips to obtain a secondary school education?

During this period, Innis also continued his lifelong efforts to educate himself outside the formal classroom experience. He sought out the few 'thinkers' in his local community, and rode to school in the smoking car rather than the regular coach in order to have better conversations with the 'characters' on the train. In a sense, these habits confirmed a conviction in the value of the oral tradition before he was even aware of its meaning. Throughout his life, he took great delight in stories. In being daring enough to ride to school with the 'characters,' Innis established his habit of gleaning whatever information he could from people he met in the course of his travels. His later emphasis on the importance of the oral tradition was one way of recognizing the significance of the smoking-car milieu to his understanding of the world.

He knew that the next steps in his training would require leaving the farm to pursue residential education, and that it would be costly. Therefore, he took the 1912–13 school year off to build up his resources. In the fall of 1912, he started his teaching career by returning to S.S.#1 South Norwich as the sole teacher; he was eighteen years old. His father was a school trustee, which undoubtedly helped him secure a post for which he was not yet qualified. It was a less than successful experience and led him to abandon his plans for teacher training. When his teaching stint ended in December 1912, he continued to help on the farm, work in his uncle's country store in Hawtrey, and trap muskrats for supplementary income.

Given his farming background, Innis is better understood not as atypical but as a type: a young man of great expectations who makes good as a colonial or peripheral intellectual without having to complete a break with the rural community he left behind. In a way, Mary Quayle Innis is correct. Many of his generation of scholars did indeed come from a farming background, and, because this was as common for his generation of Canadians as it is rare for ours, one needs to recognize that it was a specific type of farming background. These farms were situated in a European-settler colony, a particular context in which a large influx of European settlers overwhelmed and marginalized an indigenous population, confiscating their territory and turning it into a commodity through treaties and titling technology.[25] This is quite distinct from a rural youngster 'making good' through education in a developed economy such as Britain, or in a Third World context. The divisions of class, race, and language simply did not exist at all or to the same extent for Innis as they did (and do) for similarly inclined youngsters located in the latter two contexts. Innis represented at a most general level the following character: the child from a relatively deprived background who is designated at an early age as the one person in the family who will make good. The mystery surrounding Innis' given name is relevant to this designation.

It is part of the oral tradition surrounding the life of Harold Innis that the original name given to him by his mother was 'Herald' in anticipation of the great things he was expected to accomplish. This anecdote of the name 'Herald' appears in virtually every biography of Innis without any indication of whether it is true or not (attaining, in effect, the status of an 'urban legend).' For this reason, I think that it is important to spend significant time here getting to the bottom of the story.

The most vocal proponent of the 'Herald' story has been Innis' younger brother, Sam. According to Sam, 'Harold's mother, Mary Adams Innis, was the influential one of the parents, the driving force. She doted on the *Family Herald* and named her eldest son "Herald," which was noted on his baptismal certificate.'[26] She did so because, as she was growing up, she had had a vision. 'The vision was that Mary Adams would become the mother of a famous person.'[27]

Sam Innis was a 'character,' a relatively uneducated storyteller who never left the Otterville area and never lost his strong Baptist faith. Until his death, Sam continued to insist that he 'got several hints' from Mary Adams to substantiate the 'vision' story, which many in the Innis family regard as a tall tale invented by Sam. For instance, Donald Innis, Harold's eldest son, was upset that Sam's thesis might be taken seriously by some. He wrote off the issue of 'Herald' by saying that his grandparents simply were not aware of how to spell 'Harold'; he could see no connection between his grandmother, 'a very sensible, down-to-earth person,' and this purported 'vision.'[28]

All the contemporary documents I have been able to locate seem to support Donald's view. The minute book for the school trustees of S.S.#1 South Norwich shows that William Innis, Harold's father, who served as chairman and secretary of the school trustees, had a great deal of difficulty spelling names.[29] Moreover, none of the contemporary documents I have located denote Innis as 'Herald' (although several render his family name as Innes). For instance, his original 1894 birth registration and 1916 army-enlistment forms both render his first name correctly.[30]

And yet Innis himself gives the story credence. Writing in his 'Autobiography' in 1952, he says, 'Both my parents had apparently from the beginning thought of my education and I have always been curious as to why they should have registered my name as Herald and that the change to Harold was not made by my Mother until some time in the First World War.'[31] This is an odd sentence for, on the one hand, he clearly recognizes the link between 'Herald' and the idea of getting ahead through education by putting them together in the sentence. Why then is he 'curious' as to the rationale behind the use of 'Herald'?

I can think of two explanations. There is perhaps no better mechanism to expose a child to deep discomfort at an early age than to give him an odd name. Although the issue is trivial in retrospect, it was probably taken seriously by Innis as a child. The extent to which the Innis family was considered unsophisticated by others would have been forcefully underlined to the young boy if his school teachers had corrected, unconsciously or not, the name that his parents had given him. Maintaining that he did not even understand the logic behind their choice of 'Herald' may have been Innis's way of asserting that he was well removed from such unsophisticated roots.

A far more likely explanation is that the 'Herald' moniker did indeed mystify Innis. It is quite possible that Mary Adams chose the name because she had quite definite expectations (or 'a vision,' if you will) of what her son might turn out to be. I suspect that she had the clergy in mind. She would not have told her son the details of these expectations because, as a Baptist, she would have wanted him to make his own decisions as to the pursuit of a career. All that Mary Adams could do was to orchestrate a supportive environment for her son. She did so until the First World War aborted her hopes for her son's future by offering Innis an alternative; this would explain why she did not relinquish her attachment to 'Herald' until Innis went overseas.

Mary Adams was still living in 1952 when Innis wrote his 'Autobiography.' Why did she not simply put the matter straight? She came from a time when personal matters were kept private. Her discretion is evident in Innis's observation that he learned why his mother had married his father only through a comment she made fifty years later to Innis's youngest daughter: William Adams was chosen because he was the most devout of her suitors. This meant that he was not only the best bet as a husband in a time when alcohol was a serious problem but, presumably, the most appropriate choice as a partner for the realization of her vision.[32] Innis's father effectively reinforced her plans for their son's advancement, but less for visionary reasons than for practical ones. Innis says of him, 'My father expressed his general views by saying that I should try to be a teacher and in this way avoid the arduous labour to which he had been exposed all his life.'[33]

The conflicting interpretations by Innis's brother and son of the 'Herald' story reflect their different backgrounds – Sam the story-telling farmer and Donald the rational urbanite. Both attitudes simplify the character of Mary Adams. I do not see any contradiction in a strong-minded, down-to-earth woman articulating her hopes for her son in terms of a personal vision; it is in the best Protestant tradition of a 'calling' or a 'mission' in life predestined by God. However, her 'sensible' approach in passing on the impetus to personal betterment, rather than the 'vision' itself, to her son meant that the issue remained puzzling, even to Innis.

Donald's explanation that his grandparents simply misspelled 'Harold' seems unsatisfactory. Both the Innis and the Adams families were aware of family genealogies going back many generations. The tradition was to name male children after ancestors, with a strong preference for biblical and/or Scottish names such as Samuel, Isaac, Moses, and William. There is no 'Harold' in the family tree previous to Harold Adams Innis, and Mary and William apparently had no other motivation to use that name for their first-born son.[34] On the other hand, Innis's parents would have been aware of the meaning of 'Herald' both from biblical sources and the name of the local newspaper, the Otterville *Herald*, and would have intentionally applied that meaning to their child. On registering the birth, they would have been told that the commonly used 'Harold' was spelled differently from the name with the meaning they wished to attach to their son. They would have been embarrassed at this indication of their apparent ignorance and would not have wanted their son to be similarly embarrassed as he made his way through life with a name everyone thought was misspelled. How else are we to explain the persistence of this family myth, its lack of documentary evidence, and the sudden appearance of the name 'Harold' in the Innis family tree?

Donald was right, but in a way that inadvertently supports Sam's story. His grandparents did not waver on the meaning of the name, even though they bowed to social pressure in the spelling of it. Sam's 'vision' story is also correct in that we note Innis's repeated use, during his communications phase, of a favourite biblical paraphrase of his mother's, 'without vision, the people perish.' The banal context of childrearing in rural southwestern Ontario to which this lofty observation applied is apparent when the citation is placed in its biblical context:

> The rod and reproof give wisdom: but a child left to himself bringeth his mother to shame.
>
> When the wicked are multiplied, transgression increaseth: but the righteous shall see their fall.
>
> Correct thy son, and he shall give thee rest; yea, he shall give delight unto thy soul.
>
> Where there is no vision, the people perish; but he that keepeth the law, happy is he.[35]

In short, from before he was born, Innis was cast as a 'precocious child character' by his parents – a young man from a poor background but of great expectations – a 'herald' of better days to come.

Academic excellence is the most prevalent means by which such a character type makes good. In a positive sense, the person on whom great expectations are put exhibits a high degree of native intelligence, a tenacity of purpose, and a propensity for unremittingly hard work. On the negative side, the circumstances of such a life dictate a cloying sense of inferiority stemming from the impression of forever having to make up for the lost time of a disadvantaged past. This is a complex that, as a matter of self-defence, is usually well hidden but only rarely overcome.

Compared with young scholars who had come of age in the metropole, all the while adapting to existing paradigms, Innis was able to write things of significance in his early years of scholarship because he was writing about himself or about his actual experiences and observations. One suspects that the background to great writing is far more concrete than is usually credited. While visionary or mystical writing may fall outside this observation, philosophical work does not. By this I mean that, although its substance may be abstract, the long process of familiarization with an existing paradigm before a genuine contribution can be made is anything but. Innis, largely because he was not brought up in the metropole, missed the opportunity for a thorough grounding in an existing intellectual paradigm. Under these circumstances, he made a contribution in the only other way open to him. He wrote what he had come to know directly. The powerful faculty he developed for pattern recognition often masks the extent to which his abstractions were rooted in what he referred to as 'dirt' experience. For example, it is often noted that, without his experience on the traplines, Innis would not have been able to write the chapter on the characteristics of the beaver that serves as an introduction to his fur-trade study. What is less often noted is that his concentration on the importance of the characteristics of staple products necessitated a more wide-ranging insight.

This prerequisite understanding is outlined in his concluding chapter of *The Fur Trade in Canada* under the heading 'The Importance of Staple Products':

> Fundamentally, the civilization of North Americans is the civilization of Europe and [my] interest ... is primarily in the effects of a vast new land area on European civilization ... People who have become accustomed to the cultural traits of their civilization ... the social heritage – on which they subsist, find it difficult to work out new cultural traits suitable to a new environment ... The survivors live through borrowing cultural traits of a civilization suitable to the new environment, through adapting their own cultural traits to the new environment, and through heavy material borrowing from the peoples of the old land. The process of adaptation is extremely difficult ... A sudden change in cultural traits can be made only with great difficulty ...[36]

This early recognition of the importance of cultural factors is a theme that runs through Innis's work. While the theme is grounded in Innis's immediate childhood experiences, it is a profound insight with many different levels of meaning. At one level, it is a point of departure for a deep understanding of the social psychology of other 'settler states' such as Australia, Argentina, South Africa, Israel, the United States, and Canada. At another level, it is an introduction to Canada's fur trade. A third level is less often recognized: the intensely personal autobiographical statement of a farm boy from southwestern Ontario.

Yet Innis's point of departure for the theory behind the staples approach is precisely his point of departure as a young man with great expectations. His boyhood was one in which the general and the individual experience were found in close proximity. The idea of getting ahead, represented by his character, was understood by the majority of the community in which he lived. To put it another way, the 'precocious child character' assumes the dimensions of an 'everyman' figure only in a European-settler colony.[37] Nowhere else does the majority of the community share the belief in the imminence and inevitability of upward social mobility.

By and large, the migration of individuals to a new country took place as a result of coercion rather than preference. Or, perhaps more accurately, the immigrants did not 'belong' to their new milieu in the way the autochthonous populations of the metropolitan countries and the colonies belonged to theirs. The settlers did not prefer the New World to the Old so much as they did their position and potential in the New World compared to their experience in the Old. The strategy they adopted was to replace their economic marginalization in the metropolitan society with a cultural marginalization (of temporary duration) at the periphery of contemporary empires. They were the survivors of Europe who opted for immigration and a belief, not just in individual social mobility, but in group or societal mobility. The alternative with which they were faced was that of being crushed into a lumpenproletarian state if they remained in the metropolitan societies. For many of these people, to stay in the Old World entailed absolute psychological and material impoverishment, with early death as the likely consequence.

Because of the close coincidence between the aims of individuals displaying the precocious child character and those of displaced Europeans, the character took on an 'everyman' quality only in European-settler states. In metropolitan countries, the impulse for a precocious child character almost invariably comes from an anomalous personality within the immediate family. In these older societies, the advent of this character is not well understood and therefore is often actively opposed by the class in which it occurs. For example, one can imagine the virtual impossibility of its developing in the European peasantry.[38] While a strong close relative (such as Innis's mother) may serve as the origin of the impulse in a European-settler context, the important difference is that her efforts will be generally comprehended, accepted, and assisted by the community as a whole.

The extent to which the precocious child character was generalized in Innis's milieu is evident in the extent to which resources were directed to the construction of schools as soon as the problem of subsistence was overcome.[39] (Indeed, it was recognized in the setting aside of land for schools during the original surveying exercise.) The results are apparent in the truly remarkable number of scholars who emerged from this rural Ontario background. When we trace why this is particularly so in rural Ontario, we must follow the skein back to the Old World cultural propensities of the Scots.[40] Nowhere has the idea of social betterment through education become such a national passion as it did in Scotland. Certainly, at the time of the great Scots migrations of the early nineteenth century, the Scottish public school system was one of the best in Europe. Although it would take another book to examine the issue, the case of Scotland and education seems to support the suggestion that the character we are discussing becomes a social myth in marginal societies containing large numbers of people who have been forced through a long period of attrition into the position of becoming emigrants/immigrants. Innis was of pioneer Scots stock on both sides of his family and this, coupled with a general sympathetic community milieu, cannot but have helped to reinforce his mother's ambitions for him. Without the support of his 'colony,' Innis almost certainly would have given up his university career. We are used today to scholarships and state support for education. Innis, however, faced daunting obstacles to realizing his potential in the first quarter of the twentieth century. At the time, the only means of support were his family and his employment savings. During the period in which he was attending McMaster (1913–16), the Innis farm was finally generating enough surplus to help with his support; however, this was always supplementary to his own earnings. He would probably have abandoned his studies (out of his sense of pride, if not out of necessity) had he not been thrifty with his 1912–13 teacher's wages.

Typically, when it occurs in a metropolitan environment, the precocious child character needs support from a patron, usually found in the person of someone from the upper class who takes an interest in the advancement of the precocious youth. Financial backing is required, but what is more important, in a highly class-delineated society, is that the upwardly mobile require an introduction to the manners and personalities of the upper class from those already initiated into its circles. Innis and many other similar cases in the European-settler colonies did not have this type of patron. Having succeeded on their own or with the aid of their immediate family, they were far more likely to demonstrate a populist frame of mind. The character in the metropolitan environment develops in proximity to the rich cultural traditions and institutions of the centre. The problem in this case is not the lack of social resources but the unlikelihood of those from a lower-class background gaining access to these resources.

In the European-settler colonies, the school was often the first public institution to be established by general subscription after the farms had begun generating a surplus. In Upper Canada, land was reserved for schools in the surveying stage before the arrival of the first generation of pioneers; this is what one would expect of a people rebuilding their European traditions. Obviously, the general state of educational resources on the periphery is far more precarious than in Europe. While a particular paradigm or world-view may have an entirely self-contained institution dedicated to it in the metropole, it may be represented only as a 'trace element,' such as an idea in the head of one individual, in the peripheral European-settler community. An established country's or metropole's trade union movement with its associated cooperatives, self-improvement societies, and cultural organizations, for instance, may be reduced in the periphery to one individual who is viewed by the rest of the community as having eccentric beliefs. This individual's only connection with the original paradigm and its institutions may be a subscription to a left-wing journal published far away. Nevertheless, he remains an educational resource of a type not available in a colony whose population is not composed of European settlers.

Moreover, given the vacuity on the periphery of European civilization, it may be that resources available for any one paradigm are not sufficient to sustain the interest of an inquiring mind, even if the same paradigm is rich enough to stifle an equally precocious youth at the centre. The character at the centre may, from an early age, be preoccupied with 'sorting out' the intricacies of a paradigm. Yet the same type in a European-settler colony will spend all his time 'searching out' trace elements of the great paradigms in his milieu (in the form of individuals, groups, clubs, reviews, and the like). In other words, the basis for Innis's lifelong approach to his scholarship was laid early in the eccentric characters and nascent institutions of Oxford County.

Pattern Recognition

In this context, one of the key elements involved in the early education of the precocious child is the development of the ability to recognize patterns. This is particularly so in an agricultural community, and especially one buffeted by severe seasonal change. Innis's mature methodology should be seen, therefore, as being overdetermined by the circumstances present during his formative years in a European-settler colony. The germane characteristics of this methodology are: a style of research that covered a wide range of metropolitan paradigms; an intense scepticism towards the conclusions of any particular paradigm; and a reliance on the recognition of patterns not restricted to any particular paradigm. He could have developed this approach only because a wide variety of elements of the great paradigms of Western civilization had been present in trace form in the milieux of Otterville, Woodstock, and McMaster.

The Innis family farm, in particular, provided Harold with a solid perspective on imperial dynamics and an exposure, at least in small doses, to some of the great paradigms of Western culture. It sharpened an aptitude for pattern recognition. For instance, Innis recalls: 'We were struck with the differences in the size and shape of the nuts of different trees and, strange as it may appear, by looking at a hickory nut one could almost say what tree it came from. The same difference was evident in the chestnut trees, and indeed, in the sap which was collected in the spring. By tasting the sap it was possible to say in rough sense what tree it came from. But these details were simply an indication of the character of the work which compelled one to recognize such differences.'[41]

There were two quite distinct tasks calling for pattern recognition around the farm. One had to do with understanding the interrelation of all the major components bearing on the farm's main products. This was a complex system, which, in the case of the changeover to dairy farming, involved not only cows but pigs, turnips, mangels, new field-cultivation techniques, increased demand for labour, cutting machines, corn, soil types, pumpkins, silo construction, and winter dairying, among others things. It is not unreasonable to suggest a relationship between Innis's holistic viewing of the myriad details of farm life and his later propensity for identifying underlying tendencies in the massive details of the staples trades or the media histories. Once the dominant mode of agricultural production was worked out, the second pattern-recognition task was the spotting of what Innis would later call the 'unused capacity' of the farm. This was primarily his mother's work and would lead to innovations designed to raise, though marginally, the productivity of the farm. The making of maple syrup using cleared stumps is one example. The use of the whey by-product of cheese production as pig feed and the introduction of poultry to forage on barnyard debris are two others. On a much larger scale, Innis would later use the same concept of the unused capacity of a process to trace the reasons for the advent of new innovations in technology and production.

Above and beyond any specific ability, the farm background provided a different conception of the sensate. The indelible mark left by the farm at such a fundamental level determined that Innis would remain basically a farm boy, awkward and shy, despite the acceptance of his work in more sophisticated milieux. It was also this background that allowed Innis to comprehend that human appreciation of such a fundamental thing as the passage of time was not absolute but would change according to the principles of organization of a society. This is one attribute that Innis, as the precious child character, could not leave behind. Despite his great accomplishments, he was still at ease with country people in a way that was not possible with his colleagues or even with his children. He would visit the farm once every six months or so, arriving unannounced, and the farm people would reciprocate by visiting Toronto about once a year. Innis's youngest son recalls: 'To me it was like Mars. I could never understand the people. They never thought that his unannounced visits were odd. They just knew 'when the snow melts Harold will arrive'. They had a completely different sense of time.'[42] Having a different sense of time should not be thought of in too abstract a manner. In day-to-day life in the city (outside academic routine), it meant that Innis found it difficult to function smoothly. His own children felt embarrassed by his tremendous awkwardness in public. He did not spend much time with his children, so it was a special occasion when, following doctor's orders to relax, he took them out to the cinema. But the outing was marred in the eyes of his son by his fumbling with his wallet in the line-up at the box office. He was an accomplished intellectual at the peak of international recognition and yet, at the same time, he remained 'a country boy who had not solved the problem of making change in the big city.'[43]

Innis was fortunate in escaping a regular pathology of the European-settler colonies: a tendency to denigrate the values of non-European traditions.[44] In Innis's case, there was, if anything, a bias towards valorizing the cultural contributions of indigenous peoples. There were two reasons for this. First, the autochthonous population offered no threat, being entirely absent (except in the form of abandoned artefacts) from his boyhood milieu. Second, the history of the milieu was such that the European settlers picked up many indigenous skills in the course of adjusting to the new environment.

The main difficulty for the character in the European-settler state is the phenomenon of deracination that was the consequence of the cultural vacuity of these colonies. Training of the individual involved his moving to colonial metropoles and eventually the metropolitan countries themselves. For many, there was no return trip either geographically or psychologically. The minimization of personal strain often led directly to the joining of a metropolitan institute and paradigm. The only option was to return to the European-settler colony in order to contribute to the long, slow, and collective process of strengthening the cultural traditions and institutions of the periphery.

Two elements should be noted in the case of the precocious child character that goes away for training but returns to his home community. First, the individual who feels less at ease with the metropolitan paradigms is more likely to adopt this strategy. Second, this decision does not avoid the problem of psychological deracination. It is true that, in the more socially levelled environment of the New World, the home community may be able to remain more open to the return of the successful character. However, except for visits, no real return is possible. Even in the European-settler colony, the nature of intellectual or academic work is such that it becomes centred in the colonial metropoles. It is to these halfway houses that the individual returns. For the character maturing under these conditions, then, the transfer from a rural to urban milieu remains a more important axis to his success than the change from working class to bourgeoisie. This change from country to city was one of the most profound influences on the personality of Harold Innis.

Self-Improvement

Besides the overall perspective on empire to be gained from Otterville, there were myriad extracurricular opportunities through which an individual might improve himself. By the nature of the community (semi-literate), much of these opportunities involved oral communication. In church and politics, the early years of the century were an era in which grand oratory was much praised. Such oratory must have made an early impression on Innis since it became one of his teenage obsessions; later in life, he could recall election campaigns going as far back as 1900. But oratory was only one aspect of the oral tradition permeating the environment; during an era predating radio and television, conversation was relatively more important. Innis seems to have been intrigued by eccentric personalities in the community and it is likely that, by pursuing discussions with characters such as the two local agnostics, he was able to pick up a far broader range of ideas than would otherwise have been possible.

As a boy he also participated in discussions of a more formal nature at the Baptist Young People's Union and the Scouts. Many of these were no doubt insignificant, but others lead me to believe that this sort of function played a far more important role in the socialization of young people than currently is the case. We know, for instance, that one of the Scout troop meetings involved the presentation of a mock Declaration of Independence, which led to the replacement of the 'dictatorship of the Scout Master' by a 'parliamentary' form of government: a rather bizarre but typically Canadian combination of 1650, 1776, and 1917.[45]

Even the printed resources available at the time would serve far more frequently as a supplement to the oral tradition than is now the case. The Bible was more often recalled and recited from memory by Innis's father in the course of his day's work than it was read. Knowledge of the book meant the ability to repeat stories or maxims word-for-word in the appropriate situation. The role of Mao's Red Book during the Cultural Revolution or the Koran among more conservative Muslims is closer to that played by the Bible in Innis's youth than any example from Canadian society today. Not surprisingly, therefore, teaching also depended largely on the oral techniques of rote learning. Innis even invested some of his hard-earned income in a subscription to the *Globe*, a Toronto newspaper, primarily to gain access to printed facsimiles of the magnificent speeches of Sir Wilfrid Laurier, the Canadian prime minister.

It is impossible to obtain an accurate account of the written, non-academic material that would have been in the Innis household during Harold's childhood. However, an overall impression can be obtained through his statements and the bits and pieces of material that have been saved and deposited in the archives. Certainly, the Bible was the fundamental book. Even in his youth, Innis seems to have been especially interested in its historical aspects. His copy had a particularly fine set of maps indicating 'The Distribution of Nations after the Deluge,' 'Nations of the Old Testament,' 'The Plan of Solomon's Temple,' 'St. Paul's Journeys,' and 'Prevailing Religions of the World.' The significance of these to Innis may be indicated by the fact that, although the Bible has disappeared, its map pages are found in his papers.[46]

Beyond the Bible, the secular literature available to Innis would have been heavy on nature stories, judging by his extensive field notes. Also, Innis cites his reading of the 'realistic' nature stories of Ernest Thompson Seton and their portrayal of the suffering of wild animals as the reason for his abandonment of muskrat trapping. His later mania for the universalist historians can in some sense be traced back to another literature type that was popular on the farm: the universal almanac. At the turn of the century, the *Family Herald* was more than just a farm newspaper – it was *the* east-west cultural link in Canada. Through institutions such as the Maple Leaf Correspondence Club, it served the function of the Canadian Broadcasting Corporation long before that institution (or, for that matter, its underlying media technologies) existed.[47]

This variety of material was found in the farm milieu of the 1890s quite naturally, not as an exotic element. Even unpopular ideas were present in written form in such a way that they *belonged* in the community. Innis spent much time with a local agnostic whose family nevertheless remained staunch supporters of the Baptist Church. One imagines that this character had to be quite forthright defending his eccentricity in the face of local religious fundamentalism. In this situation, he seems to have become a kind of evangelist of progressive thought who placed great stock in the work of Ernest Haeckel, the chief German defender and propagator of Darwinism.[48] Innis was also introduced to radical socialist ideals by this same individual through a publication called *Cotton's Weekly*.[49] The history and content of this news sheet is a concrete example of the Gad Horowitz/Levis Hartz thesis of the proximity of the conservative and socialist traditions. It also helps us to fathom how such material could be present in the milieu of Innis's youth.

Originally, *Cotton's Weekly* started as the regional newspaper for the Brome County area of Quebec's Eastern Townships. Its *raison d'être* was covering the local news of small towns such as Sutton, Cowansville, and Farnham. This was supplemented by material of a religious nature and by a full page devoted to temperance news. It was Christian and prohibitionist but became increasingly socialist in its orientation, carrying articles on the latest successes of Eugene Debs and the American socialists, on German socialism, on farmer's unions, and on child labour (this last by Jack London.) It was militantly in favour of women's suffrage. The popularization of Marxist economics was also one of its regular subjects, and it would often produce what would seem to us unorthodox combinations of topics such as 'Adam Smith on

Strong Drink' or 'Christ the Socialist.' It was vitriolic in its criticism of the free-enterprise system and saw government intervention in the economy as a panacea. 'Government ownership, when properly managed, means increased service at diminished rates with good prices for labour.'[50]

What is most striking about the material is the popular format in which the ideas are presented. It was just another small-town newspaper, except that it contained items such as the following, which Innis would have found unique and interesting:

> Why Study Economics?
>
> Now Political Economy may not amuse you, but if you want to enjoy life, if you want to get the full value of what you produce, if you want to act intelligently in all things that pertain to yourself, your family, your class, if you do not want to be eternally fooled by the politicians, you must study out this subject, study it just as diligently as you would study book-keeping or shorthand or medicine, for not until you do, not until your fellow workers do, you and they will continue to be wage slaves.[51]

Thus, even before Innis left the farm, he had been exposed to what was on the way to becoming the official paper of the Socialist Party of Canada.[52] From this example, we can see how Innis sought out those trace elements of great paradigms that existed in the relatively culturally rarefied milieu of the Canadian hinterland. In this case, one individual out of the entire community carried and disseminated both Darwinism and the heritage of the intellectual left.

Church and School

The church had a great impact on Innis's early life. His grandfather had converted to the strict Baptist faith, and Mary had chosen to marry William largely because he demonstrated the zeal of the newly converted and would, over the long term, exhibit the values of faithfulness, hard work, and thrift.

The Baptist Church flourished in farm communities because its radically decentralized congregational structure provided a sense of community to widely scattered and relatively isolated farm families. It was one of the few organizations of what we now call civil society to exist in the rural context. Innis grew up in the church, attending services every Sunday and interacting with other youngsters through the Baptist youth organization. From his childhood notes, we know that from an early age he struggled with a strict moral code in a manner that seems foreign to our urban society over a century later.

The choice by the mature and informed individual of salvation through Christ was a key element of Baptist belief. This was interpreted to mean that, while children might be reared in the church, they could not make a legitimate profession of faith and be baptized until they reached maturity and actively chose the faith. To this end, Innis and other young Baptists were encouraged to expose themselves to alternative (that is, non-Baptist) ways of looking at the great questions of life in order to be in a better position to make a genuine profession of faith as an adult.

It would be impossible to separate the religious from the scholarly elements of the hopes for betterment instilled in Innis by his devout mother. A righteous and disciplined upbringing was looked on as the key to individual social advancement and salvation, as well as the secular and spiritual renewal of the community as a whole. The link between the individual and communal aspects of this project was the Protestant conception of being 'called.' The maturing individual, in responding to such a call, was not only deciding his own life's vocation but accepting a mission within God's vision of what was good for human society. In this sense, any career could be a godly career, the key being the correct recognition of and response to the call. The terrifying responsibility inherent in these beliefs stems from the radical indeterminacy of the calling. Neither the hidden potential nor the future mission of any particular individual was necessarily apparent to the community in which he lived. This posed a particular problem for those few individuals called to be the thinkers or visionaries of the community. The risk was always that either they would not recognize their mission (it being so exotic) or that the community itself would reject their efforts for the same reason. The consequences would be the material and spiritual stagnation of the community. In his later years, this mode of thought became transformed into Innis's deep concern with the extension of the educational system to the farthest reaches of society in order to identify and foster the abilities of gifted individuals wherever they occurred.

Backwoods religion was first and foremost a means of getting together, of breaking down the isolation of the frontier. The sociologist S.D. Clark, who worked under Innis, describes the context: 'The fundamental problem of religious organization in pioneer Canada was the problem of meeting the needs of the scattered backwoods settlements, the traditional attachments of the old world ties of folk and class broke down in face of powerful forces of individualization, and new attachments had to be established in terms of a new sense of social purpose. It was the failure of the traditional churches that they offered no effective support of forces of social reorganization in the Canadian backwoods society.'[53] This observation is supported by the general fickleness of the backwoods population. Regardless of previous denominational inclinations, the settlers flocked to churches that could provide preachers to the peripheral communities. On this score, the Baptist Church was one of those evangelical institutions that successfully challenged the more established denominations for the patronage of the hinterland farms. '[Their] evangelical principle of relying simply upon a "call" to the ministry rather than upon educational qualifications – or qualifications of social class – eased the problem, already greatly facilitated through close geographical ties, of building up in the country an effective preaching force.'[54]

An earlier Innis would have been destined never to leave the peripheral rural areas. He would have been a Bible-wise but essentially ignorant individual. Yet, in other respects, Innis's personality was marked by the frontier church. Certainly, he exhibited a conservative anti-establishment bias similar to that of the backwoods church vis-à-vis the urban and materially rich denominations such as the Anglicans. These establishment denominations, with their financially and politically influential congregations, represented faith as the proper balance of spiritual and secular interests. Such an orientation could not compete in the backwoods milieu where secular concerns, by and large, had not yet developed. The paradox is that, by being uncompromisingly religious or otherworldly, the evangelical movements met more effectively the primitive 'secular' need of the backwoods, almost an emotional need, to break down the isolation of the farm and lay the basis for community. This being the function of the rural churches, it is not surprising that their ultimate power for enforcing secular affairs came from shunning, the practice of excluding individuals from the religious community. From today's perspective, this form of Protestant excommunication seems excessively vicious given the deprivation of social intercourse it involved.[55]

Innis's revulsion at this type of petty cruelty formed the basis of his lifelong aversion to strategies of institutional ostracization. Yet, these

Draconian practices of the pioneer Baptist Church were the natural outcome of an intense belief in the responsibility (and accountability) of the individual for his or her decisions. In rejecting the application of institutional pressure, Innis exaggerated, rather than abandoned, this belief in the primacy of the individual. Nowhere was this more apparent than in his approach to his own work, which was driven by the sense of deep personal responsibility he felt. His lifelong complaints of exhaustion through overwork are more correctly described as his application to himself of this exaggerated Baptist view of the individual.

Again, we see Innis in his early years caught in the middle of the periphery as the Baptist churches of his early experience evolved from individual community-based centres to local manifestations of a national denomination. This period of transition in church life should not be thought of as an abstract background to Innis's development. It implied very immediate consequences for his project. For example, the self-taught backwoods preacher was replaced as an ideal, in the minds of people such as his mother, by the university-trained theologian whose strength was not only faith but knowledge. On a congregational level, these new ideals took institutional form in the Canadian Literary Institute (later Woodstock Collegiate Institute) and the Toronto Baptist College (TBC), which would eventually become McMaster University. As with all significant social changes, the transformation of the Canadian Baptist Church did not take place without conflict, in this case between the more traditional, rural-based fundamentalists and the liberal and urban theologian-scholars. In fact, Innis's home church broke with its urban-based, university-linked counterparts in the late 1920s and retained its fundamentalist character as a 'fellowship' Baptist church. It continues to do so to the present day.

Innis's religious experiences put him in the middle of the margin more generally, considering the way the Baptist churches were pulled between influences of the British Church and the American movement. Southwestern Ontario Baptists in particular were subject to dual influences reflecting the biases of the British and American mother organizations. Local churches incorporated this cultural schizophrenia in their acceptance of a relatively conservative, non-emotional, Sunday-sermon format, interspersed with special revivalist meetings led by visiting evangelists at which entry into the church would follow outbursts of emotional frenzy.

The structural organization of the Baptist Church ensured that this tension would be biased in favour of the rural congregations. As S.D.

Clark puts it, 'by virtue of the fact that the Baptist principle of church autonomy weakened denominational organization, the Baptist ministers located in the larger cities ... enjoyed little more prestige than their fellow preachers in the country and consequently they were not able to exert a dominant influence in determining the general outlook and appeal of the Baptist churches.'[56] This meant that, while education and scholarship became accepted within the Baptist Church, the scholar-theologian never gained the political hegemony that characterized his position in other denominations. An uneasy truce existed in which the scholarly Baptists always had to be aware of the reaction that their work or pronouncements might call forth from their fundamentalist (and generally) rural brethren. It was probably the existence of this group of Baptist scholars that allowed Innis to equate, and then replace, his traditional faith with the scholarly search for truth. It is also likely that the interference by the denomination in the academic pursuits of these scholars established Innis's long-term revulsion at infringements of academic freedom.

Beyond this, the church was largely responsible for moulding Innis's attitude towards basic political concepts. Early experience with the Baptist ideal of the autonomous, democratic, local congregation helped to ensure that he would remain a lifelong radical democrat. Just as the basis of the highly decentralized Baptist Church was conservative, so was Innis's attitude towards politics. His sympathy for local decision making in the church and, at the same time, his realization of the necessity of more centrally coordinated goals (such as missions and education) provided him with an example of the difficulties of democratic federalism over which he would muse for the rest of his life.

Although Innis grew up in a rural community in which religious orientations were taken seriously, it was a pluralistic community that had worked out practices to ensure that these distinctions did not lead to destructive tensions over common efforts such as the local one-room school. As he began school, it was class rather than religion that made an impression on the young Innis. 'The problem of religious division seldom arose in the appointment of teachers, particularly as conventions were built up assuring there would be alternate appointment of Roman Catholic and Protestant teachers. There was much more concern ... in the question of distinction between ordinary farmers and people [referred to as] "low class"... This term ... was probably used to describe who had head lice and who gave them to other pupils in the school.'[57]

The rural phenomenon of the single-classroom school provided the ideal setting for the rapid advancement of the bright and disciplined child. Innis cites it as the non-specialist environment par excellence, and his positive experience in such a classroom at S.S. #1 South Norwich would later reinforce his opposition to specialization in education. This environment honed Innis's ability to jump rapidly from one subject area to another, since various lessons and subjects would be pursued simultaneously (theoretically, with each individual following only one grade and lesson at a time). He also lived the experience from the perspective of the teacher at this same school in Norwich (a relatively established community) and in the pioneer prairie settlement of Landonville, Alberta, where he served as the first teacher to its frontier school. He was never to forget the hopes for the education of their children demonstrated by the people of these places or the difficulties faced by rural people in an educational system biased towards the more elaborate and specialized approach of the city schools.

In reviewing the material remaining from Innis's years at Woodstock Collegiate Institute, one cannot help but be impressed by the quality of the instruction. For example, essays were marked far more closely on grammar and style than they would be today. Innis followed a full range of subjects, including composition, literature, grammar, history, geography, arithmetic, algebra, geometry, art, Latin, botany, zoology, and physics. He seems to have passed up only one subject: French. The high quality of the teaching and of his own work at WCI is demonstrated by the fact that he won the second prize for a short-story competition in his final year as a McMaster undergraduate. The story had been written years before at Woodstock.[58]

The superior teaching at WCI reflected the extent to which it was a special educational institution. Upon entering the institute, Innis found himself in the Canadian Baptist educational milieu, one in which he would remain until completing his master's degree. Besides providing effective basic training for Innis, these Baptist educational experiments were also a source of his later attitudes towards the theory and practice of higher education in Canada.

The paradox of WCI and McMaster is that, although they were the projects of a religious denomination, that denomination's beliefs held strongly to the principle of secular education. One has the impression therefore, that the prime motivation behind them was not religion so much as it was 'practical' nationalism. The Baptists were grappling with the problems posed within their denomination by a rapidly maturing nation. 'Their "incipient nationalism" ... expressed as a "common feeling of identification with Canada" ... arose naturally out of the experience of the churches in creating acceptable syntheses from imported and local materials.'[59] In terms of the ministry, this meant a gradual turning away from untrained 'called' preachers in favour of 'man-made' (university-educated) ministers. This in turn entailed a slow change from the conception of the church as an evangelical movement towards a conception of it as a denomination. It further implied that the training of man-made preachers for the Canadian church would not be dependent on the institutions of the British and American Baptists, but would be done by new institutions in Canada capable of handling the particular needs of the Canadian church.

Key among these new institutions was Woodstock Collegiate Institute. The church was still overwhelmingly based in rural farm communities and the training of autochthonous personnel had to take into account the nature of this raw material. The founders of the institute were of the 'opinion that it had always been unrealistic to expect semi-literate farmers' sons to be rushed into theological studies without adequate preparation in 'literary and practical' subjects. Moreover, a sound preparation [at WCI] would afford Baptist youth with more rewarding instruction than ... was then ... believed ... available in the publicly supported schools of Western Ontario.'[60] The breadth of vision behind the establishment of the new school was captured in its original name: the Canadian Literary Institute (CLI), founded in 1857. 'This school ... was not for [Woodstock] alone, nor even for its immediate hinterland, but for the whole country.'[61]

The scope of this idea went far beyond passing on polished farm boys to the Baptist theological school. WCI was a secular school run by the Baptists. As such, its curriculum was not religious, and neither its student body nor its faculty was exclusively Baptist. It was, in short, the ideal preparatory school that identified promising rural students, refined them, and gave them a grounding that allowed them to compete in the university setting with students from far more privileged backgrounds. It was designed to give rural, mainly Baptist, youth a fighting chance at gaining access to avenues of higher education, leading not only to the university-trained ministry but to other professions and trades. This should not be thought of as implying strictly academic training: 'Newcomers were subjected to even more basic [subjects] than the formal curriculum. One graduate recalled that many red-faced youngsters had to be taught the rudiments of proper table manners and given persistent instruction in the use and care of a flush toilet.'[62]

Whether or not the relative cultural deprivation of Innis's rural background required this sort of supplementary training, two things are certain. First, WCI was an ideal institution for a farm boy to pursue his hopes for social advancement through education. Second, Innis was deeply impressed by the necessity of constant attention to the development of this type of institutional support if the intellectual talent of the rural areas was to be recognized and fostered in a higher education system fundamentally biased in favour of individuals from a more privileged urban background.

The same recognition of the need for a special institution that would compensate for the relatively deprived background of the Baptist rank and file led to the advent of a denominational university, despite the Baptists' support for secular, state-financed higher education. When the constitution of the provincial university was debated in Ontario, the Baptists had argued for a government-funded, non-sectarian 'college of literature, science and art' to which the theological colleges of the various denominations would be able to affiliate. The Canadian Literary Institute had been established only when it appeared that the Church of England's King's College would be converted into the provincial university (the University of Toronto). In other words, the CLI represented the Baptists' decision, in the face of imminent political defeat on provincial university policy, to 'go it alone' in implementing this idea of a general arts and science program open to *all* denominations. Yet the Baptists held so closely to their ideal of the separation of church and state in the field of education that, even though the new institution was undertaking what they considered to be the legitimate responsibilities of the state, as a denominational initiative it was not allowed to seek government subsidies. As a result of this strict application of principle, the new school experienced chronic financial difficulties.

The relevance of these passionately held beliefs on the organization of education to Innis's later attitudes towards the university becomes clear if we consider their historical roots. The historian Charles M. Johnston writes:

> The principles behind the Baptist concept of a university could be traced to the Reformation, when the Anabaptists, the radicals of that spiritual revolt, had, unlike the followers of Martin Luther and John Calvin, deliberately set the temporal off from the spiritual out of a revulsion for the excesses of the Renaissance papacy. Whatever the terms might have meant to other sectarians, 'church' and 'state' had a special meaning for the Anabaptists' ideological descendants. For them the church was not an

> elaborately structured entity comparable to the polity of the state. Rather, its distinctive feature was a loosely organized congregation of believers who were not manipulated by a higher power and who stressed the idea of complete religious liberty. It then followed that a church so conceived could not readily consort with the more complex arrangements that the state had devised to manage its affairs. But spiritually as well as structurally the Baptists had reason to shun an intimate association with the public authority. By its very nature the state was the antithesis of Christianity and ought not to be encouraged to mix with the sacred to the extent of supporting one sect to the exclusion of all others.[63]

If we substitute here 'academic' for 'religious,' 'community of scholars' for 'church,' 'search for truth' for 'sacred' and 'Christianity,' 'intellectual' for 'believer,' and 'school' for 'sect,' we find we are left with a close parallel to the mature Innis's educational policy.

Innis also seems to have been influenced by the Baptist educationalists' manner of seeing the provincial university as the cornerstone of independent national development. They opposed the Church of England approach, not only because it was denominational, but also because it was excessively British and colonialist. In contrast, the Baptists hoped that a non-denominational secular institution 'would be invaluable as a nation-building force. "I cannot but lament a state of affairs in this country," [wrote the founder of the CLI]. "Men come from Ireland, England, and Scotland retaining all their peculiarities, even their nationalities. We ought to look upon ourselves as Canadians and earnestly enquire by what means we can advance the interests of the country ... Now the great Provincial Institution ... would tend to remove this evil. It would form the nucleus for a national feeling in Canada."'[64] Innis's nationalist orientation in his later work at the University of Toronto was in the mainstream of traditional Baptist educational thought.

Had the Baptists been working out these principles under the different conditions applying in the United States, the school at Woodstock might have developed into a high-quality, small-town, liberal-arts college. However, the structure of the country militated against this eventuality, a fact reflected on the life of the great patron of Baptist educational efforts, Senator William McMaster. McMaster had made his fortune in those sectors in which Canadian business has traditionally excelled: banking, resource exploitation, transportation, and distribution. As a natural consequence, he was strongly metropolitanist in his philosophy. This meant that the senator subtly but effectively threw his influence behind initiatives to move the central institution of Baptist training from Woodstock to Toronto. He did so not only to strengthen Toronto in its competition with Montreal for the domination of the Canadian hinterland, but also because he genuinely believed that an urban environment was more conducive to higher education. Senator McMaster was thinking of the training of Baptist professionals useful in industry and finance as much as he was that of ministers with a solid background in classics.

For political and economic reasons, the change of location was accomplished gradually. First, the theology school was separated from the general arts and science program at Woodstock and relocated under the new name of the Toronto Baptist College. However, it now became impractical for theology students with weak backgrounds to do make-up work at Woodstock while enrolled at TBC. As result, it was necessary to begin teaching a few general arts and science courses at Toronto. From this inauspicious beginning, McMaster University would grow as a Toronto-based general educational programme that would soon dwarf both the divinity school[65] and the Woodstock school. (Eventually, in 1930, the university would be moved to Hamilton, Ontario.)

The university-level liberal-arts program had been established at McMaster Hall in Toronto in 1890. So, before Innis entered McMaster, the Woodstock school had become a high-quality secondary school that supplied students to the new university. The continued existence of the general arts and science faculty at McMaster, rather than its federation with the University of Toronto, seems to have been due to a number of factors. The most important of these was the continued necessity of providing appropriate courses for students who were arriving at the university directly from a rural educational background without passing through WCI. (The University of Toronto, even at this time, tended to focus its attention on honours programs biased in favour of students coming from the more developed educational infrastructure of the city.) This difference in orientation between McMaster and the University of Toronto manifested itself in curriculum differences. Those Baptists who now argued against federation with the provincial university pointed out that the University of Toronto was ignoring certain subjects that were essential to the cultural development of the country, for instance, English literature and political economy. This criticism underlined the relative sophistication of University of Toronto and its student body, rather than the reverse, for these curricular deficiencies were not oversights but conscious policies resulting from the more colonial British attitude of the institution. 'The notion of introducing English studies as a formal discipline on its own would have appalled those who regarded a knowledge of one's literary heritage as part and parcel of a cultivated gentleman's cultural equipment, as an adornment not to be reduced to the indignity of a "classroom" subject.'[66] This was obviously not an attitude conducive to the effective schooling of farmers' sons. Similarly, for years after the establishment of the University of Toronto, the subject of political economy was avoided for fear of introducing party politics into the classroom. Thus, a general difference in the needs of Baptist students, related in turn to the different socio-economic background of the Baptist church, led inexorably to the abandonment of federation plans with the University of Toronto and the adoption of a more generalist approach in the curriculum of the Baptist institution.[67]

As a consequence of these developments, Innis was able to benefit from surprisingly high-quality institutions both at the secondary-school and at the university level. Moreover, these institutions were specifically concerned with overcoming the educational problems faced by rural youth. Innis absorbed not only the content but also the methodology of training at WCI and McMaster, both of which underlay his commitment years later to a generalist and open-access approach to educational policy.[68]

Innis's move to McMaster represented more than just another step in his educational odyssey. He was leaving the farm that had been his home since birth. He did so without making a public profession of faith and being baptized, as was the Baptist tradition. As it turned out, he would never undergo the ceremony of religious baptism. Instead, at McMaster he faced baptism in a sea of ideas. This baptism led him on to a more sinister baptism in the trenches of the First World War. The experience would mark Innis's personality for the rest of his life and lead him away from the Baptist faith that had served as the cornerstone of his youth.

CHAPTER TWO

The Great War, 1914–1918

The prevailing opinion in the trenches was that anything might be true, except what was printed. [From this scepticism about anything official there arose] a prodigious renewal of the oral tradition, the ancient mother of myths and legends.

Marc Bloch[1]

For a young man from the margin, attendance at McMaster University in Toronto was a revelation. The university was run by the Baptist Church, but this did not mean that issues and claims of faith went unexamined. In fact, the theology professors at this school were intellectually driven to articulate a philosophical basis for faith.

When Innis entered McMaster in October 1913, it was a lonely time for him. He was given advanced standing based on his examination results at Woodstock; entering at the second-year level meant that he missed the first-year activities so important in developing undergraduate friendship networks. This lonely existence was compounded by him living on his own in a boarding house, a less costly option than the student residence. Even at that, he skimped on food to save money, and by Christmas he was significantly emaciated and ready to quit. The quick intervention of his mother, who brought in his Uncle William (a medical student) to counsel him, and his acceptance of increased monetary assistance from his parents, led to his return to studies in January 1914.

The university had just passed through a period of intense controversy in which the more conservative (mainly rural) congregations and preachers had expressed serious concern that some of the more progressive professors had abandoned true spirituality and the revealed truth of the Bible. This controversy was sparked by the approach of two particular staff members in the pre-war period: I.G. Matthews and George Cross. Both men had been trained at the University of Chicago, an institution also founded by the Baptists, and, although the campaign of the conservatives focused mainly on the 'heresies' taught by Matthews, it was really targeted at the approach to theology developed in Chicago.[2]

During the 'muckraking' progressive era, the University of Chicago had been in the vanguard of attempting to adapt the church to modern times. It had championed the subjecting of the Scriptures to intense scientific, historical, and literary analysis and the establishment of a philosophical basis for the Baptist faith. This approach included a comparative study of other religions and gave rise to the 'social gospel' movement that linked Christian belief to direct action in social reform on a number of issues.

Baptist traditionalists objected to these new approaches because they believed that they undercut the spirituality that lay at the heart of Christianity. They charged that the new approaches covertly questioned the revealed truth of the Scriptures, brought to the world by Christ and literally set down in the Bible. The traditionalists mounted their attacks by laying out the 'fundamentals,' which formed the basis of Christian faith, and insisting that all teachings at Christian institutions comply with these basic elements. (In an era in which 'fundamentalist' is usually applied to Muslim conservatives, it is interesting to note that the term was first coined during this period to describe, in a positive fashion, the position of these traditionalist Christians.)

The intense campaign to discipline Matthews had subsided in 1911, just before Innis arrived at McMaster, but had seriously split the Baptist congregations and had come close to undermining any notion of academic freedom at the university. It was a bitter battle that coloured the atmosphere in the university and led Innis to a lifelong interest in some of the themes involved, the most obvious being the importance of academic freedom.

The effect of this atmosphere on the devout farm boy was profound. His plans were still based on the hopes of his parents. His father's wish that he avoid the hard life of a farmer by becoming a teacher had been abandoned when he decided to attend McMaster instead of going to normal school (teachers' college). He was now considering the career path favoured by his mother: training for the Baptist clergy. At McMaster, he did not feel at ease with the more conservative rural clergy who stressed revealed truth to such a degree that they were willing to attack the principle of academic freedom and restrict scholarly inquiry. But neither did he feel entirely comfortable with the position of the progressives. He was still a conservative rural Baptist whose favourite university entertainment was going out to hear visiting evangelical preachers. For instance, he wrote to his family on one occasion: 'Was over Tuesday night and heard Rowell and last night John MacNeill "The Presbyter." He is about the best in Toronto and he is certainly pretty good. Heard Dr. Easton Monday of New York and he is certainly the foremost Baptist minister in America although I later saw him going down the street smoking ... Since Christmas I have run across a lot of fellows in McMaster who tend towards Materialism or who believe there is no God which was an astonishing fact to me ... am going to hear Dr. John R. Mott tomorrow "the greatest missionary since St. Paul."'[3]

Innis must have felt some merit in the fundamentalists' criticism of the 'present-mindedness' of the progressives – their tendency to discount the achievements of the past (in this case the revealed truth of the Christian Scriptures and their oral tradition) in an effort to find a perspective more suitable to addressing the problems of the present.[4] Later, using some of the concerns that had been raised by the conservatives in this bitter pre-war debate, Innis would resolutely argue against the activism of the social gospellers during the Depression – this, despite his loss of faith during the First World War. In short, his mother's hopes that he become a clergyman suffered because he did not feel totally at ease with either side of the debate: as a clergyman, he would not fit easily with either the doctrinal certainty of the rural brethren or the present-mindedness and social activism of their university-educated, urban-based colleagues.

The McMaster that Innis entered in the fall of 1913 was a tiny institution by today's standards, the total undergraduate population being composed of 204 men and 44 women. The whole university – classrooms, offices, and student residence – was housed in a single building on Bloor Street just west of University Avenue that today serves as the home of the Royal Conservatory of Music. Yet its small staff was beginning to produce significant scholarly works. W.S. Wallace, who taught history, was just at the start of his long scholarly career; he published *The United Empire Loyalists: A Chronicle of the Great Migration* in Innis's first year at the university, and his *Family Compact: A Chronicle of the Rebellion of 1837* would be published in 1915. That same year, W.J.A.

Donald, who had trained at Chicago and taught political economy, published *The Canadian Iron and Steel Industry: A Study in the Economic History of a Protected Industry*. James Ten Broeke taught philosophy and authored *A Constructive Basis of Theology* in 1914. A.L. McCrimmon, who had a long history in Baptist educational institutions (he had been classics master and eventually principal at Woodstock), had just become chancellor. He had taught political economy at McMaster since 1904, and, as chancellor, had established the subject as a separate course of study just before Innis's arrival. He also taught education and sociology. Above all, he was an educationalist who kept the curriculum up to date. In short, McMaster was a small but intellectually impressive institution.

Baptism by Intellectual Fire

The very strength of the Baptists' faith seems to have allowed for a militantly secular approach to the ideas that were present in the McMaster classrooms. Their faith demanded that agnostic and heretical modes of thought be examined, though, to be sure, for the purposes of refutation.[5] There were, as a result, no dominant intellectual paradigms present at McMaster. Like the farm milieu, but in a more concentrated form, McMaster exposed students to a wide variety of trace elements of various disciplines and paradigms of contemporary Western civilization. Innis would come away from McMaster not with a catechism but with a grab bag of notions.

The modernity of the curriculum was impressive for the times. Innis had been introduced to the theory of evolution by one of the few agnostics in the local Otterville community. It was, after all, a time in which the topic was being attacked by fundamentalists everywhere as anti-biblical and heretical. Yet, when he arrived at McMaster, one of the new professors (the first to have completed his undergraduate work at McMaster) was being praised as 'living proof that one may believe in God and Evolution, [for] ... by his wise mediations of the truths of Science he strengthens the Christian faith of those who sit at his feet.'[6] Similarly, the works of Sigmund Freud were being discussed in McMaster classes as early as 1909. The political economy department was likewise quite progressive, having become separated from the more traditional subject of constitutional history shortly before Innis became an undergraduate. 'The new course reflected some of the great social and economic concerns of that generation, trade unionism and labour economics, the emergence of trusts and other forms of modern capitalism, and the doctrines of socialism. Beatrice and Sidney Webb, J.A. Hobson, the leading anti-imperialist of that age ... and F.W. Taussig, the student of American tariff structure: these were some of the impressive names on [the political economy course] ... reading lists.'[7] It is highly probable that Innis was initially introduced to the work of the American economist and social scientist Thorstein Veblen at McMaster rather than later at Chicago, where he would receive his doctorate. Innis followed an enriched program in this progressive setting, working towards a double honours degree in political economy and philosophy.[8]

Given Innis's lifelong concern with developing a 'philosophical approach,' we might expect to find a clue to his understanding of this term in the content of the classes he followed at McMaster.[9] Philosophy at the time was a one-man department. That man, James Ten Broeke, was a Baptist minister trained at Rochester and Yale (PhD) who had done post-doctoral research at Oxford and in Germany. Ten Broeke viewed philosophy as 'the mind approaching its materials to relate them all, according to its own modes of action, into a consistent whole in relation to some ultimate, unitary ground of all that is. Its synthetic task is to arrive at a whole of knowledge that will relate rationally 'the collective knowledge of a given age.'[10]

For Ten Broeke, of course, the unifying principle for this wide-ranging syncretism was God. Innis's later obsession with the philosophical approach might be seen as a consequence of pursuing the same wide-ranging syncretism but without Ten Brooke's conviction as to the underlying unifying principle.

The philosophy course was astounding in its scope, covering everything from the history of philosophy and psychology to ethics, metaphysics, logic, and the psychology of the religious experience. Ten Broeke's approach was primarily historical; he was especially concerned with theories of individual consciousness and how these related to collective historical phenomena such as religions.

Ten Broeke was a transitional character: as Innis put it, he was 'primarily concerned with the problem of working out a philosophical basis for theology, when to the orthodox no such problem existed.'[11] Innis might have added that, to the agnostic, Ten Broeke's approach was equally irrelevant. In pursuing his goal, Ten Broeke displayed a very liberal attitude, exposing his students and himself to all sides of contemporary scientific debate, whether these points of view tended to subvert his own religious position or not. Despite his tremendous faith in the ultimate unity of scientific endeavour and Christianity, this method of teaching, which stressed pushing the search for truth to the limits, inevitably implanted doubts in the minds of young Baptists like Innis.[12] Probably the three most radical sets of ideas that Innis ran across at McMaster pertain to 'mechanistic' (physiological) psychology, evolution, and relativity. We do not know how Innis viewed the arguments of the supporters and opponents of these theories during his undergraduate years. But we do know that he was exposed to them and that they all carried an agnostic bias. Innis clearly recognized the unconsciously subversive influence of the devout professor on his Baptist students. Later, in his *Autobiography*, he would write: 'He must have been by far the most heretical thinker in the university.'[13]

The relatively liberal attitude on the part of the majority of the faculty at McMaster was largely the result of feedback from the University of Chicago. In the field of theology, the Chicago approach was unhesitatingly to subject the Bible and contemporary Christian beliefs to the improved methods of verification that became available with the development of new techniques in archaeology, linguistics, psychology, and the social sciences in general. Christian scholars influenced by Chicago operated on the conviction that, whatever the immediate consequences of their research, there would be no long-term contradiction between faith and science. Their total dedication to the search for truth was founded on the basis of the belief that such 'truth' would, by definition, be found to be Christian truth. A faculty member at McMaster summed up this attitude by saying that his university 'must welcome truth from whatever quarter, and never be guilty of binding the spirit of free enquiry. As a Christian school of learning under Baptist auspices, it stands for the fullest and freest investigation, not only in the scientific realm but also in the realm of Biblical scholarship.'[14]

Unfortunately for the relatively sophisticated faculty of the university, the base of the denomination was still overwhelmingly rural and fundamentalist in belief. Furthermore, since the Board of Governors was elected by, and directly responsible to, the congregations' representatives voting at Convention, there was always the possibility of serious interference in what was being taught. There was a direct monitoring link, after all, between the classroom and the farm through the sons and daughters being trained at the university. In this context there was always a danger of backlash to the liberal spirit at McMaster, which existed alongside attitude such as the following: 'We are not now denying that there is truth to pursue, but we do most confidently and solemnly affirm that there is truth to teach. However vast may be the domain of the unexplored and the unknown, it is yet true that something is known [and is] ours by the ... attestations of the ages [or] by the unequivocal revelation of God ... Before such truths as have been abundantly proven or clearly revealed we dare not take the attitude of the ... doubter and the agnostic.'[15] Given the prevalence of these attitudes in the denomination as well as its structural control of the university, it is not surprising that there were a series of bitter confrontations concerning teaching at McMaster. These were especially serious, for they had direct and immediate financial implications for the university. They also had implications for Innis's views on the structure of higher education.

Once the economic difficulties of his first term had been resolved, Innis' undergraduate days were among the happiest and most carefree of his life. One senses, in his correspondence and notes, a tremendous fascination with the interplay of ideas and opinions as well as a growing sense of confidence as the farm boy adjusted to urban and university life. Nowhere was this exuberance more apparent than in his success in the Debating Society. (The group itself was the descendant of Ten Broeke's Philosophy Club.) From the point of view of socialization, involvement in this form of competitive public speaking, with celebrated dignitaries in the audience, could not have been a more effective antidote to Innis's reserved nature. The essentially liberal methodology of the debate, involving the effective presentation of both sides of a question without reference to the debaters' personal sentiments, provided Innis with a lifelong respect for the anti-fanatical tendencies of the common law tradition. Arguing the case may not have necessitated a personal commitment to an issue, but it did require an understanding of the topics at hand: one debate assigned him the challenge of arguing the merits of socialism as the only genuinely Christian form of government! During this period, we again find that the themes that entered Innis's consciousness would percolate in his mind over time before being rearticulated in the firmly held opinions of his later years.

The beginnings of Innis's approach to scholarship may be seen in his debate notes from 1916. Concerning a proposal to conscript all undergraduates, he writes:

> McMaster University Exists Not for the Sake of
> Physical Strength

> Our ideal is higher than that, and if our ideal must amount to something it is in leading nations by precept and example to the better, nobler paths of peace ... It is to the men who stand fast to the work of strengthening the nation by leadership that the glory of victory will go as assuredly as it goes to the man in the trenches.[16]

In a different debate, he refers to the relationship between commerce, transportation, and communications:

> Before people can exchange goods they must have a media of intercommunication. That is, they must understand each other. Before goods can be exchanged there must be a means of efficient transportation.
>
> Through man's ingenuity the steamship evolved, railways were built, the telegram invented, cables laid down and the penny postage effected. Every one of these brought about a tremendous expansion of commerce. As Prof. Mavor has said the ocean telegraph cable has made the world one city and given the world one market.[17]

In these debates, Innis seems to have learned to use a flexible perspective that kept him from accepting the terms of any particular paradigm; he would never lose this characteristic.

Baptism by Military Fire

Things were going better for Innis but worse for the world. The outbreak of the First World War in 1914 had a profound effect on Baptists, who were traditionally exceedingly wary of militarism. McMaster had been one of the few universities during the pre-war period to decline to establish a Canadian Officers Training Corps (COTC) program.[18] However, the fate of Belgium and the propaganda on German atrocities quickly led Baptists to view the conflict as a just war in defence of Christianity, democracy, and civilization.

The university was greatly affected by the war. The COTC program was promptly established, and social activities were curtailed in favour of rifle practice. COTC military training was encouraged in two ways: by counting it as the equivalent of one course (of the student's choice), or by granting an automatic 'bonus of a certain percentage, for instance 10%,' on the final marks of every course for which the student sat the examination. Innis participated in the COTC and was entitled to choose from these incentives.[19] Enlistment in the regular army was even more encouraged, by giving students automatic credit for all their courses in the year in which they enlisted.[20] In one respect the situation probably benefited Innis, since he had to overcome his chronic shyness in taking over many of the responsibilities left open by the departure of others. As he puts in his 'Autobiography': 'Those of us who had not enlisted were compelled to carry the additional load [left by the departing students]. At one time I was acting editor of the McMaster Monthly as well as general poo bah for all sorts of other organizations.'[21]

What seems to have bothered Innis most about the war was the effect it had on standards. As a precocious child character, he had worked incredibly hard in pursuit of his education. Now, when he was in striking distance of earning a university degree, it was as if his lofty goal had been irrevocably devalued by the reward for enlistment, namely, the automatic awarding of such a degree during the final year of studies. Innis tried to avoid this devaluation by not enlisting until after graduation, but he was never convinced that his degree was up to standard. These nagging doubts bothered him and later served to reinforce his objections to university disruption during the Second World War. If a veteran or a man about to enlist could not take full advantage of the traditional cultural institutions of their society, how could the human sacrifices of the front be justified?

McMaster provided, in both relative and absolute terms, a zealous environment for the war spirit. Before the war, McMaster's sentiment had been almost pacifist in nature. Yet when the war broke out the atmosphere changed overnight. In Charles M. Johnston's words, 'the zeal and dedication released on the McMaster campus seemed to be an extension of the traditional evangelical militancy of the Baptists. Combat against anti-Christian forces at home and on foreign mission fields in time of peace was now to encompass a campaign to destroy the enemies of the British Empire, the acknowledged safeguard of civilization and of Christianity itself.'[22] One justification for war to which the Baptists were particularly susceptible was that of 'making the world safe for democracy.' To quote Johnston again: 'This theme elicited a warm response in the humble churchgoer of the back townships who could boast that every Baptist congregation in the land was a democratic or autonomous community and that "the sum total of these small democracies was a mighty democracy willed and ruled by God." That mighty democracy ... must gather up its strength to defeat prussianism.'[23]

The Great War was a prime mover for Innis, a scarcely visible force that set his personality, imposed the great problematics that he would later tackle, and forged the complex link between his sense of individualism and his nationalism. Yet Innis's experience of soldiering was quite brief. He went overseas as an artillery private in the fall of 1916 and, as a result of a serious wound received at Vimy Ridge in July 1917 (three months after the ridge was taken), was invalided home in March 1918.

Innis never exhibited the frequent fascination of men for things military; in photographs, he invariably looks awkward and out of place in his uniform compared to his colleagues. His enrolment in the COTC was undertaken as a necessary duty, without passion. He indicated on his forms that he wished to train only to a certification level of 'efficiency' rather than 'proficiency' and that he was not willing to attend 'camp.'[24] He had no desire to become an officer, although that was the whole point of the COTC program; he went to war simply to play his small part in the great struggle that was taking place.

Above all, he wanted to come back alive and for that reason chose to enlist as a gunner. With his limited knowledge of military affairs, he thought that this would entail working well behind the front lines. Yet in England, perhaps because of his obvious intelligence, he was trained as a signaller. He wrote home happily: 'This morning they put me at the signalling bunch so I shall not have much to do with the big guns after all and will likely be safer still.'[25] Given his later communications work, it is ironic that the event that finally cleared Innis for posting to France was an examination testing his proficiency in various media of (military) communication: '12 on the buzzer, 8 on the flag, 10 on the semaphore, and six words per minute on the lamp.'[26] He was still very much the upbeat young farm boy out to see the world; on a six-day pass before going to the trenches, he rapidly toured the sights of London, Glasgow, and Edinburgh.

Unfortunately, Innis badly miscalculated on positioning himself for relative safety during the war. Artillery signallers (spotters) were used by gunners to observe landing shells and send back aiming corrections so that the shells could be more accurately 'walked' onto their targets. These spotters had to make their observations from forward positions, often in front of the first lines of trenches. On 7 July 1917 Innis was doing duty as a spotter under cover of darkness when his position was noted from an enemy observation balloon by flare light. The balloon called down German artillery fire and Innis received a serious shrapnel wound through his right thigh. He spent eight months in various military hospitals in England. During this time, he would complete his

McMaster MA, even before leaving the army. It would take his physical wound seven years to heal. His psychological wounds would last a lifetime.

To begin to understand this phenomenon of profound personal change over a short time period, we must come to grips with the personal situation and the motivation of the young Innis who presented himself before the recruiting officer in the spring of 1916. On 4 April 1916 Innis wrote home announcing his decision to enlist in these terms:

> You will wonder what has struck me but the fact of the matter is I have been thinking. It isn't so much because these other fellows [volunteers from McMaster] went through that made me think but it is because if the Christian religion is worth anything to me [enlisting] is the only thing I can do. If you were not Christians I don't know how I should write to you. If I shouldn't go I could [not] content myself with the fact that I had not lived up to my duty, that Christ had asked me once to take up his cross and follow him and I hadn't been able to do it ...
>
> [Many others from McMaster have already gone] everyone of them smarter than I ever dared to be and it is the least I can do to give [them] my support ...
>
> Germany started in this war by breaking a treaty, by breaking her sealed word. Not only did she do that but she trampled over a helpless people with no warning and with no excuse. If any nation and if any person can break their word with no notice whatever, then, what is the world coming to.[27]

These sentiments are so foreign to our overwhelmingly secular minds that one is tempted to believe that they were somehow insincere – a rationalization that Innis produced in terms that his still devout family could understand.[28] Yet, only in recognizing Innis's complete sincerity in expressing these beliefs, can we appreciate his crushing disillusionment with traditional religion and other institutions following his war experience. After the war, 'preaching' took on the meaning of an obscenity in Innis's vocabulary.

In fact, before enlisting, Innis had nearly gone along with his mother's wish that he join the clergy. The strength of his pre-war faith is evident in notes made for his eyes only in his McMaster diary: 'Do not go into the presence of temptation ... Be forearmed. Link up with other men. Associate yourself sufficiently with Christ. Keep your eyes in the right direction ... Take Christ as the great solvent of your doubts ... Be decisive. Spend time in study of life of Christ. Meditate, instruct. Spend time unhurriedly in daily prayer ... A dominant purpose to live the right life. Christianity a life spirit and method of living.'[29]

He was still, after all, the farm boy from a poor background whose training was taking him in the direction of the ministry. The Christian faith of his youth was so intense that it would admit, through the influence of individuals such as Ten Broeke, examination of doubts of the most profound nature, in the belief that overcoming these doubts would lead to an ever more perfect faith. Thus, the notes to himself continue: 'Do not look upon state of doubt as the final stage but as a transitional stage. Approach these questions as far as possible without presuppositions.'[30]

Through Ten Broeke and other professors at McMaster, Innis came to believe in the underlying unity of Christianity and the critical spirit. From this perspective, doubt became of central importance in the genuine articulation of faith, and subverted his mother's hopes for his ministry. In replying, decades later, to his son Donald's question as to how he had managed to escape from his rural, Baptist background, Innis said, with his characteristic terseness about things personal, 'They pushed me too hard and I got suspicious.'[31]

His family and friends were so keen to complete their plans for Harold that they arranged for him to be given a rural charge (parish) as a lay preacher upon graduation. They did this in spite of the fact that Innis, because of his continuing doubts, had not yet accepted to make a profession of faith and be baptized. The plan backfired. In taking these exceptional measures, they had devalued, in Innis's eyes, their conception of faith. They were willing to accept his preaching to others while knowing that he still was unsure of his own beliefs.

At this crucial moment, the war presented itself as an option that was articulated in intensely Christian terms without demanding suspension of personal religious doubts. One could be certain that Germany's actions were anti-Christian without necessarily being any more at ease with one's own faith. In Innis's case, the war offered him a way out – a way that allowed him to escape his family's plans for him while still being in keeping with the intensely religious milieu which had bounded his progress so far.

Innis's motives for going off to the war were the motives of his generation. Living in more sophisticated and cynical times, we are accustomed to think of soldiers' motives as irrelevant and pretentious rationalizations for the real forces leading to military conflict. It is easy for us to comprehend Innis's going to war as a way of avoiding the ministry, but the bias of our times makes it hard for us to take seriously the Christian rhetoric in which his decision was couched. And yet these were the sentiments of a generation: Innis was not alone. Moreover, whatever the era, motives are of central importance for the decisions taken by individuals. They are, after all, life as it is lived. And when, as in the case of the First World War, there is a tremendous commonality of motive, they are significant in the general psycho-history of an era – the typical becomes the archetypical. The confrontation of these noble expectations with the reality of the conflict (or, to put it another way, the legitimate counter-motive of the enemy) created a complex trauma that went far beyond the success or failure of the individual battles.

The Uniqueness of the First World War

Innis was not unique in feeling a disconnect between his motives for going to war and his experiences of the war itself; it was a general sentiment of the veterans of trench warfare. During the Great War, there was an absolute chasm between expectations and reality. It would be difficult to find another example of war in which the motivation was expressed in nobler ideals and the fighting done in a baser manner. The depressing recognition of the contradiction permanently set off the heroes of the war (the veterans) from their worshippers (the civilian population). The trauma of this war involved an alienation for the front soldier so extreme that it could not be easily dissipated.

Innis's bitterness at the chasm between the front and home indicates that this war was unlike any other. Its features were the result of a particularly catastrophic alignment of ideological and technological factors in military affairs. On the ideological side, the military establishments of the countries entering the war had adopted the doctrine of *offensive à outrance* ('offensive to the limit'). The will to win was thought to be *the* essential factor determining defeat or victory. Hence, 'the defensive is forgotten, abandoned, discarded; its only possible justification is an occasional "economizing of forces at certain points, with a view to adding them to the attack."'[32]

On the technological side, this military ideology of the offensive came to be dominant at a time when militaries had inadvertently developed an overwhelming superiority of defensive firepower. Chief among these developments were the perfection of the rifle in terms both of range and of rapidity of fire; the mass production of machine-guns and long-range artillery; and the use of barbed wire and highly effective signalling systems. These factors, when combined with the underdevelopment of air planes and tanks, led quickly to stalemate – something not anticipated by anyone. Trench warfare arrived unheralded, and brought with it an unprecedented transformation of the traditional characters of the 'soldier,' the 'enemy,' the 'staff officer,' and 'civilian society.'[33] The domination of the war by defensive technology meant that the enemy became invisible and impersonal, with the odd result that the stereotypical idea of the aggressive soldier gave way to that of figure who was, above all, defensive. For every operation in which men went over the top in a bayonet charge, they spent months dug in, surviving the rain of metal that was part of life on the front. It was a war that demilitarized the men who fought it. As Innis would write later, 'the bandoleers of artillerymen filled with rifle ammunition were gradually emptied and the ammunition thrown out by the side of the railroad. The troops were not interested in carrying a heavy load of rifle ammunition and did not hesitate to throw it out at the earliest opportunity.'[34]

In *No Man's Land*, Eric J. Leed summarizes the effect of this technological stalemate on the personality of those who fought. The continual hiding underground from the artillery barrage led to 'the dismantling of the sense of self as offensive executor of a national will upon a quasi-human enemy.'[35] The precarious collective safety of the dugout also led to 'increased narcissism [and] expanded self-love' among the soldiery. This 'provided one of the strongest and most positive bonds that survived the war, the bond of the individual to the men in his unit. This regression also produced a body of men with an enormous need for care and reassurance, a need combined with anger and hostility towards a society that had placed them in the position of victims.'[36] The defensive personality soon expressed itself in the 'language' of warfare that was developed at the front. There was an almost immediate recognition that the enemy was experiencing the same frustrations of military stalemate. Trench soldiers soon recognized that routine offensive actions by one side triggered retaliatory efforts from the other lines without any change in position. As a result, there was a continual tendency for the fighting to quiet down when it was left to the routine efforts of line soldiers in a stable sector.[37]

Special offensives planned by staff officers in chateaux well behind the lines and the shifting of units in the line were meant to counteract this dissipation of the offensive spirit. It was an almost universal comment of the veterans that they were motivated to go 'over the top' without hesitation, not out of animosity and aggression towards the enemy but because in so doing they would cut down on the chances that other men in their unit would be hit.[38]

The impersonal nature of the machine-gun and barrage, and their high efficiency in killing, meant that the resentment of the line soldier in trench warfare became directed, above all, to his own staff officers. These were the human agents who had the most obvious role in determining his death. This inversion of standard military and societal stereotypes marked Innis indelibly. Almost no anti-German statements can be found in his writing or correspondence during his time at war; rather, his ire was expressed in a deep-seated bitterness towards those in authority.

Bizarre things happened in the trenches. Describing the barrage before Vimy, one veteran said, 'If I put my finger up I would have touched a ceiling of sound. Sound had solidity.'[39] In like fashion, death lost the abstract clarity it had in civilian life – a clarity that was based on the passage of time: the moment between life and non-life. At the front, 'it ceased to be an abstraction and became a term defining the growing distance from which the combatant viewed his home.' Time changed from a metronome to an accordion. 'The exclusive attention upon events that threatened death expanded time.'[40] All of these experiences became part of Innis's invisible personality as a veteran. They were to be called on during the communications period as he developed apparently abstract concepts concerning the relative malleability of time and space.

The final and most painful inversion brought about by the structure of the war was the destruction of the ideal of 'home' – the very ideal for which the soldiers were supposedly fighting. That this destruction took place 'behind the backs' of the veterans gave it the taste of treachery. In fact, it was the result of the realization that a modern war of attrition required the orchestration of jingoism to an unprecedented extent, and that the means for this profound manipulation of public sentiment were at hand in the increasingly perfected mass-communications media. The returning veteran, therefore, was obliged to face, in an exaggerated form, those same naive and ugly sentiments that he had just discarded in his first-hand experience with the reality of the front.

In brief, during the First World War, the ideology of the offensive coupled with the preponderance of defensive military technology led to a radical inversion of stereotypes. The community became the veterans, the men of the trenches, while the home societies were transformed into chauvinistic, militarized machines. The staff officers became, in a sense, an enemy ordering one to certain death. And the enemy often was recognized as a comrade who also suffered under the despotism of the barrage.

Leed examined countless veterans' commentaries on the war in order to determine its effects on the personalities of the young men who fought it.[41] The underlying theme of these commentaries is a view of the war as a rite of initiation taking place on the extreme margins of humanity: the trenches. The common impression was one of having lived in two completely distinct worlds: peace and war. This impression was accompanied by 'a profound sense of personal discontinuity ... on every level of consciousness.' The intensity and exotic nature of the experience made it, like the rites of a cult, difficult to recognize or understand for those who themselves had not passed through it. Like cultists, the veterans could not or would not convey a sense of the rites of the front: 'An examination of the identities formed in war must come to terms with the fact that these identities were formed beyond the margins of normal social experience. This was precisely what made them so long lasting, so immune to erosion by the routines of post-war social and economic life, and so difficult to grasp with the traditional tools of sociological and psychological analysis.'[42] This is also why it is so difficult to understand the long-term impact of the war on Innis.

Innis's Invisible War

The very fact that Innis wrote little about the war indicates the profound effect the experience of trench warfare had on him. It was an experience that only those who had lived through it could comprehend. To try to described it in words alone to those who had not been there could only demean the lives (often lost lives) of those who had been there. It was a topic best avoided.

For instance, given the horror of the war, Innis's letters home have a strikingly nonchalant air to them. Few contain references to the fighting any more direct than this: 'The war is still on and we still see some of its horrors as well as some of its humorous aspects.'[43] The bulk of the letters are made up of sarcastic remarks on the weather and of commentary on letters coming from home; there is virtually no mention of acquaintances he has made at the front.[44] There are several obvious reasons for this. Principally, Innis would have been concerned with not stimulating anxiety among his family as to his welfare.[45] Also, censorship was strict. Once, even a reference to the size of the *incoming* shells was removed by the censor's scissors before Innis's letter reached Canada. But beyond this, the content of the letters can be fully explained only by assuming that the scenes at the front were beginning to disturb Innis himself at a profound level. Overtime, there were increasing references to a general weariness that seems to go beyond mere physical strain. The humour is forced and the tone dull because the alternatives are too painful to reduce to writing.

We know that this is the case because Innis kept a clandestine diary during his time in the army. (The practice was officially illegal for intelligence reasons.) While the diary does deal with warfare, it does so with the same dulled, detached air. Innis writes about his friends at the front being killed or wounded with the level of emotion of someone noting penalties at a sporting event. The mechanism at work here appears to be one typical of individuals placed in long-term horrific situations. It involves self-preservation through an auto-anaesthetization of normal sensibilities. It is an unconscious mechanism that cannot easily be turned off after the dangers have passed.

Innis's treatment of the war in his 'Autobiography' is an example of the long-term effects of this phenomenon. While he notes the events he was engaged in, he does so euphemistically. For instance, 'after an hour's shelling our Battery could be said to be out of action.' The horrors of war had become bearable by being experienced as the 'ordinary' for that period. No further probing was bearable even after an interval of thirty-five years. While Innis recognizes the bias of his war letters, he draws back from a more profound description and opts for an autobiography that is essentially a gloss for these documents. 'The letters were, in the main, pieces of paper filled with inconsequential information such as would escape the censor's attention.'[46]

Outside the circle of fellow veterans, the soldiers talked about the war as rarely as they wrote about it. As one put it:

> [We] were all 'old timers,' the men of the trenches ... We were prisoners, prisoners who could never escape. I had been trying to imagine how I would express my feelings when I got home, and now I knew I never could, none of us could. We could no more make ourselves articulate than could those who would not return; we were in a world apart, prisoners, in chains that would never loosen till death freed us.

> And I knew that those at home would never understand. They would be impatient, wondering why we were so dumb, unable to put our experiences into words; and there would be many of the boys who would be surly, taciturn, moody ... We, of the brotherhood, could understand the soldier, but never explain him. All of us would remain a separate, definite people, as if branded by a monstrous despotism.[47]

Nor did Innis talk much about the war; the subject was virtually untouched in all the interviews with him and in all the commentaries on his life and work. Alexander Brady remembers not only Innis's reticence to talk about the war but also his air of resentment when others discussed the subject in his presence. As far as I have been able to determine, Innis made only one Armistice Day address (1933), and it dealt primarily with contemporary problems. This fact itself underlines Innis's sense that the sacrifices of the war were to be honoured not in the rituals of memorial but in the resolution of the problems of the peace. Direct reminiscence was insupportable: 'It has not been long since most of us have been awakened by nightmares of intense shell-fire, and even now the military bands played with such enthusiasm by young men are intolerable, and Armistice Day celebrations are emotionally impossible.'[48] The war experience could not be articulated because it was of such a singularly horrific nature that it could be fully comprehended only by those who lived it. This is the message Innis passed on to his son Donald. It also explains why Innis – uncharacteristically, some might think – kept in regular touch with the veterans' association of his old battery.

The process of auto-anaesthetization that went on in the trenches left psychic wounds among many of the veterans; these were far more lasting than physical wounds. Upon demobilization, the shock of returning to a changed domestic society and the longevity of the various neuroses acquired during deployment meant that many men who had not succumbed to nervous collapse in the trenches did so at home, in the post-war era. Indeed, as the war years receded, an increasing percentage of the veterans on disability allowances were victims of nervous disorders rather than physical disabilities. 'Paradoxically enough, "war" neurosis was a condition more prevalent in "peace" than in war.'[49] These cases were just the tip of the iceberg, however. Many veterans simply coped with their neuroses in the course of their everyday life.

Innis seems to have been one of those for whom the difficulty of dealing with the war experience grew rather than abated with the passing years. Not surprisingly, one of the only records we have of his talking about his feelings regarding the war was written by a fellow veteran and close friend, George Ferguson. Even in this case, it took Innis's approaching death to make the war a subject of conversation.

> Innis ... had an increasing horror of his memories of [the war]. I'm not sure of his thinking. My impression was that he regarded it as the ultimate obscenity.
>
> The veterans of his generation and mine looked back upon their service with an amused tolerance and nostalgia. What days they had been! The fear and the horror and the endless casualties faded. Not so with Innis. He never spoke of it except briefly and with loathing. When he was dying and could hardly work any more, we spoke of it and I said I had some of the old ordinance maps of the areas in which the corps had fought. I asked him if he would like to look at them and, rather to my surprise, his face lit up and he said he would.
>
> He talked to me in those last months for the first time in his life about the war ... All he could say about it was the horror of the performance. It was not that he himself had been wounded, that wasn't it. It was the more he thought about the thing, the more he thought of young men being destroyed, who might have been so valuably useful. And he would speak with real bitterness ... bitterness I've never seen in another man about the stupidity of the whole performance which he had embarked on *himself*! *Everybody* did in those days! ... But by God he had come to some pretty violent conclusions about it ... about the idea of war.[50]

One of the only non-veterans Innis talked to about the war was his son Donald, who tells us:

> The war was something else again. What it meant to him I can only express in a word I have heard used by men when speaking about someone that has been at the Front or has escaped death in a certain fashion that they have been 'kissed' and no other word that I know expresses it so well. The war was a test of his physical and mental resources such as he found nowhere else and could have found nowhere else. That he passed this test meant a great deal to him and I think would be difficult to understand for anyone that had not gone through it. In passing it he joined a fraternity that is unseen and unspoken but exists nevertheless. When we went over Vimy Ridge his reactions were like those of someone visiting an old school, a place not liked but a part of his life.[51]

A Hard Patron[52]

Although Innis shied away from the ministry because of doubts concerning his faith, enlistment in no way implied his readjustment of long-term goals away from an essentially Christian calling. If Innis had learned anything from Ten Broeke, it was that 'in Christian doctrine, disillusionment had [to be] regarded as a positive experience ... Through their disillusionment those following the path of Christian enlightenment acquired wisdom.'[53]

The First World War, however, tested such ideas. In that conflict, as Eric J. Leed writes,

> disillusionment describe[d] not an ascent towards grace but a spiritual and social descent ... [Men] were compelled to resign themselves to the omnipotence of those material realities that were already familiar to the industrial working classes – realities that were described as 'industrialized' or 'technological.' In the war disillusionment signif[ied] a loss of social and existential status or, more exactly, a process of self-definition through realities that had a significantly lower moral and existential value than abandoned expectations.[54]

The war acted as Innis's patron but in a decidedly secular sense, directing him towards economics and away from religion, philosophy, and law. Even though his habits of churchgoing carried over sporadically into the early 1920s, his efforts at self-improvement and self-definition were of an exclusively secular nature after the war experience. As well as changing the orientation of his career, the war acted as his patron by providing him with the means to go on with his studies. For instance, Khaki University, a military extension-education program, was a Canadian initiative involving the Young Men's Christian Association (YMCA), the Chaplains' Corps, and several universities including McMaster. The program made available higher education facilities, namely instructors, books, and courses of study, to reserve and convalescing troops. Fortuitously, it was just beginning operations as Innis was recovering from his wound in England, and it enabled him, as a long-distance student, to complete his McMaster MA before being discharged from the army.

Finances had always been a key problem for Innis, and here, too, the war stepped in to help. Innis received a tuition allowance and a vocational-training allowance. He had also accumulated a small amount of savings from his war pay. As a private, he received $1 per day in regimental pay plus field allowances of ten cents daily. Despite the parsimony of army remuneration, Innis saved twenty dollars every month to send home to his mother.[55] Most significantly, his status as a returned veteran likely opened up teaching and other employment opportunities at Chicago, the income from which allowed him to pursue his work without interruption.

In his 'Autobiography,' Innis describes the three career options that faced him on his return:

> [First] to join the church and then the ministry ... The second alternative which had been very much on my mind was that of law, and I had decided in a general sort of way to attend Osgoode Hall in the fall, but I still had an uneasy conscience that I knew very little about the subject of economics and that I ought in fairness to myself make some effort to gain a more intensive acquaintance. I decided, therefore, that I would use the funds that I had received from the army to take the summer school at the University of Chicago and thus to remedy to some extent my defective knowledge. This uneasy feeling of conscience unhappily or happily continued to bother me. I found that after a summer's work at the University of Chicago that I had greatly underestimated the difficulties of breaking from army life to academic life, in spite of my Master's degree. I was therefore compelled again to face my conscience and to decide to remain in Chicago and complete my work for the PhD. before starting my studies in law. The uneasy conscience, however, has continued to worry me and I never felt completely equipped to go into the profession of law.[56]

This revelatory passage merits closer examination, for it tells us that Innis's career in economic history was simply a detour on the way to his major goal (that is, after the idea of the ministry was abandoned): a career in law.

It is easy to see why law was appealing to him. Oratory was respected, even in the farm context, in a way that is utterly foreign to us today. Furthermore, Innis had excelled in this field in the McMaster Debating Society. The abandonment of one profession, the ministry, to which oratory was central, quite naturally led to the picking up of another to which it was equally central. In fact, the Debating Society format, with hypothetical questions argued technically without reference to the actual beliefs of the speakers, fits the legal rather than the clerical style of oration. Innis's disillusionment with religion and the fundamental questions of 'justice' posed by the war experience would have further fixed his hopes on a legal career.

However, his patron dictated that this was not to be. We will return shortly to this issue of 'an uneasy conscience' – a certain insecurity about his ability to complete law. It is sufficient to say here that this self-distrust was a widespread phenomenon among veterans of the war. Neither, for Innis, was it a temporary affair. It should be noted that, when writing in 1952, he used the present perfect tense in stating: 'The uneasy conscience has continued to bother me.' It would not be until the last few years of his life that he would feel adequately prepared to return to deal directly with the subject of law.

There were other less subtle ways in which the war and its aftermath turned him away from law. On the pecuniary level, he was informed, in no uncertain terms, by the Invalided Soldiers' Commission that he would receive greater benefits from the government veterans' programs if he continued at university rather than immediately pursuing a career in law.[57] Another factor may have been his unsuccessful performance as a volunteer Liberty Bond spokesman selling government war bonds in Chicago. In the Debating Society, he had no qualms defending any point of view no matter how contrary to his own beliefs. After all, the swaying of adjudicators through the building of a rational and stylish argument had been the point of the exercise (as it was in the courts of law). However, in Chicago, the oratory involved in selling war bonds imposed impossible emotional demands on him. He was caught in a bind. He abhorred talking about the war, but he realized that his presentations could help the efforts of those still at the front. Then again, this type of oratory was most effective in the jingoistic environment of the time, when it was at its most chauvinistic and propagandistic. Having seen the front, however, Innis must have felt incapable of glorifying it in the requisite manner.

He worked for an organization that sent out returned men to give short 'pep' talks before public gatherings such as cinema showings. Given his later work, it is ironic to find that he was introduced by the organization's literature in the following manner:

I am a Four-Minute Man.
I am the Mouth-Piece of Democracy.
I make men THINK.
I wield the most potent power of Human
Endeavor –

THE SPOKEN WORD.[58]

Despite this build-up, Innis's presentations were low-key. He invariably made three points: a successful attack took much preparation and back-up (as at Vimy); this was the best way to keep casualties down; the audience's role in the preparation was to give money. He never attempted to stir up anti-German sentiment. Yet even this low-key approach must have bothered him for it involved a large degree of misrepresentation. Publicly, he assured his audience that effective preparation led to victory. Privately he knew, as he would recall in his 'Autobiography,' that this was not so: 'Our task ... was that of bringing material to the front line so that a bridge could be thrown over the line immediately after the attack and that the guns would be moved over these bridges to a new front. This was, of course, highly speculative and dependent on the weather and became a complete fiasco because of a heavy snow fall on the night of the attack and the impossibility of moving the guns forward in the mud.'[59] After the armistice on 11 November 1918, when his speeches no longer involved life-and-death support for those still at the front, Innis made an unremittingly bleak attempt to describe the chasm that existed between the expectations of enlistment and the reality of the front. This had a depressing effect on his audiences and the invitations to speak rapidly ceased.

Prior to the armistic, at a gathering in Chicago in the late spring of 1918, Innis had been asked to accept the toast for the returned men. Knowing the jingoism of the audience and the disgust that Innis felt about the war, the chairman suggested that he simply say, 'I am glad to be back,' and sit down. Innis, enraged, took the opportunity instead to try to get across the horrific nature of the gap between the the common portrayal of the war and the reality. His speech notes are a form of black poetry:

As we had imagined it:
afraid war would end
chasing Germans with bayonets
Pleasure of going over the top
Loaded down with German helmets [trophies]

As we found it:
Bully beef and hard tack
[railway] cars marked 8 horses or 40 men,
ditching ammunition,
bayonets used to toast bread and cut wood
Filling of sandbags.

Helmets to wash in
damned dull, damned duty and damned monotony of it
continual mud
continual reading of sheets [maps]
continual bread marmalade and tea
continual shelling
hide and seek warfare
With the monotony came fear
Instinctive location of deep dug-outs
Mathematical probability of shells
landing in the same place twice
Flattening against the trench wall
Drinking poisoned water
How long we were [at] battery
eating cordite
Gradual longing
for blighty
Before these influences all men are alike
Canadians
English
Scotch
Americans[60]

The war devalued oratory by reducing it to propaganda. The mobilization of the civilian population through the rhetoric of religion and justice sickened Innis, both because he had seen the carnage at the front that this propaganda was designed to obscure, and because he himself had unwittingly been mobilized originally by the same sort of high-sounding sentiment. The experience may well have been a key one in turning him away from law and the ministry towards a career in economics. It also helps to explain his utter disgust when economists were called on to take their turn playing the role of propagandists during the Depression.

War and Nationalism

The war defined Innis's nationalism. Vimy Ridge, where Innis was wounded (albeit three months after the famous battle in which he participated), had tremendous significance in the development of a Canadian national identity. This significance was recognized not only after the fact on the home front but also by the soldiers taking part in the campaign. It was the first time that all four Canadian divisions participated in a joint operation under Canadian command. Furthermore, they were well aware that they were attempting to overrun an objective that the troops of various nationalities had failed repeatedly to take. For these reasons, it was looked on as a national test and is still considered the most celebrated battle in which Canadians participated – in the Canadian psyche, something on the order of Gallipoli for Australia or Gettysburg for the United States.

The nationalism of the veterans, however, was ambivalent. After all, it arose not in the relative calm and the self-confident cultural milieu of the metropolitan countries but in 'an isolated cut-off world of marginalized men among whom all correlates of identity had turned inwards.'[61] This is the reason why Innis's 'nationalism' has been presented in such a confused way by his commentators, whose basic mechanism has been to identify Innis's attitude as either unalloyed nationalism or anti-nationalism, depending on the bias of the commentator, and then look to his writings for textual proof. For instance, if we limit ourselves to the wartime correspondence, we might select the following passage on English conscientious objectors in Canada to substantiate a virulently nationalist position: '[It] makes my blood boil when I think of Englishmen in Canada who have not the backbone to enlist ... They are over in Canada making all kinds of money ... while our lads are out here suffering heaven knows what and for what is England's cause if it is anybody's ... [They] are living off the fat of the land ... These people stand back while men die and suffer – [they] wax fat and refuse to lend a hand. They may be conscientious objectors, well so am I and I object to that.[62]

On the other hand, we might choose to demonstrate his anti-nationalist sentiment through this diatribe on American nationalism: 'The Americans never get tired of talking about the things they do or the things they are going to do ... I never heard such a line of bragging and boasting in my life. It was really disgusting at least to Canadians ... Some of them think they have won the war. Of course all this will change as their casualties get larger. That fact will sober them just as it has sobered every nation. So much for July 4th.'[63]

Given these last comments, we should not expect to find Innis's own sense of nationalism articulated in similar jingoistic terms. In his correspondence, references to the perfidy of the 'Hun' drop off dramatically with the recognition at the Front that the 'enemy' was, in fact, a companion in suffering. Similarly, little reference is made to 'home' in the sense of nation, largely because he detested jingoism and because his concept of 'nation' had been forged with the community of Canadian soldiers that endured the holocaust at the front. Yet, whatever label we apply to it, Innis developed a tremendous sense of his 'own' that viewed the following as positive: the efforts of the Canadian soldiers at the front; the quiet but wholehearted support for these efforts on the part of the domestic community; and the attempts to keep the home society carrying on in its usual manner despite special wartime measures. The negative foils to this sense of nationalism were phenomena such as jingoism, profiteering, and conscientious objection.

The concentration of Canadians at the front had the effect of decolonizing their concept of nation. By all accounts, the placing of Canadian troops under the supervision of British officers or NCOs (non-commissioned officers) generated an immediate and intense sense of nationalism. As Innis put it in the understated prose of his 'Autobiography' more than three decades later: 'The treatment of Canadians and all others by officers sent out from Great Britain must have been an important factor hastening the demands for autonomy throughout the Commonwealth. Their insolence and brutality were such that Canadian recruits could scarcely overlook, indeed they condemned them in the most hearty fashion.'[64]

Changes in the military structure to rectify this became one of the chief political gains made by Canada during the war. The wholly Canadian nature of the reorganized units further strengthened the sense of nation. But perhaps more significant was the irate reaction of Canadian troops to the assumptions of the English as to their motives for fighting the war. Men who thought of themselves as Canadians going to war for Christian and democratic principles found that they were viewed as colonials whose arrival on the scene was motivated by a desire to defend the 'mother country' and empire.[65] This realization was made all the more striking by the recognition among the soldiers that many 'Canadians' themselves, especially at home, viewed the war in the same fashion.

Innis and other returning veterans retained elements of this wartime resurgent nationalism and subsequently applied them, in a complex manner, to a peacetime environment. In Innis's case, the result was not only a rejection but an abhorrence of the jingoistic chauvinism that had characterized the war. This explains his refusal to define openly his nationalism either belligerently, in terms of an enemy (the Hun), or self-righteously, in terms of Canadian superiority in certain fields. Rather, it is developed almost in an invisible manner through criticism of the British and American cases, and is to be equated neither with American jingoism nor with British colonialism.

Innis's conception of nation was an existential one: the problem in Canada was not one of mouthing a national identity, but of constructing one. This required a long-term strategy to exert the hegemony of Canadian interests in all social sectors – politics, the economy, and culture. Only when self-confidence had been assured in all these areas could a quiet, non-jingoistic sense of 'one's own' be assured. The Canadian military effort in the First World War came to be viewed as the price of entry to this process; efforts after the war in sectors other than the military were judged in terms of the extent to which they lived up to the great sacrifices made by the 'nation' qua front line soldiery. For the veterans who returned to Canada infused with the notion of the tightly knit community-of-the-trenches-as-nation, this was not just Remembrance Day rhetoric.

Yet it was an impossible basis on which to consolidate a healthy, measured sense of one's own. This passage from George Grant graphically depicts why:

> Just read how English-speaking Canadians from all areas and all economic classes went off to that war hopefully and honestly believing that they were thereby guaranteeing freedom and justice in the world ... When one thinks what that war was in fact being fought about, and the slaughter of decent men of decent motive which ensued, the imagination boggles ... It killed many of the best English-speaking Canadians and left the survivors cynical and tired. I once asked a man of that generation why it was that between the war of 1914 and 1939 Canada was allowed to slip into the slough of despond in which its national hope was frittered away to the U.S. ... He answered graphically: 'We had our guts shot away in France.'[66]

The price paid for the new sense of identity had been too high, both physically and psychologically. We can gain an inkling of the physical sacrifice by imagining what the death of 240,000 and the wounding of 600,000 of the younger men of this generation of Canadians would mean to our society today.[67] But above and beyond this, the generation that returned from the war was one psychologically crippled by the experience of the front.

A great sense of undirected anger existed among the returned men, who found no equivalent in domestic society to the tremendously supportive sense of community at the front.[68] Moreover, Innis was among those who recognized, as they had never recognized before, that a British colonial spirit continued to pervade the country. The most obvious manifestation of this was the presence of British and British-born officials, often with an active bias against Canadian personnel, in key positions in some of the central institutions of the country. Over this issue, Innis's veteran's nationalism comes to the fore in its most strident form. (Yet it is precisely here that commentators on Innis are silent or at their most apologetic, probably because Innis views do not accord with the present-day liberal ideal of the scholar.)

Perhaps the most volatile psychological scars that the war inflicted on veterans concerned their attitude towards authority. The trench soldier lived under the sway of an ultimate despotism. He hated it, and yet he internalized it. The war experience did not tend to produce anarchists (or democrats for that matter), for the overwhelming oppression of the chain of command taught one first and foremost that there was no escape. The best one could hope for was that the gap between authority and responsibility in decision making could be minimized, or, in the language of the later Innis, that monopolies of force and knowledge could be avoided as much as possible. In practical terms, it did make a difference who was in command, and the best commanders, in the wartime situation, were those who retained a thorough knowledge of the position of the private soldier. Individual personalities notwithstanding, this meant a preference for line officers over staff officers; for those who had come 'up through the ranks' over those who received an automatic commission; and, in the case of Canadian troops, for Canadian over British officers. Within the authority structure of the university, similar principles applied: in general, the scholar was preferable to the administrator; the scholar serving an administrative function while still pursuing his research was preferable to the professionally trained, full-time administrator; and Canadian faculty (especially in administrative positions) were preferable to non-Canadian personnel.

Yet the experience of the front had taught many veterans that there was something fundamentally evil in authority (power) per se. Innis's decision to follow a career in scholarship, and his militant efforts to define the university as an institution operating outside the realm of the powers of the day, reflected this veterans' attitude. The conception of power as evil was incorporated into his work (in particular the communications studies) and in his manner of carrying out administration within the university. In viewing the pursuit of scholarship as the antithesis of power, Innis judged suitability for administrative duties within the university largely in terms of scholarly accomplishment.

War and Scholarship

The lessons learned in the 'school' of the trenches would be important in Innis's later scholarly work. For instance, the Front exposed him and the others to circumstances in which 'the bizarre becomes the normal, and through the loosening of connections between elements customarily bound together in certain combinations, their scrambling and recombining in monstrous, fantastic, and unnatural shapes, [the soldiers were] induced to think (and think hard) about cultural experiences they had hitherto taken for granted.'[69] This was an aptitude that Innis carried far beyond the bounds of the war. Yet its manifestations in his communications works are seldom recognized as being connected to his battlefield experience. For an example of the everyday 'transgression of categories' in Innis's war experience, consider this comment on billeting on a French farm: 'We finally settled ... with our guns in the cellar of the house and our horses in what had been the parlours and living rooms of the first floor.' Shades of Otterville gone mad![70]

In a similar manner, his attitude towards technology can be traced back to 1914–18. The sacred force to which the novices of the Great War were exposed was, above all, technology, but technology in a diabolical rather than its traditionally benevolent guise: 'Technology ... was 'resituationed' into a context of destruction, work, and terror, where it made human dignity inconceivable and survival problematical. In this process of 'resituationing,' the neutrality of technology seemed to fade and certain features – heretofore unsuspected – of the means that industrial civilization had developed to control nature and transcend human limitations became obvious.'[71] This painful realization of the ambivalence of technology would mark Innis's work profoundly, to the extent of setting it off from the basically optimistic tenor of McLuhan and others working in the fields in which he was active. In staples, it led to his being concerned primarily with the problem of cyclonics: of the exhaustion of resources that follows the application of new technologies to virgin materials. (This concern with cyclonics was also fostered by his first-hand experience of them in his own farming background.) In communications, the central function of the press in creating the prerequisite jingoism underlying the slaughter was obvious. But, more subtly, the experience of the war allowed Innis to focus his work on the underlying perspectives or myopias that technologies of communication imposed on human consciousness.[72] This latter idea was also related to the war experience, for the sensory dislocations produced by the front were important antecedents to his communications studies: 'The invisibility of the enemy put a premium upon auditory signals and seemed to make the war experience peculiarly subjective and intangible. One could not see the incoming shell but one could hear and learn to understand the language of the artillery. Surviving required it.'[73]

The extent to which his war experience motivated Innis's scholarly endeavours has not been adequately appreciated, largely because he wrote so little on the subject of the war – the phenomenon of silence that characterized the veteran's reaction. The war imprinted his thinking with its symbolism, evidenced by how the suppressed subject comes to the fore. This is shown particularly in private correspondence when he was expressing a passionate concern about some issue. Invariably we find Innis at these points articulating his 'nationalism,' a veteran's sentiment that was much more ambivalent than wartime jingoism. This sentiment amounted to the collapsing of his point of view on the issue at hand into that of the 'soldier' in general, and of the young generation. For instance, in defending the continuing operation of the university during the Second World War, he writes in a confidential memo to the president of the university:

> Here it may be said that I am still a psychological as well as physical casualty of the last war.
>
> It may be that my army training compels me to take battles as they come ... I feel strongly that we must carry on as we have been doing and making the best contribution we can with an intact and integrated but badly weakened organization. To do less than this is to be unfair to a generation and to ourselves.
>
> I am sorry to inject a personal element in all this and my excuse is that the individual soldier has been lost sight of.[74]

The scholar is equated to the trench soldier holding the line despite battle losses. The concreteness of this equation for Innis is apparent in his harsh remarks on those who left the university to do 'war work' of a non-combat nature: 'After eight months of the mud and lice and rats of France, in which much of the time was spent cursing government officials in Ottawa ... I have never had the slightest interest ... in people who were helping in the war with a job in Ottawa or in London. The contrast between their method of living and France simply made it impossible for me to regard them as having anything to do with the war and I continued to look on them with contempt.'[75]

For Innis, war and the search for knowledge had been merged; had the war occurred a few years earlier or later in his career, this identification might not have been made. In the experience of the veteran, one of the most disturbing psychological shocks is the return to 'home,' a place that has become hopelessly idealized during the combat years. The sense of treachery at this moment is common, for the 'home' that is expected has been replaced by a sorry version of a pre-war society that no longer exists. There are a thousand personal versions of the bitter story of the veteran's return: disgust at those who did not go, missing those who did not come back, the adulterous wife, the peaceful society turned jingoistic.

Innis left for war in the middle of a project for self-improvement through education. In consequence, the bitterness he felt on his return related to the decimation of the university infrastructure during his absence.

> When I came back in the spring of 1918 I found the universities depleted of staff ... because people were bustling back and forth winning the war, they said, or their friends said. This meant of course that after taking the dirt in France I was expected to take more dirt when I came back from people whom I regarded with contempt ... [At this] point ... I determined never to have any part in letting men down who had been in the Front lines. It is that attitude and the attitude of men in the government when Canada entered the war who had had similar experience that led them to insist that as far as possible educational institutions must carry on.[76]

The passage underscores not only the tremendous psychological difficulties that Innis experienced as a veteran in the post-war world but also his understanding of what war was fundamentally about. For Innis, it was and remained a moral issue. If the traditions, and their institutional repository, the university, for which these wars were fought were simply being frittered away behind the lines, then nothing less than treachery was involved. Furthermore, if critical, autochthonous institutions did not continue to exist and to probe the inner significance of these events, how could one be certain that the massacres were not meaningless? How could war be avoided if not by thoroughly understanding it? In this sense, the soldier and the scholar occupied the same front lines.

Innis's MA thesis, 'The Returned Soldier,' which he began researching and writing while convalescing in England, manifests a wide variety of the effects of the war on its participants. A reticence or inability to probe the war's profound influence on individual consciousness is evident in him switching his topic from 'Psychology of the Canadian Soldier' to a more general approach outlining various elements of a 'just' public policy for veterans' affairs. A strong sense of nationalism, portraying collective effort and diligence as the only way to vindicate wartime losses, is apparent everywhere: 'Work, work, of brain and of brawn, cooperation, organization, and determination to heal the sores occasioned by war, and to start again along the lines of sound national progress, is the hope of the Canadian people ... that she may take her place among the nations of the world for the privilege of which her best blood has been shed.'[77] Innis also repeatedly points out the intensity of feeling of the veterans' brotherhood and the problems it can entail: 'Again the army is far from lonesome. There are few moments in the army when one is alone ... This fact is of more importance than at first seems possible. It is to a great extent responsible for the failure of the Government to settle [veterans] on homesteads ... The furnace of war has melted men into a brotherhood in which the frills of humanity have disappeared.'[78]

Far from romanticizing the war experience, Innis sees its results as overwhelmingly negative: 'There remains the fact that no one who has been wounded, or no one for that matter who has seen a great deal of front line service, is physically or mentally better for the experience[79] ... A long period in the trenches varying from six to twenty-six days without relief is usually followed by a reaction from which only the strongest survive.'[80] Yet, while the connection between the barrage and psychological problems is recognized, Innis is reticent to discuss the resulting psychopathologies directly. Instead, he outlines the detrimental habits they throw up: alcohol, tobacco, and women. There is still much of the young Baptist's puritan morality intact: 'A heavy bombardment of an hour's duration immeasurably increases the consumption of cigarettes.'[81]

Innis does identify the key pathology of the veteran but seems puzzled by it, linking it loosely with an environment of danger and discipline.

> The fact is always forced upon the soldier that he must obey orders. The result is more or less an indifference to what happens ... This cramping of individuality and the enforced idleness that usually accompanies it, have eradicated many of the characteristics that marked the ordinary civilian.

> It has introduced, and created a lassitude, an indifference to surroundings which is not in his best interests. A man who has been over the top taking chances with life and death has become carelessly indifferent to the mere happenings of everyday life.[82]

This is the key contradiction in Innis's presentation of the post-war period. On the one hand, battlefield deaths are accepted as the price of entry to independent status for Canada; on the other, the war sapped his generation of the numbers and the will essential to carrying forward the national project. His thesis was a detailed description of the public-policy measures that were necessary, not only to provide a supportive milieu to help veterans get over the effects of the war, but also to move on with national reconstruction.

Innis arrived in Canada in March 1918, having completed his course work and thesis through the Khaki University program. On 19 April he took his final MA examinations, and on the 30th he was awarded the degree at commencement ceremonies in Toronto. As a returning veteran, he received a standing ovation. On 3 May 1918, at a medical-board hearing in London, Ontario, he finally received his discharge 'in consequence of being medically unfit due to wounds, although fit for employment in civilian life.' He had served for '1 year, 351 days.'[83]

Of his McMaster colleagues, 237 had served during the war. (To put this in perspective, only 105 men were still enrolled at McMaster at the close of 1918.) Twenty-three of these had died, and approximately three times that number were wounded.[84] Innis would spend the rest of his life trying to deal with the impact of such losses.

For Innis, and for his generation of veterans, it was a time in which a vastly increased sense of responsibility had to be reconciled with a diminished sense of confidence. Years later, he would summarize the impossible situation: '[There are] certain elemental facts about the post-war. And the first fact is the loss of life and the loss of the younger, abler and more vigorous part of life. In every sphere of activity, Canadian life since the last war has been impoverished. In politics, law, the church, medicine, business, there is a preponderance of older men or of very young men, with the generation of the last war without representation. That generation has been lost and its scantier numbers have not the strength to maintain their position.'[85]

CHAPTER THREE

One of the Veterans, 1919–1923

You will find, as those of us who survived the last war found, that the loss of your comrades may make it impossible for you to withstand the pressure from later generations and indeed from earlier generations. Both youth and age of other generations will crowd in on your generation as they did on ours. But at least you have had the advantage of warning from the experience of an earlier war and you will have the very particular interest and sympathy of a generation which had not such warning or experience to direct them. There is 'an incommunicable bond' between those who have faced death when young.

H.A. Innis, 1945[1]

In 1918 Innis moved to the University of Chicago to continue his studies. Given the encouragement of his professors, their openness to a Canadian research topic – the history of the Canadian Pacific Railway (CPR) – and the income he could obtain through teaching, residence duties, and marking, Innis decided to pursue a PhD in political economy. And there, too, he met the woman who would become his lifelong partner – Mary Quayle. It is difficult to imagine that Innis could have succeeded in his career without the unfailing encouragement and psychological support that Mary Quayle provided to her physically and mentally damaged veteran over the next three decades. He was awarded his doctorate in August 1920, and they married in May 1921.

His thesis was published as *A History of the Canadian Pacific Railway* in 1923, but Innis was not happy with the result. At the time, the history of Canada had been written as a heroic struggle for constitutional government that had overcome, through human efforts, the natural pull of continentalism. The building of a railway across the country readily fitted this interpretation. However, Innis noticed in his research that there were underlying geographical features that had given rise to the country. These were the drainage basins of the great river systems that had defined the space of Canada during the earlier era of the fur trade.

Innis's scholarly life can be divided into three decade-long blocks during which he focused on different areas of research and academic activity. Yet one theme was common to all three periods. This was his belief that a new perspective, developed by the collective effort of individual scholars at Canadian institutions, and with an unprecedented first-hand knowledge of the country's past and present, would lead to true understanding of the country and eventually make a unique contribution to Western civilization.

On the surface, the years from 1918 to 1923 were, for Innis, ones of great success. Under the surface, however, things were quite different. In referring to himself as a psychological as well as physical casualty of the Great War, Innis was not using the term 'psychological casualty' as a mere figure of speech.

It is difficult to imagine the pressures soldiers faced on the front. To be living in pits in a field bombarded regularly by high explosives was bad enough. To be asked from time to time to leave that modicum of safety to advance across the barbed-wire strewn, mud-churned 'No Man's Land,' exposed to flying shrapnel and rifle and machine-gun fire for no obvious military reason, was the height of madness. This type of behaviour, which went against all instincts of self-preservation, had to be reinforced by extreme measures of discipline. Refusal to go forward often meant execution by one's own side – either immediately by one's own officers advancing behind each wave of men, or later by a firing squad composed of one's own comrades. Minor infractions of regulations such as drinking called forth such punishment as 'the cruelty of cartwheel crucifixion ..."field punishment No.1" ... men, volunteers, spread-eagled to cartwheels, tied there for hours in a biting, bone-chilling wind all because the fellow had not shined a button or given some snobby officer a proper deference ... And these men had left good jobs and homes and had come, as the orators said, to fight for right and loved ones.'[2] It was an authority whose only limit was a clandestine bullet in the back of the overly strict officer in the chaos of battle.

Innis himself did not suffer such severe mortification as a soldier, and the majority of officers were not sadists. However, the war could not have been fought the way it was without the corollary of severe repression. Furthermore, whether or not each soldier had directly experienced extreme punishments, they all knew the worst stories. Indeed, these stories became exaggerated in the retelling. All of the soldiers knew that there was no escape from the despotism under which they fought. Once the rules were set in motion, there was little hope of stopping their application.[3] And they all experienced this type of mortification on at least a minor scale. The particularly high incidence of psychological disorders occasioned by this war among the front-line soldiers resulted from their inability to take any action to avoid either the random danger of the barrage or the absolute tyranny of military authority. Only during the Nazi Holocaust of the Second World War has a great mass of individuals been put under such pressure for such a long period of time.

The principal horror of the context is that personal strategies were severely limited. We have no statistical record of suicides, since these cases were, by and large, covered up to preserve both morale and the individual's reputation. Recollections of veterans and war literature, however, indicate that a significant number of soldiers chose this way out. In some ways, this is less depressing than the universally accepted view among the soldiers at the front that the best solution to the unbearable situation was 'a blighty one': a wound serious enough to ensure that one would never be fit to return to the trenches, but not serious enough to kill or permanently immobilize. Innis, for instance, was congratulated by his comrades on being hit by shrapnel in this moderate way. The extent of alienation was such that 'blighty' wounds were often self-inflicted.

Psychological Scarring

For the first time in military history, hysterical reactions become the dominant pathology accompanying the campaign. The uniqueness of these syndromes is attested to by the term by which they became known – 'shell shock' – and by the reaction of authorities to their advent. Although unconscious and involuntary reactions, they were largely treated as a disciplinary problem. Shell-shock victims were, quite literally, tortured back to 'normal' behaviour.[4]

Under these circumstances, by far the most general and effective reaction of the individual soldier involved an unconscious, partial withdrawal into himself – a kind of auto-anaesthetization to the reality of the trenches. The name for this syndrome, neurasthenia, covers a general range of reactions. It can be applied to the front soldiery of the First

World War as a body. Robert Graves 'insisted that every man who spent more than three months under fire could legitimately be considered neurasthenic.'[5] Neurasthenia was a functional pathology that kept men in the fighting by desensitizing them somewhat. Accordingly, it was thought at the time that 'there are no major distortions of thinking and behaviour evident in individuals suffering from this condition.' As we might expect from an anaesthetizing syndrome, it 'is characterized by complaints of chronic weakness, easy fatigability, and sometimes exhaustion.' It continues after the cessation of the horrific reality, just as memories of that reality remain with the individual. Its precipitating factors have traditionally been seen as mental and physical overwork, especially when occurring in situations conducive to emotional tensions. It has a 'tendency to chronicity and frequently severe incapacitation to which it subject[s] its victims.' There are a number of related symptoms, including morbid fears, a sense of hopelessness, depression, anxiety, and an inability to make decisions.[6]

When Innis talked of his being a psychological casualty of the war and of the difficulty of making the readjustment from army to student life, I believe that he was referring to the problems he faced in picking his project for self-improvement through education. These were exacerbated by the traumatic reversal of social roles that he had experienced at the front, as well as by the neurasthenic state imposed on his personality by the trenches. Yet three factors served to mask the severity of his psychological problems. First, ostensibly at least, he remained the precocious southwestern Ontario farm boy, albeit one with less optimism than before. Second, his career success overshadowed his mental difficulties. Was it not natural for someone who worked as diligently as Innis to seem overworked? Third, there is a bias within existing archival documentation that tends to preserve the invisibility of this darker side of Innis's personality. If we conduct a careful examination of that documentation, we find that, prior to the First World War, there is no evidence of recurring depression and lack of energy. Nor do we find the scathing remarks about groups or individuals that characterized his later correspondence. Yet topics that were important during the pre-war years – for example, commentary on religious events and clerical personalities – drop off drastically after 1918. Church had become a family rather than a personal obligation. He wrote to his mother from Chicago shortly after his arrival: 'Sunday in obedience to your commands I went to church and as to the sermon I can say very little. The Bishop of Chicago preached the sermon but I am no judge of music.'[7] Changes in his religious attitudes and practices deserve further comment.

From the scraps of original documentation that remain from this period, we can see the extent to which Innis was imbued with a deeply religious spirit prior to the war. In numerous notes to himself scattered through his papers, he reminds himself of the need to maintain his faith in the midst of his daily activities. His entry into the war was couched in terms of Christian principles. Similarly, his letters home from McMaster indicate an enormous naivety as far as matters of faith are concerned.[8] In the manner in which today's undergraduates might talk about a rock concert, the young Innis was excited about the prospect of going to see Dr John R. Mott, 'the greatest missionary since St. Paul.' Even if we make allowances for the bias of the audience to which Innis was writing, it seems that his chief extracurricular activity involved going to hear evangelical preachers who were visiting Toronto.

Given the obvious importance of religion in his background, the virtual absence of the subject in his 'Autobiography' is striking. This was largely the result of the deep bitterness that Innis felt towards those social forces which had underwritten the slaughter that he experienced in the trenches of Flanders. What he does focus on in the 'Autobiography' is precisely the transitional tension between the fundamentalist and the less emotional, modernist pastors, and between the camp meeting and the more staid Sunday sermon and Sunday school.

> In the Baptist denomination, there was a sharp distinction between those who held strongly to certain items of doctrine, such as adult baptism and closed communion. The storms which raged over the community of the Baptist denomination depended on the fervour of individual ministers. Particularly active ministers had no hesitation in excommunicating members from the church for dancing and other practices. In the main, religious instruction meant attendance at the Sunday school from week to week and the following of a routine of lessons provided for the teacher and student of different age groups ... Periodically revival meetings were held ... The pressure of these meetings was generally followed by what was called 'conversion' and in turn admission to membership.[9]

Innis's great reserve and general feeling of bitterness towards the church is evident when he couches his personal quandary during his adolescent years in the third person: 'Since the denomination emphasized adult baptism, and since this meant a deliberate public testimonial of faith, the Baptists were always faced with the prospect of losing individuals who reached adult age and were reluctant to submit to public testimony of this character ... In spite of all these efforts [to apply pressure] there still remained individuals who refused to make any public statement of their faith and who were the object of arguments advanced in sermon after sermon.'[10]

If the Baptist denomination had permitted Innis's religious doubts and convictions to remain a private affair, he may have remained within the church. But, because he took his faith and his doubts seriously, he felt obliged to move in a direction that would free him from the need for a public profession of faith while at the same time allowing him to remain consistent with his inner Christian convictions. The war provided such an outlet, but in a way that would sabotage his faith entirely.

Yet there was a basic continuity between his religious and non-religious phases. We have already mentioned some Baptist carry-overs in his later work. Even while still an adolescent within the church, Innis was demonstrating the propensities that would mark his adult life. His interests were in the political and institutional underpinnings of the church rather than strictly scriptural or moral questions. While others chose such topics as 'Who Is on the Lord's Side?' or 'Joys of the Christian Life' for their presentation to the Baptist Young People's Union, Innis commented on 'Our Denominational Boards and Our Relation to Them,' providing a description of the role of the coordinating bodies in the radically decentralized decision-making structure of the Baptist Church.[11]

The Baptist influence on Innis was especially evident in his continued aversion to power structures. George Ferguson recalls: 'He was the only man I ever knew who customarily referred to persons of the Roman Catholic faith as "Romans." This expression must have come straight out of the Woodstock back concessions, and it was quite clear to me that "Romans" scared him. This was not doctrinal. I came to the conclusion ... that it was based on his hatred of power again. The Church represented itself to him as one more great agglomeration of power and, as such, [was] to be profoundly distrusted.'[12] After the horrors of the First World War, this distrust applied as well to the Baptist Church despite its more democratic structure.

The subject of church, faith, and religion drops entirely out of his correspondence in the early 1920s. By 1926, he is writing to his old alma mater, 'I am not a Baptist.'[13] The apparent loss of faith was accompanied by an increased pessimism. We can well imagine the tone of the

Innis letter that drew forth this sarcastic response from a hospitalized veteran-friend:

> I really think I am a good Samaritan after having seen through your plot of wickedness.
>
> I have filed two charges of manslaughter against you but have through leniency asked that judgement be reserved until the 'Great Judgement Day.' You made an attempt on my life by trying to hit me such a blow that the shock would overcome me ... Harold, I give you great credit for such a deeply planned plot of execution, however, it failed ... Say old boy! Please ... have a heart and don't kid me that way ... I know you're not serious, for were you serious the University of Chicago would be hunting [for] another instructor in elementary economics.[14]

The depression of spirits brought on by the war simply exaggerated certain anxieties already present in Innis's pre-war thinking. For instance, with his hopes for the future based exclusively on academic success, Innis understandably felt a certain vulnerability because of the relatively deprived background of his rural youth; this preoccupation, always present, became more pronounced after the war. Indeed, the feeling of inadequacy was something that Innis never managed to overcome completely, despite his later successes. Consider the following example from his 'Autobiography': 'At Woodstock Collegiate Institute ... I was not a brilliant student.'[15] Yet a report card preserved from these years shows him coming second in a class of forty-four! Similarly, of his last two years at high school, Innis comments: 'I managed to secure all subjects ... though I doubt whether many of them were of high standing.'[16]

As for having been selected to teach in Alberta before he began his studies at McMaster, Innis did not feel that it was related to his marks or experience. Rather, in referring to the head of the local school board, he says: 'I have always suspected that a Scottish name had its own attractions for a Scot.'[17] Of his 1915 success in the intercollegiate debate, he admits grudgingly: the judges' 'decision was not unanimous but was given in our favour.'

There are many other such examples. In 1916 Innis refused the opportunity to receive an automatic BA for enlistment, putting it off in order to assure himself that he had earned his degree on its own merit. This strategy did little to assuage his doubts, however, for he always retained the suspicion that the degree had been devalued by the enlistment incentives. He was equally disparaging about his MA work: 'I am afraid that my thesis could not be regarded as an outstanding performance and I have always felt a debt of gratitude to McMaster University in accepting my application ... [On my final comprehensive I] wrote a very fair examination, though I did not do well on it.'[18]

While in hospital in England, Innis entered an essay competition as part of his correspondence course with Khaki University. The head of the university, Dr Henry Marshall Tory, awarded him the prize for his entry, 'The Press.' Despite this, Innis says: 'But I have never been certain this was due to the essay or to the fact that there were few entries.' As to his overall performance in Khaki University under difficult circumstances, he informs us that he 'had an average standing of 81%, whatever that might mean.'

I have already cited the passage in which Innis relates that his 'uneasy conscience' concerning his preparation for law at Osgoode Hall and an underestimation of the 'difficulties of breaking from army life to academic life' caused him to enrol in economics at the University of Chicago. He continued to have problems with self-confidence throughout his stay at Chicago. It is not enough for Innis to say, 'Professor Frank Knight ... did much to rouse my interest in economics,' – he feels compelled to continue with the words, 'as well as to make me feel how little I knew.'[19] Regarding his final examinations, he says, 'These I passed successfully except for the single subject of cost accounting and the instructor concerned was good enough not to hold me up on the grounds of my failure in that course.' Of his oral examinations, he tells us: 'I had been warned by Professor Wright that I was not in good physical condition and that I should limit myself to two hours of study per day. In view of my abysmal ignorance I found this impossible, with the result that I was worn out when I came up for my final. But again it was an indication of a never failing kindness of the members of the staff of the Department to me that my mistakes were overlooked and that I was allowed to take my degree at the end of the quarter.'[20]

Nor did Innis's lack of confidence end here. Even after the publication of his doctoral thesis, *A History of the Canadian Pacific Railway*, he reports: 'I had a further uneasy feeling that my thesis was inadequate and indeed one or two reviews had pointed out this fact. I must, therefore, satisfy an uneasy conscience by continuing along lines which would offset its defects.'[21] We should remember that, partly as a result of the continued vitality of the British Empire and the importance of railways within it, Innis's CPR study was more widely reviewed than any of his later efforts. (It was published by one of the leading British presses on economic and political subjects, P.S. King and Son, in 1923. Innis had been introduced to the publisher by Gilbert Jackson, one of his British colleagues in the department of political economy at the University of Toronto.) Furthermore, these reviews were, on the whole, very positive – an enviable record for a young academician's first publication.

Several points should be made here. First, Innis's comments about his education are made at the end of his life and project a more negative view of his earlier accomplishments (up to the BA degree) than what is found in contemporary documentation. Second, the comments reflect a deep sense of inadequacy, of getting ahead on the favours of others, which bears little resemblance to the diligence of his performance – although it is true that his doctorate was accomplished in haste and he was given generous consideration for his MA work.[22] Third, the directing principle behind his career was this very sense of inadequacy that led him to concentrate first on what he saw as basic subjects before considering more complex ones. In other words, his change in focus from religion, culture, and philosophy to law and then to economics should be viewed as conscious retrogression (not a progression) that would be reversed once he felt that he had established an adequate academic background. In a similar fashion, he did not have the impression that he was moving *on* from the CPR study, but rather that its fundamental inadequacies were requiring that he move *back* to a more rudimentary examination of the economic history of Canada.

The fact that his original project was deformed by the war had left him with a terrible sense of inadequacy, often barely controlled, that was to provide the motivating force for his career. He lived under the triple pressures of an unsophisticated rural background; an overwhelming workload resulting, as he perceived it, from the absence of the many promising young Canadian scholars the war had claimed; and a chronic sense of overwork and depression that was the long-term result of the psychological torture of the trenches. The intensity of these consequences of the guilt carried by Innis as a survivor of the First World War could well have led to his complete disintegration.

We are fortunate in having an odd document that gives us an insight into Innis's self-esteem at the time he was studying in Chicago: a psychological test that one of his acquaintances asked him to fill out in the summer of 1920.[23] Innis does not by any means assess himself totally negatively. As we would expect, he sees himself as 'cautious,' 'ambitious,' 'punctual,' 'modest,' and 'thrifty.' Yet, aside from a 'good memory,' he views himself negatively in the areas that have to do most directly with his scholarly work. He finds himself 'careless,' 'inaccurate,' 'lazy,' 'impulsive' (as opposed to 'deliberate'), 'disorderly,' 'impatient,' and of 'inferior intelligence.' As we would expect from a veteran bearing the psychological scars of the trenches, he feels 'gloomy,' 'self-distrustful,' 'indifferent,' and of 'inferior character.' Innis was and continued to be an individual who needed support if these negative emotions were not to reach a point where they would destroy rather than motivate his scholarly work. He got that kind of continuous support from Mary Quayle. One of the most propitious events in his life was her arrival on the scene during his Chicago years.

Mary Quayle Innis

Mary Quayle was not a cure for Innis's psychological wounds but a crutch that he would use right up to his death. Innis, a lecturer since 1918, was one of her teachers when she returned to Chicago to complete her undergraduate degree after a stint of wartime work in Washington. She shared his intellectual interests and went on to have a scholarly career of her own, her main contributions being as an editor. Her publications include the novel *Stand on a Rainbow* (1944), *Mrs. Simcoe's Diary* (1965), *The Clear Spirit: Canadian Women and Their Times* (1966), and *Unfold the Years*, a history of the Young Women's Christian Association (1949). In 1935 she wrote *An Economic History of Canada*, which summarized much of the progress that Innis and his fellow economic historians had made to that time. After Innis's death, she also served as his literary executor, editing his *Essays in Canadian Economic History* (1956) and revising *Empire and Communications* for republication in 1972.[24]

Mary Quayle was six years younger than Innis. Her father was not a farmer but a technician who had immigrated from Britain and made his living installing telephone systems in rural and small-town America. The family moved with the father's work, finally settling in Wilmette, Illinois, a short distance from Chicago. After completing high school, Mary pursued a degree in English at the University of Chicago, where she met Innis. The Quayles were a small-town family and more prosperous than the Innises. Overall, however, the couple shared a similar background: rural, evangelical Protestant (the Quayles were devout Methodists), and fixated on education as a means of getting ahead.

Mary had a personality that was ideally suited to supporting Innis. From her novel, which is largely autobiographical, we have the strong sense of a mother figure who believes that the essence of this role is to suffer the impositions of others: to encourage and mobilize people when they are 'down,' and to put up with their insensitivity when they are active and in high spirits.[25] The complementarity between Innis's needs and her personality deeply attracted them to each other. It was a union based on mutual appreciation of their 'troubles' in readjusting from 'war work' to 'university work,' as Innis so bluntly put it in a letter to her father asking permission for their marriage.[26]

That Mary Quayle was the outlet to which Innis directed his pent-up feelings of depression and loneliness and from which he received assurance is apparent in their correspondence of the 1920s, including the following selections:

> My ignorance is so astounding that I can hardly see the force in this admonition [by Professor Chester Wright to rest before the final exams] ... I am terribly sorry that I wrote the last letter but everything was pitch black at the time ... Please forgive me. I couldn't help it – you were the only one I could go to and as you can imagine things were more or less upset ... I wish I were so built that I would afford you less trouble.[27]

> This Sunday hasn't been a bit better than any of the others and probably worse.
>
> In the first place I haven't been to any kind of church ... I [met] one of the originals of the Fourth Battery ... He was with them all through and ... he happened to be the first man I have met since leaving France ... We talked on the street corner for almost an hour and he gave me the approximate location of nearly everyone. Several of course had died or been killed whom I had thought came through alright. Others were successfully re-established or were 'on the road'. The unemployment is getting steadily worse and a good many of them have been caught. It was a real treat to meet him ...
>
> The result was that I arrived late for dinner ... There were ... two of my aunt's sisters and their respective 'beaus' ... It made me feel terribly lonesome. It was almost like being an unwelcome guest.[28]

> It is terribly lonesome as you can imagine. I had grave doubts about the advisability of getting up at all this morning.[29]

> Yesterday I was as usual very disgusted with myself at not having accom-

plished anything ... Friday is such an impossible day ... I went over to McMaster and endured the unveiling [of the War] memorial ceremony.

It was not a success. It was an occasion where words were altogether inadequate and yet the Chancellor ranted and raved about the Huns ... There was such a chance to have a really great occasion and they fell so far short of it.

... I do hate being alone and I hate being with people. I wish you were here. I never could be so lonesome. But why should I bother you again ... The house is so lonesome.[30]

Here I start whining again. I know this will be the last weekend and I shall get a letter from you in the morning. I am thoroughly fed-up though I have no reason for such feelings.

... The place gets under my skin. It is driving me mad. I could yell or do anything to get the awful loneliness off the place.[31]

Don't worry about me in the least. Don't bother about what I write. I am feeling dead and everything but that doesn't matter and it can't be helped. I am sorry my feelings keep creeping into these letters but don't pay any attention to them. You have enough to look after now without bothering about me. I shall probably live through it. I have lived so far ... It would probably be better if I didn't write all this gloom but your letters mean so much ... If there was only some escape from the monotony of it all – but it doesn't matter. I must close this. It doesn't do anybody any good.[32]

I am afraid I wrote a nasty reply to your horrid letter. I wasn't feeling well that day but I have since got back on my feet.[33]

I am getting awfully fed up with this continual grind of morning, noon, and night. Nothing ever happens and I have to keep constantly going or I should be so terribly lonesome for both of you [Mary and their infant son, Donald]. That is the only way I can fight it off.[34]

Even more striking is Innis's deep-seated fear, in 1922, that another world war was breaking out over the Chanak peninsula in Turkey:

It has been a long and dreary day but I have got it over as best I could. As you probably know War has been declared. Everyone is very apathetic here but with time probably some enthusiasm will be worked up. I hope it quits before it gets going that long.[35]

These excerpts may be gloomy enough, but consider the fact that none of the 'black' letters referred to in these passages has been preserved in Innis's archival papers.[36]

Archival Censorship

A basic problem with existing assessments of Innis is that they have taken their orientation primarily from what has been retained as archival material. No effort has been made to determine what material has *not* been included in the archives and the reasons why.

The norm for Innis was to keep everything. Virtually all of his former acquaintances testified to his squirrel-like habits. Mary Quayle Innis herself noted: 'I don't think he ever threw anything away. He kept not only all notes, letters, reports, magazines, the manuscripts and galley proofs of all his books but work books and essays from public school and high school, all the letters ever written to him, etc.'[37] The secretary who took dictation for his autobiography was amazed to find that he composed it by rummaging through his remarkable collection of correspondence and memorabilia contemporary to the time about which he was writing. This included everything down to the theatre ticket stubs for the London shows he had seen during the First World War![38]

The 'Autobiography' offers us a point of departure for assessing the mass of Innis's archival material. The book was cut short, at the year 1922, by Innis's death. It is an extraordinarily reserved document. There are no conscious revelations concerning confidential elements in his life. Rather, there is a shying away from anything having to do with emotion. For instance, his passionate love for Mary Quayle is dealt with in half a sentence: 'My wife and I were married on May 10th, 1921, and the summer was spent chiefly in finding a temporary apartment at 696 Markham Street and in continuing work on my thesis.'[39]

The 'Autobiography' has more of the flavour of a biography, resulting from its basic structure as a gloss on the original documentation saved from the period. Correspondence is either repeated verbatim or summarized. Nothing personal is added except in a most guarded fashion in order to situate the documentation. Compared with other autobiographical treatments, personal reminiscences, except those recorded at the time, are negligible. When Innis is dealing with the early period of his life, for which relatively little documentation exists, he retrojects his later economic-history work by way of explanation. His childhood is firmly set, as we have seen above, in the context of a series of staple products. It is a caricaturized presentation that has little to say about family personalities, or his personal attitude towards fundamental issues such as religion. The 'Autobiography' is, like the unpublished 'A History of Communications,' a set of reading notes based on the written material, in this case material related to his life. We know this is so from the reports of the departmental secretary who worked on it while Innis lay on his deathbed. The archives show that the original pre-1922 correspondence carries the crabbed notation of Innis's later handwriting. This indicates in summary the contents of each letter as it was reviewed in 1952.

Mary Quayle Innis became the custodian of these papers after Innis's death. Her summary notes replaced Innis's on his post-1922 documentation. Over the years, she continued this process of reviewing the papers with as much attention to detail as that evident in her work on the incorporation of Innis's marginal notations into the text of the second edition of *Empire and Communications*.

Some might say that there is no need to introduce Innis's increased sense of responsibility or the psychological damage caused by the war into the treatment of his depression – that this depression can be seen as a perfectly natural reaction of a young man deeply in love but separated from his partner, or of a young scholar struggling to establish an academic reputation under the burden of a heavy teaching load. I would argue, however, that the obvious nature of this explanation tells us more about why the documents relating to these matters were saved than it does about why they were written. They are love letters, after all, preserved by the woman to whom they were addressed. Love is a laudable emotion that Mary Quayle especially would have wanted passed on as a memory of Innis. The accompanying negative sentiments are unimportant relative to this overwhelming sense of infatuation. But, as time went on, the more the bleak emotional quality came to the fore, and we find that the documentation for this later period has largely been 'lost.' Without this judgment, how are we to explain the anomaly that virtually all other material that would reveal the personal or emotional side of Innis has not been included in the archival files?

Preserved letters to friends and family drop off drastically after the 1920s despite Innis's extensive travelling and correspondence. One can safely assume that most of this would have been handwritten. It is therefore unlikely that Innis kept copies. However, we still cannot account for the absence of incoming correspondence that would give us an insight into Innis's preoccupations. As a general rule, the closer the friendship, the less likely was the correspondence to be preserved. Yet each friend is represented by a limited number of letters (often a single letter) in the files. It is as if these letters are to serve as a 'sample' indicating the general tenor of the much more voluminous complete correspondence.

Virtually none of the correspondence relating to the serious crises that Innis went through at the University of Toronto (especially those concerning his promotions) has been preserved. Similarly, correspondence relating to support or criticism of individuals or institutions as far as research money is concerned is almost wholly absent. Yet we know from the extensively documented general discussions with officials of the Ford, Rockefeller, and Carnegie foundations, from the thank-you letters from successful candidates, and from the oral history of the time, that Innis exerted a preponderant influence in this field. We are left with the tip of the iceberg in the form of non-confidential general letters of reference for either inconsequential purposes or minor personalities.

Finally, we note a significant discrepancy between Innis's written work and the archival files. In his essay 'The Church in Canada,' there is one of those extremely strident passages that crept into the last years of his communications work.

> The Department of Political Economy, if I may judge from personal experience, is under constant surveillance by a wide range of individuals. If in the course of an article I make a reference to a large government department or a large business organization, I will receive in an incredibly short time after the article has been published, a personal letter, possibly directly from the public relations officer of the organization concerned or indirectly from the president or head of the organization, explaining that my remarks are liable to misinterpretation and inferring that the head of such an influential department in a large university should be very careful about the way in which his views are expressed. *I plan to leave in my estate a valuable collection of autographs of prominent men in this country.*[40]

We know that this reference is not merely an example of Innis's rather odd sense of humour, for he repeated the statement privately to his son.[41] However, no such collection of autographs is to be found in his papers.

A review of the smaller collections of Innis's papers located outside the main archival holdings at the University of Toronto tends to reinforce the impression that the principal collection has been subjected to a selective editing process. Letters and memos from Innis in the University of Toronto Presidents' Papers, as well as correspondence with Irene Biss and George Ferguson, all provide evidence of someone functioning at a high emotional pitch. A picture emerges of a tortured personality characterized by those elements mentioned above: an unbearably heavy sense of responsibility for individual scholarship as well as institution building and national politics; an antipathy to authority, evident in a recurrent 'crankiness' (for example, a tendency to negotiation by means of threatened resignation); recurring bouts of depression and overwork to the extent of incapacitation; and, in general, a 'siege mentality' – a feeling of being part of a vulnerable minority on nearly all issues.

Mary Quayle Innis seems to have controlled these negative forces in Innis's make-up during his lifetime and to have minimized their significance after his death. For instance, we know that in March 1937 Innis was incapacitated by neurasthenia.[42] Owing to this bout of 'nervous exhaustion,'[43] he remained bedridden for most of the month, cared for by his wife. However, none of his colleagues at the University of Toronto, when interviewed years later, could recall such an incident. Instead, Professor S.D. Clark remembers that Innis was convalescent around this time on account of, as he recalls it, a war wound that was 'acting up.'[44] Furthermore, in June of that year, Innis was to take over from E.J. Urwick as head of the department of political economy. It is understandable that in a more conservative age, and at such a crucial time in his career, a confusion of physical and psychological war wounds was welcomed.

The personality of Mary Quayle Innis seems to accord with the confidant/editorial role suggested here. In fact, she remains an even greater enigma than Innis. She destroyed the bulk of her own papers, including all of her personal correspondence and the manuscripts of several complete but unpublished novels, before depositing the residue with the University of Toronto archives. It is significant that she appears not to have allowed a *single* item of her correspondence with Innis to be preserved in the Innis papers deposited in the archives. (The only revelations concerning her husband that we have from her are second hand, provided through a few short recollections of statements by Harold.) Mary Quayle played acolyte to Innis's academic priest. As such, she helped orchestrate the remembrance of the man while at the same time guarding her own invisibility. For instance, she cooperated wholeheartedly in the work leading to the production of the 1972 radio documentary on Innis yet never allowed herself to be interviewed.[45] She was an intensely private person, even more so than Innis himself.

In suggesting that she is partly responsible for reinforcing the statue-like image of Innis as the archetypical scholar, I am not saying that he would have had her do otherwise. Indeed, she may well have received a mandate to complete a process already begun by Innis as he reviewed his papers with a view to producing an autobiography on his deathbed. What should be recognized is that, arriving on the scene when she did, she provided Innis with the psychological support he needed and would continue to need given the tasks imposed by his exaggerated sense of responsibility, as well as his recurring bouts of depression and exhaustion. The tremendous positive impulsion this gave to the recently returned veteran is evident in the following passage:

> Just why I should be so happy ... is more than I can explain ... I am at the third phase of the cycle. 'One extreme has followed the other.' There are several reasons why I suppose I should be grouchy but all of them avail not ... But the mystery remains as to whatever there is in me that you seem to think is worthwhile. And still more how I am ever to reach up to your confidence. To think that you believe in me so wholeheartedly. With such confidence I shall certainly do my best and I think we can do anything! There is no job too big ...[46]

Chicago

It had been the practice at McMaster for some time to direct those students going on to graduate work towards the University of Chicago. Originally, this had been related to the Baptist antecedents of the institution, but soon the successful record of the McMaster students reinforced the trend. Distance must have been another factor. In any case, there was a rapport between the institutions that had been established before Innis's arrival. For this reason, it was quite natural that Innis, in searching for a thesis topic, should have been given one regarded as a follow-up to a study by his recent professor at McMaster, W.J.A. Donald. The CPR thesis was a logical response to questions posed by Donald's *Canadian Iron and Steel Industry*.

Both Innis's 'Autobiography' and Robin Neill's *A New Theory of Value* contain significant descriptions of the staff members who influenced Innis during his stints at Chicago from 1918 to 1920. These included C.W. Wright, who allowed Innis to work on a Canadian thesis and taught him American economic history; F.H. Knight, who taught economic theory and was of a sufficiently philosophical cast of mind that he stayed in correspondence with Innis even throughout the communications period; Professor James Alfred Field, who apparently demystified statistics as a subject; J.M. Clark, who introduced the importance of the subject of overhead costs; and C.S. Duncan, whose work relating market structure to product characteristics served as a point of departure for the later development of Innis's staples approach. Innis also pursued several law courses through the commerce school, which led to his further respect for the 'logic and force' of the subject, to the extent that he seems then to have fixed on the goal of getting to the bottom of economics as a prerequisite to returning to a consideration of law.[47]

Here again, it is the context surrounding Innis rather than individual influences that has been inadequately dealt with in previous commentary. As we might expect, the general climate of intellectual discussion at Chicago was to mark his later career to a greater extent than the influence of specific instructors. To begin with, the recent departure of Professor Thorstein Veblen, who became a seminal influence on Innis during this period, did not mean that the atmosphere had ceased to be permeated with his unorthodox ideas. In fact, Veblen's influence was so important that it is discussed separately in the next chapter.

One of the most significant effects on Innis at Chicago arose out of his activities as a lecturer under the supervision of Chester Wright. Following on Veblen's principles, Wright and others were experimenting in the reform of the methodology of teaching economics.[48] This, in turn, had grown out of a fundamental criticism shared by many of the faculty at Chicago of the primarily deductive methods that lay behind neo-classical economic theory. They maintained that economic policy could not be based on a static model, for the idea of equilibrium was an abstraction never attained in the real world. Instead, they argued for the basing of economic policy on a methodology that examined particular cases as a prelude to the formulation of general rules. Their inductive approach implied the preparation of a new set of teaching materials stressing actual case studies. Innis must have been impressed by these discussions; both his teaching and his research methodology were to be characterized by this approach on his return to Canada.

The intellectual climate of Chicago had implications for Innis at a more abstract level, too, since it reinforced the basic perspective he had experienced on the farm. On an intellectual plane, Chicago represented the point of view from the 'middle of the periphery.' James Carey has managed to isolate the main elements of the context,[49] and, drawing on his work, we can define this conception of the middle of the periphery in both geographical and ideological terms. Geographically, the majority of the intellectuals who went to Chicago came from small, relatively rural, Protestant towns, a tradition that had been established with the founding of the university in the 1890s. These intellectuals – who collectively became known as the Chicago School of Social Thought – had their personalities formed between the mass urban agglomerations of the Atlantic seaboard and the outposts of the frontier. Accordingly, they tended to replicate the experience and values of these communities in their work in various academic disciplines. In ideological terms, their task was to homogenize the small-town perspective in order to produce a national ideology that offered an effective alternative to those social paradigms put forward by Marxism, frontierism, and the liberalism of Harvard and Columbia. Those intellectual movements valorized, respectively, the European industrial experience, the American frontier, and the universalist systems-theory approach generated in the eastern seaboard schools of the United States. The Chicago School, in opposition to this view, posited the small town as a model of social organization. Its members contradicted frontierism by proposing that the essential dynamic in American civilization was not the extension of its limits through space but the creation of democracy through the collective action of small communities as they developed through time. The building of common institutions – schools, churches, community halls – where none had been before, and the correlative levelling of distinctions of class and ethnicity, furnished the central archetypes through which this approach transcended frontierism. It was not the edge of the frontier but those areas between the edge and the larger metropoles that were hegemonic in the American experience. Here were to be found the points of focus at which a new way of life, a new national culture based on 'consensus, communications and co-operation,' was being generated.[50]

Possibly because of a more intimate connection with the European experience, both culturally and as a result of their experiences with the factory system, the intellectual establishments of the eastern seaboard had countered Marxist ideology by the development of a new universalist approach to social analysis. The result was an aggressive paradigm that incorporated a strongly individualistic position. This system-analysis approach claimed validity of application to any sociocultural formation. Marxism was, in comparison, denigrated as a culture-bound paradigm that related only to the experience of early European industrialization.

The Chicago School likewise attacked the Marxian paradigm of class struggle as European-culture-bound and, therefore, inapplicable to the New World, where the relative classlessness of the small-town milieu was dominant. But, as opposed to the approach developed at Harvard and Columbia, and in reaction to the severely individualistic nature of their Protestant background, the Chicago School stressed the effect of the group on the formation of personality. In so doing, they rejected the universalist pretensions of the social thought of the eastern centres, and concentrated on a historical approach that admitted some of the collectivist elements of Marxism. However, they replaced class and conflict as the underlying principles of these elements with 'neighbourhood and consensus.' The main problem to be resolved in maintaining the plausibility of this approach was to demonstrate how – contrary to Marxist theory – small-town dynamics could remain hegemonic in an era of increased urbanization, industrialization, and an apparently 'European-style' exploitation and conflict. The Chicago School attempted to do so by bringing forward a particular conception of 'communications.'

Communications in the small-town context had always been viewed in a cultural rather than a technical sense. It was a general term referring to all those means by which the small town facilitated the melting away of class and ethnicity and formed a consensus. The challenge for the Chicago School was to determine how these processes could continue in healthy communities or neighbourhoods at the same time that an overall national culture and economy (with concurrent industrialization and urbanization) was being established. It was proposed that the new means of mass communications would allow for the continued existence of the local neighbourhood so essential to the American spirit, while also ensuring the integration of this neighbourhood into the national culture and economy of the modern era. This process was not viewed as inevitable but as one possible alternative. However, it was a necessary alternative if the essence of the American culture was to be preserved, free from the stifling control of new centralized national organizations and bureaucracies.

Another school of thought was present at Chicago in the post–First World War period that, at an infrastructural level, dovetailed well with the Chicago School's implicit critique of Marxism. This intellectual movement centred on the idea of an 'imminent benevolent technology' that would reverse the negative results of industrialization while still incorporating higher productivity. This train of thought had been passed on from the Old World to the New through Prince Peter Kropotkin to Patrick Geddes to Lewis Mumford. In summary, it posited that the exploitative centralizing and environmentally destructive features of early industrialism were not inherent to the industrial process itself but rather were the characteristics of technology in its initial phase only. This Palaeotechnic technology, as it was termed by Geddes, was based on coal, steam, and mechanical production. It was destined to be replaced by 'Neotechnic' technology based on oil, electrical power, and electrical machines. The new technology would be cleaner, more humane, and decentralizing in nature. It was argued that Marxism was mistaken, therefore, to have raised the effects of one stage of technology to the level of universal applicability. As a line of argument, these ideas tended to support the contention of the Chicago School that the 'community' could remain the central principle of American culture even in the modern industrial era.

Clearly, many of the notions that were in the air at Chicago were absorbed by Innis,[51] though some took years of germination before they were incorporated in his work. He retained an interest in the communications media as a focal point of social change; in the oral tradition (the idea of the neighbourhood) as the essential basis for the formation of personality and culture; and in the dynamics of 'Palaeotechnic' and 'Neotechnic' technology. Chicago provided the intellectual ammunition to reinforce his lifelong lack of empathy both with the frontier thesis and with liberal or Marxian blueprints for a rational new society.

However, Innis's thought should be seen as a fundamental critique, rather than a revised version, of the work of the Chicago School. His position as an artillery signaller undercut any naive utopianism concerning the ability of 'Neotechnic' technology to lead to a reversal of the centralizing tendencies in society. The Great War, in general, with its mass mobilization of the domestic population through the application of electricity to the media, and the unprecedented massacre at the front through huge assaults coordinated by electrical signalling, had left him with a rather sceptical view of the potential of these new techniques for social progress.

Innis subjected this thought to assessment in terms of the Canadian rather than the American experience. The central difference, of course, was that Canada had not cut its colonial ties with Europe in the way the United States had. For Innis, therefore, it was more evident that the

New World communities were essentially an extension of the society of Europe. Their development was seen far less as an autochthonous dynamic than as an extension of European culture and metropolitan demands for resources. Oddly enough, the very fact that Innis came from a background that was more provincial than that of American intellectuals allowed him to develop a perspective on communications which was much less parochial. As Carey puts it: 'the significance of Innis is that he took the concerns of the Chicago School and, with the unvarnished eye of one peering across the 49th Parallel, corrected and completed these concerns, marvellously widened their range and precision, and created a conception and a historically grounded theory of communications that was purged of the inherited romanticism of the Chicago School and that led to a far more adequate view of the role of communications and communications technology in American life.'[52]

Ideas of a more general nature to which Innis was exposed at Chicago lay in his mind, developing until their release twenty-five years later in the communications works. From the beginning, he was concerned with reconciliation of the larger dynamics of Western culture with the detailed historical data of his research. In this sense, his thesis, *A History of the Canadian Pacific Railway*, sets a pattern for his later work. Understandably, as a first work, it is the least successful of Innis's publications in elaborating this pattern. It contains vast quantities of statistical information bracketed by the briefest of conclusions from a man renowned for his brief conclusions. Here, he is content with a simple assertion that the study has dealt with 'the spread of Western civilization over the northern half of North America.'

The thesis led him quite logically back from the railways to consideration of the antecedent waterways. What is less often noted is that it led him to recognize the necessity of understanding, as a point of departure, the cultural dynamics that were correlated to the development of any economy. It was this realization that gave *The Fur Trade in Canada* a coherence that *A History of the Canadian Pacific Railway* lacked. It was also this realization that gave his work, from the start, a self-critical and political perspective, for it implied coming to terms with the cultural poverty of the intellectual tradition in Canada. His work, by force of necessity, became part of a concerted effort aimed at the overall strengthening of the cultural environment that produced him.

In his 'Autobiography,' Innis tells us that he was 'determined by two powerful influences to concentrate on Canada. One was the tremendous affection which Canadian soldiers who had been abroad came to have for their native land. I was very much struck by a lecture given by Professor G.M. Wrong at the University of Chicago which he stated that the War had completely changed the attitude of Canada toward Great Britain.'[53] Wrong had championed the idea of sending a group of scholars to accompany the Canadian delegation to the Paris Peace Conference. Although the Americans had adopted this practice, Wrong's idea was rejected on the basis that the Canadian delegation would be serviced adequately with information from British sources. Wrong's reaction was blunt: 'Canada goes to the Peace Conference with no opinion of her own on these matters. That is to say she goes as a colony and not as a nation.'[54]

Innis responded to this context by melding into his career the concerns of individual research, academic institutions, and national policy. What appeared to be matters of minor importance in the eyes of many of his colleagues often took on for him a much more general significance. From the beginning of his career, therefore, he was viewed as a controversial figure with a habit of unreasonable persistence in pursuing his goals. Both the close linking of individual research, academic infrastructure, and national policy in Innis's mind and the disturbance that this vision caused within the university were evident in the follow-up to his doctoral thesis.

> I had gradually come to realize that the Canadian Pacific Railway had linked up a land unit which was basically dependent at an earlier stage on water navigation. It was important, therefore, to have maps which would show how the rivers interlocked from Eastern Canada to the Pacific coast, and to restore the idea of unity which had been destroyed by maps relating chiefly to the railroad. In this I had infinite trouble, since map-makers were very reluctant to cut new plates bringing out the inter-relation by water between different regions and since they felt that such new plates would have a limited demand, and it was only after considerable expense, and with high-priced maps that firms could be induced to undertake the tasks and even then endless controversies developed between the book stores, the map producers, the importers and myself. It is perhaps this controversy which led to the general feeling in the University that too much time was being taken so far as the student was concerned and that some more precise arrangement should be made. I am reasonably certain that I was defended in all of this by my old friend and teacher, Mr. W.S. Wallace. In any case early in October I was approached by Professor MacIver in a

> confidential talk and told that Sir Robert Falconer, President of the University, was anxious to appoint a man in Geography.[55]

Innis's strong vision of his role as a Canadian intellectual thus led, at the outset, to a switch in the direction of his scholarly career that only in retrospect is less startling than his later move into the field of communications.

CHAPTER FOUR

The Search for a New Paradigm, 1920–1929

Internationalism, in this sense, ought not to consist in lying prostrate before the ... theorists of our choice, or in seeking to imitate their modes of discourse. The reasons why this kind of imitation can never produce more than a sickly native growth are complex. Mimesis, for some reason, can copy but cannot originate or create. The 'adoption' of other traditions – that is, adoption which has not been fully worked through, interrogated, and translated into the terms of our own traditions – can very often mean no more than the evacuation of the real places of conflict within our own intellectual culture, as well as the loss of real political relations with our own people.

E.P. Thompson[1]

During the 1920s, as a young professor at the University of Toronto, Innis concentrated on grounding himself in a knowledge of Canada. Focusing on the fur trade and the river systems along which it was pursued, he came to view Canada as a country that was developed because of – not in spite of – geography. He spent long hours in the archives reviewing historical documents. Beyond this, he undertook many trips to often remote parts of the country, frequently travelling under difficult circumstances. Throughout his travels, he made use of the skills he had developed in the smoking car on the Otterville to Norwich train. He learned much from the 'characters' he met along the way, not only about the fur trade but about all the staples industries that formed the economic basis of Canada.

External activity and success masked an inner turmoil. He accomplished a tremendous amount of research and travel during this decade courtesy of the psychological and moral support of his new partner,

Mary Quayle. The shrapnel that had punched a hole in his right thigh during the Great War had led to an infection, causing long-term swelling of his knee and lower leg as well as poor circulation.[2] The psychological effects were delayed, and appear to have included depression and exhaustion alternating with periods of elation and hyperactivity. In the early years of the decade, he struggled with a sense of panic that war was about to break out again. His sense of brotherhood with other veterans and the war dead was accompanied by a distancing from everyday society as he recognized how easily benign common-sense could be deformed into jingoistic militarism through propaganda techniques playing on high ideals. Mary Quayle served the role of Innis's trauma counsellor in an era when trauma counselling did not exist and when many veterans' lives spun out of control.

Part of re-establishing a 'normal' life was the starting of a family, and here, too, Mary Quayle's role in providing a supportive base for his exhausting travelling and scholarship was crucial. They had four children – Donald (1924), Mary (1927), Hugh (1930), and Ann (1933). Innis clearly loved his children, but, even by the standards of the day, he invested relatively little time in the duties of fatherhood. In later years he would occasionally take one of his children on a work-related trip. But, during the early period, he travelled without family, and Mary Quayle made it possible for him to do so without the home front becoming dysfunctional. Even his European honeymoon in the summer of 1922 was a working trip involving the confirmation of a British publisher for his thesis and the checking of references in the national library in Paris. Much of the social side of the trip was dedicated to the development of what would become an unprecedented network of contacts and friendships in the international academic community. He also made a cathartic visit to the scene of his wartime service, accompanied by Mary Quayle.

His wartime loss of faith, or rather the transformation of his calling from a religious to a secular project, was confirmed when he accepted an appointment starting in the fall term of 1920 as lecturer in the department of political economy at the University of Toronto over an offer of a post at Brandon College (at the time, another Baptist educational institution).[3] At Toronto, he encountered a situation not unlike the one he had faced during the military training he had undergone at Shorncliffe in England. Instead of British officers, he found a largely British-born and Oxbridge-educated elite dominating leadership roles and favouring, quite naturally, young British nationals or Canadians with a British educational background. From the start, he engaged heavily in academic politics, arguing for the placement of Canadian personnel – or at least senior academics from other 'new lands' such as Australia – in new openings. He did so for scholarly rather than strictly nationalistic reasons. He believed that these individuals were more likely to invest time in the 'dirt' research necessary to know the facts of Canada, and would be able to develop a paradigm more appropriate for understanding both the country and later the world at large.

His desperate pace of travel, research, and writing in the 1920s was in large measure motivated by his desire to prove this contention through the example of his own life. By the late 1920s, he had largely done so. He won over many British-educated superiors to his logic, but, while they often supported his recommendations for staffing, they were also continually exasperated by the intensity with which he lobbied for his cause. By the end of the decade, his reputation had become established, and he used it, through the threat of resignation (starting in 1929), to get his way in the discussions on the staffing and structuring of higher education in Canada. It was to become a lifelong tactic.

Innis's accomplishments of the 1920s are impressive by any standard. By the early 1930s, he had published four books: *The History of the Canadian Pacific Railway* (1923), *Selected Documents in Canadian Economic History, 1497–1783* (1929), *Peter Pond: Fur Trader and Adventurer* (1930), and his magnum opus *The Fur Trade in Canada* (1930). He had progressed from lecturer to associate professor. He had shown that a geographic and economic approach to understanding the history of Canada could lead to breakthroughs that would engage a whole generation of scholars in groundbreaking studies of individual staples industries. He had arranged for the production of new maps of Canada, attended the International Geographical Congress (1928), and essentially laid the foundation for the establishment of geography as a discrete discipline of study at the University of Toronto. He had helped establish the Canadian Political Science Association in 1929 and contributed numerous essays and reviews to academic journals. He became the first president of the Commerce Club at the University of Toronto and found time to lecture to such widely diverse audiences as the Workers' Educational Association and the Bankers' Educational Association.

When Innis returned to Canada in 1920, it was with a sense of mission firmly bound to the idea of an intellectual project. He was animated not so much by the content of what he had been taught at

Chicago as by the ideal of what was being attempted there: the collective construction of a new paradigm. Thus, he avoided being a disciple (even of Veblen) by becoming convinced of the need for an intellectual movement of comparable scope, but differing concerns, in Canada. The fusion of the ideal behind the Chicago School of Social Thought with his veteran's nationalism and sense of responsibility to the war dead became the basis of the project that would both animate and burden him for the rest of his life.

The key lesson learned by Innis at Chicago was that knowledge is always socially situated. Chicago was critical to his intellectual development because of its relation to a set of social problems and a milieu that were not being addressed by the traditional centres of learning. Innis took from the Chicago School the view that, while genuine but incremental advances in knowledge could be made using the perspective implicit in an existing paradigm, the really significant advances in knowledge took place when these older paradigms were confronted with new general perspectives emanating from previously marginalized social groups. This process might occur within traditional societies – the rise of the bourgeoisie would be an example – or, as was the case at the University of Chicago itself, between societies new and old: a Midwestern university versus the established institutions of the eastern seaboard.

There are three postulates implicit in this vision of the advancement of knowledge through a supercession of paradigms. First, each system of knowledge must depend on a social hinterland that fosters it. Second, a mature system of knowledge or paradigm becomes a self-referring realm that can be effectively criticized (or renovated) only by being subjected to a basically different 'philosophical approach' that in turn is dependent on another social hinterland. Third, this means by extension that the colonial intellectual, his institutions, and his society are central to the continued cultural vitality of Western civilization.

This view of the effect of Chicago on Innis is diametrically opposed to the orthodox one of Innis as a cloistered and naive scholar recoiling from any suggestion that the university and its staff should intervene in the day-to-day world or vice versa. The orthodox view does not recognize the relative political sophistication of Innis's project, in particular his awareness of the danger of cultural imperialism. Innis was horrified not by the prospect of intervention of university personnel in outside activities but by the tendency this created within the university to abandon long-term attempts to articulate a new paradigm. Innis felt that premature outside involvement led academics towards the intellectual short cut of borrowing ready-made paradigms having nothing to do with the Canadian situation. This created problems not only for the national interest but also for the advancement of knowledge, since it was only in intellectually peripheral contexts that one found the freedom necessary for the development of a genuinely new approach to the world. John Grierson used to complain that Innis' view of the role of the university was that of the ark on Mount Ararat. It would be closer to the truth to say that Innis viewed the university as the ark still caught in the full flood of metropolitan intellectual paradigms.

From the beginning of his career, then, Innis viewed the search for the truth and his personal research, the strengthening of the university in general and University of Toronto in particular, and the articulation of a Canadian national project as different but essential elements of a common endeavour. He had a habit of collapsing these categories with such breathtaking agility that the import of what he is saying is missed by our reading of only one level of his meaning. This is particularly so today, when a development of such a project among Canadian intellectuals seems so far-fetched as to be either preposterous or pretentious. Nevertheless, Innis entered the 1920s with just that kind of project in mind. It led him to take full advantage of any opportunity he was given, however minor.

And so, Innis's project continuously tended to reinforce itself, leading inexorably on to the next phase. As a young man, Innis had no money, and no contacts either in politics or in the world of business. His adequate but less than brilliant academic performance leading to his doctoral degree was an indicator of his colonial background. He lacked a thorough grounding in any particular contemporary school of thought. Ironically, had he gone on to work in a traditional research area already crowded with metropolitan scholars, he most likely would have produced nothing but mediocre and marginal contributions to existing works – the traditional lot of the colonial intellectual.

Faced with these prospects, Innis set out to build his reputation in an area in which he could compete with metropolitan scholarship: he developed an extensive and direct knowledge of Canada through travel and through archival research. Yet, in so doing, he never lost sight of his ultimate hope of developing a more general philosophical approach that would be recognized for its methodological genius as well as its factual content. Even in his early career, Innis pursued a wide-ranging parallel set of readings that had nothing to do with his

Table 3: Innis's Travels, 1922–33

1922	Great Britain, France, Germany, Italy
1923	Kingston, Montreal, Landonville (Alberta), Prince Rupert, Vancouver
1924	Winnipeg, Edmonton, Peace River, Lake Athabasca, Slave River, Great Slave Lake, MacKenzie River to Aklavik
1925	Ottawa (Public Archives of Canada)
1926	Hudson, Ontario – point of entry to Red Lake mines, Vancouver, Skagway, Whitehorse, Dawson, (walk overland to!) Fairbanks, Anchorage, Vancouver
1927	Three trips (not in order of visit): – Ontario: Ottawa, Sudbury, Fort William, Duluth (Minn.), Smooth Rock Falls, Cochrane, Iroquois Falls, Timmins, Kirkland Lake, New Liskeard – Quebec: Montreal, Trois-Rivières, Shawinigan, Grand Mère, Quebec, Chicoutimi, Arvida, Kenogami, Rouyn-Noranda, Malbaie – New Brunswick: Edmundston, Saint John, Newcastle, Bathurst, Campbellton – Nova Scotia: Weymouth, Yarmouth, Halifax
1928	Canada: Ontario: Haileybury, Kirkland Lake, Iroquois Falls, Cochrane; Quebec: Rouyn United States: Chicago U.K.: Liverpool, Glasgow, Edinburgh, Cambridge, Harwich, Oxford, Cornwall Europe: Berlin, Leipzig, Dresden, Prague, Vienna, Geneva, Freiburg, Frankfurt, Bonn, Cologne, Paris
1929	Manitoba: The Pas to Churchill N.W.T.: Chesterfield Inlet, Wager Inlet, Repulse Bay
1930	Newfoundland and Labrador: Port aux Basques, Corner Brook, Grand Falls, St John's, Bay Roberts, Trinity, Twillingate, St Anthony, Cartwright, Makkovik, Hopedale, Bishop's Falls Nova Scotia: Sidney, Antigonish, Pictou, Truro Prince Edward Island
1932	More extensive fieldwork in the Maritimes and Labrador, with visits to many small Quebec towns on the way (Gaspé), as well as along the Welland Canal and in northern Ontario (Chapleau)
1933	Extensive travel in Nova Scotia and Vancouver Island[4]

travels or archival work except insofar as he was trying to see beyond the basic data to a more general organizing principle.

Present and Past: Travel and the Archives

Nothing better indicates Innis's dedication to the pursuit of his project than the extensive travelling he did at the beginning of his career, funded largely, in the absence of institutional resources, out of personal earnings. None of these trips involved the speed or convenience of air travel. Consider the itinerary of his travels during the first decade of his career, as set out in table 3. Such an itinerary would be impressive for a scholar today. In pre-commercial flight days, it was astounding.

As with all of Innis's activities, we must put his travels into context if we are to understand their relationship to his project. Innis talked to people constantly in the course of his voyages. This habit was a natural consequence of low-speed travel and one on which Innis depended for much of his understanding of the economy. The view of Canada that he acquired, therefore, was built from the individual level up rather than from an academic, statistical, or theoretical level down. Furthermore, we should bear in mind that Innis's travels were undertaken in a pre-polling era in which the phenomenon of a professor arriving to ask a barrage of questions at a remote outport, northern Indian settlement, or pulp and paper mill was an utterly exotic event. Innis was continually confronted with evidence of the relevance of the direction in which his project was carrying him through the reaction of the people he met towards his work. His was a popular approach that described the overall economy largely on the basis of the particular experiences of the myriad individuals he met during his trips. He 'ransacked' the towns he visited in the same way he did the archives. He set to work immediately, searching out all the installations he considered important and striking up conversations whenever he met a character who seemed to have an interesting story to tell. He moved on equally quickly when he had exhausted the major institutions. He made copious notes along the way. The typed version of these notes from the Mackenzie River trip (1924), the Churchill trip (1929), and the Newfoundland trip (1930) run to three hundred pages.[5]

The nature of these notes reinforces our impression of his project. While the more specialized social scientist of today would tend to make notes on the particular theme he or she is currently addressing (if that scholar does descriptive fieldwork at all), Innis was describing the country in general. For instance, while researching the fur trade, he was also taking every available opportunity to look into lumber and pulp and paper operations, precious- and base-metal mining, and other resource-based industries. As a result, Innis, while he was still a young professor, gained a reputation with the public as well as his colleagues as someone who 'knew what he was talking about' because he had seen so much of it at first hand. His travels meant that he was in great demand as a public speaker in an era in which a heightened sense of Canadian nationalism led to an increased interest in the country's hinterland. It appears from the bulky file of newspaper clippings in the archives referring to these talks that Innis treated his public presentations far more seriously than his colleagues did, while also making more of them as well. Certainly, the amount of press coverage itself, as well as Innis's careful cataloguing of the publicity, indicates that he viewed this sort of work as an important part of his activities.

Direct research was one aspect of the essential grounding of Innis's work in Canadian economic history. Another key aspect was his archival research. Again we find a context in which the stereotype of the isolated scholar must be replaced by the image of a more collective endeavour, one in which a war-hardened new generation of Canadian intellectuals rewrote the history of Canada. This effort required work of many different types. On one end of the scale, it entailed the popularization of social science through new 'civics' texts such as *This Canada of Ours* (1931) by C.N. Cochrane and W.S. Wallace; a *Canadian Men of Action* series published by Macmillan, to which Innis contributed *Peter Pond;* and the provision of new maps that would show the geographical unity of Canada by highlighting river-drainage and transport systems. On the other end of the scale, the effort involved new primary research by Canadians, resulting in seminal scholarly books such as Innis's *Fur Trade in Canada*. Somewhere between popularization and original scholarly work was a need for the dissemination of key documentation in order to facilitate further scholarly work. In this regard, Innis convinced A.R.M. Lower, at the time a history professor at United College in Winnipeg, to undertake a joint project for the multi-volume publication of material from the Public Archives of Canada under the title *Selected Documents in Canadian Economic History*.

Innis recognized a need for coordination of the research being done by the new generation of graduate students and for the dissemination of information on research work in progress. He stepped in and imposed a division of labour on University of Toronto students that resulted in far more efficient research in the field of the staples trades, and undoubtedly strengthened his own scholarly work as well.[6] Innis also arranged, in an ad hoc manner, for the intra- and inter-university circulation of unpublished papers by graduate students, and, for the last four years of the 1920s, he compiled and distributed an annual bibliography of new work being undertaken on the Canadian economy, not only at Toronto but at other institutions too. In short, he was largely responsible for the groundwork that led to the establishment or revitalization of

Canadian scholarly organizations in economics, as well as of the journals of these organizations.

Yet, although the young Innis was a key animating force in this intellectual movement, he remained remote: 'I do hate to be alone and I hate being with people.' The 'joint' project with Lower did go forward, and the first volume of the *Selected Documents* effort was published in 1929. However, Lower recalls that 'we went our own way.'[7] The two scholars were never really working together at any point. There were two separate works, one edited by Innis and the other by Lower. Lower could not even follow the logic of Innis's selections. He found Innis 'close as a friend, but not as a colleague, not as a scholar.' It would be a common refrain from Innis's acquaintances throughout the course of his career.

Professor Innis

Innis's impressive activities in research, professional organization, and public speaking during the 1920s were conducted under gruelling conditions. In 1929 E.J. Urwick, the disgruntled head of the department of political economy, complained to Sir Robert Falconer, president of the university:

> The Department now contains 391 Honours students ... In addition, the Department has between 600 and 700 students from other Departments ... For the teaching of these students there is a staff of 12.
>
> ... The freshman class (147 strong) receives instruction in Economic Geography, one of the basic subjects, from a single man [Innis] ... no lecture room in Baldwin House is large enough to hold the first year, and the class has been split, each lecture being given twice.[8]
>
> ... We have no map room and hardly any maps ... There is no room in which any of the five hundred students can spend an hour between lectures to read a book or discuss difficulties etc. except a landing on the top floor which is also a classroom, and also a thoroughfare.[9]

This environment served only to exacerbate Innis's shy and rambling style of lecturing. He was, by all accounts, terrible in his delivery, to the extent that a group of students around the young Wynne Plumptre, who would become one of the first of Innis's students to join the department, repeatedly appealed to Vincent Bladen, a young Oxford-educated import, to intervene on their behalf and speak to his colleague, Innis. It is clear from reports that Innis's forte was in the non-lecture structure and smaller groups of the graduate-seminar program. Some of his teaching techniques were naively simplistic. He tried to get across the central importance of a conception of geographical space in the study of resource development by giving out maps with blank areas to be filled in as homework. What had come alive to him in his travels was unfortunately not transmitted to his undergraduate students by obliging them to slavishly recopy the related maps found in the atlas in the main library.

Innis's most successful teaching initiative was one that, not surprisingly, best combined the areas of individual research, teaching, and public policy. This was a Chicago-style seminar that Innis called 'Research in Economic and Social Studies Relating to the Dominion of Canada.' He took a case-study approach in which students were expected to undertake a smaller version of 'dirt' research, looking at a specific component of the overall fur trade. This method allowed Innis to teach, gather research material, build up university resources, and publicize the new research. An article in the Toronto Star of 5 January 1926 describes the program:

> The first tangible product of the undertaking has been a series of 16 theses dealing with the fur trade in all its details ... 16 fourth-year students in Commerce and Political Science set to work on the task of investigation under the direction of Dr. H.A. Innis. Each student engaged prepared a comprehensive thesis setting forth in detail the discoveries he had made, and in general the results of his research. The completed theses were left with the Department of University Extension, and W.J. Dunlop, Director of the Department, sent a circular to lists of fur-traders contained in Government publications, offering the free use of each work for a period of two weeks. The response to the offer was as astonishing as it was gratifying. Requests came from all parts of Canada, from the United States, from England, and even from Russia. The Hudson's Bay Company borrowed one thesis after another and made copies of them. Numerous fur firms in the United States did the same thing, as well as several companies in Canada. All sorts of letters came asking specific questions on fur farming problems. The Government Departments at Ottawa made copies of several of the theses ... Altogether it has taken from July to January, fully half of one clerk's time in the University Extension office to deal with the requirements of the fur trade awakened by the public announcement of the research work done.[10]

The role of the university's extension department in distribution appears to have been exaggerated. The few boxes of departmental papers surviving from this period indicate that Innis carried out extensive direct correspondence with furriers and other individuals outside the university who were interested in the new research. Even during this early period, Innis's correspondence, on the evidence of the surviving papers, was much heavier than that of any other member of the department. In fact, the overwhelming response to the circulation of the research papers undoubtedly facilitated the publication of *The Fur Trade in Canada*. Nevertheless, the scholarly book market turned out to be much more restricted: it took fifteen years to sell the first one thousand copies of Innis's landmark book!

In addition to public-speaking engagements and the circulation of the fur-trade essays, Innis was active outside the university community through teaching assignments with the Workers' Educational Association and the Bankers' Educational Association. He began working with the WEA immediately after his arrival from Chicago, and seems to have been tortured to no end by his feeling of inadequacy in teaching economics to mature working men. As we might suspect, this doubting of his ability related more to his own psychological state than to the reactions of the worker/students to his performance. Bladen, who took over his class the following year, commented, 'They worshipped Innis and tolerated me.'[11] Bladen also believes that the perception of Innis as having disturbing 'radical' views warranted the assignment to Innis's class of a spy from the Royal Canadian Mounted Police. The twin pursuit of such starkly different groups as the WEA and the BEA by the department of political Economy indicated a policy of social neutrality on the part of the university that Innis both believed in and, through his influence on the department, propagated. However, it should be pointed out that this was a socially activist policy rather than an 'ivory tower' one.

The 1929 Resignation Crisis

Insofar as Innis may be said to have practised departmental politics, his activities almost always were aimed at fulfilling his own mission and getting the university to fulfil what he viewed as its mission. For instance, in his early years at Toronto, Innis aggressively followed up deficiencies he had noticed in existing maps that did not show the underlying unity of Canada. His concern with the publication of new maps led to his being consulted, in 1928, on the possible appointment of the first geography lecturer at the University of Toronto. President Falconer's straightforward inquiry addressed to a relatively junior faculty member, Innis, led to a decade-long bombardment of the president's office with confidential memoranda urging the development of geography courses and suggestions of available personnel.

Innis's overall project from the early part of his career is well summarized in these memoranda. Here is a typical excerpt:

> I have perhaps no right to make any comment at this juncture but I plead the thought, time, and money spent on the whole question as an excuse for this note ... I am very much impressed after this summer's [travel] experience of the growing importance of the Hudson Bay area. We must follow up along these lines. 1. Scientific work on the Arctic and the Arctic archipelago. 2. Closer relations between the Maritimes and the West by Hudson Bay. 3. Closer relations with Newfoundland. My fear is that the Americans will precede us and that we shall lose our one chance of broadening out in our own country. This may appear beside the point but to me it makes more urgent the problem of geography. Moreover, I am convinced that from the standpoint of economics or of geography – human geography – Canada must take the lead among new countries in working out their own peculiar problems.[12]

The appointment of a lecturer, basic research in the geography of the Arctic, Canadian unity, the expansion of the economy, the intellectual and territorial counterbalancing of the American empire, the development of a perspective appropriate for countries such as Argentina, New Zealand, Australia, and South Africa – to Innis, all these elements were inextricably intertwined.

At this favourable juncture in the late spring of 1929, Innis sent a short handwritten note to Falconer, and an identical one to E.J. Urwick, his department head. 'It will probably come as a shock to you to receive my resignation which is herewith sent to you to take effect with your permission at the end of the session 1929–30. I should like to have another year to complete work for the department which I have in hand but if you insist it can take effect this current year.'[13] The immediate precursor of this bombshell was the announcement of salary increases. Hubert Kemp, who had entered the department at the same time as Innis, had been given a slightly greater increase. The matter was promptly rectified and Innis stayed at the university.

Few incidents in the life of Innis have generated such a variety of comments. The episode is most often interpreted, given the relative insignificance and lack of premeditation behind the action that triggered it, as an indication of an unreasonable petulance in Innis's character. Yet an overview of the comments made by various observers of the crisis paints a radically different picture.

Carl Berger, the historian of Canadian historians, provides us with a rendition of the affair that could be termed the orthodox view of the time. He tells us that 'Innis was not above occasionally identifying his personal interests with the cause of "Canadian scholarship." He knew that he was an indispensable figure in the scholarly associations and his university and he was not immune to the resulting temptations. His favourite device for getting his own way was the blunt tactic of resignation, or the threat of resignation. It had worked in 1929 when he wrote a letter of resignation in a huff because a colleague was promoted above his head.'[14] The strength of Berger's interpretation lies in his identification of the crisis as the first example of a tactic that would be used time and again by Innis in the course of his career. Also, Berger correctly points out that the differential salary increases had the effect of promoting Kemp above Innis. However, he trivializes both of these observations by placing them in a context that presents Innis in the role of the scholar as prima donna. It seems difficult for Berger to imagine either that such an exceptional scholar as Innis could have felt a genuine threat to his activities, as opposed to a personal slight, or that the interests of an individual could be identified objectively with the cause of Canadian scholarship. The necessity Berger feels of putting this latter term in quotation marks gives us some idea of the chasm that separates Innis's day from our own.

That someone has a tendency towards paranoia does not automatically mean that they have nothing about which to be paranoid. (Even paranoids have enemies, as it has famously been said.) During this early period, Innis was in fact under a very real threat. His radical stance on the Canadianization of faculty, on the expansion and the structuring of the department of political economy, and on the role of the university in national development meant that, from the beginning, he was in a greater measure of conflict with senior faculty than any of his junior colleagues. J.B. Brebner, in a draft of his paper 'Innis as Historian,' that is more forthright than the final version, puts it this way: 'In his early days at Toronto, some of his myopic, or perhaps merely genteel colleagues wanted to get rid of this academic maverick, but MacIver, the head of his department believed in him and defended him until his reputation was made.'[15] The full significance of the hostile campaign against Innis and the conscious premeditation by those in authority in the matter of the salary increase are bluntly outlined by Innis's arch rival of the time, Gilbert E. Jackson.

> During Urwick's Headship of the Department (I being in charge of Commerce and Finance) E.J.U. [Urwick] consulted much with me.
>
> Looking at the distant future of the Department, we could see that some day, Kemp or Innis would become Head of it. On a salary basis (by some few dollars yearly) Harold was the senior man: but he literally recoiled from administrative problems – and was quite without capacity to deal with them.
>
> By contrast, Kemp had a natural talent for that kind of thing.
>
> Once on a time (at my suggestion) Urwick recommended an all-round increase of salaries, which would have left Kemp with something like $25 per annum more than Innis. Technically, that would have made Kemp the senior of the two men and (in the silly system universities use) would at some time in the future, have made him, not Innis, Head of the Department.
>
> In that event, Innis, working full time in the field of scholarship, would have become a better scholar than in fact he was: Kemp would have run the Department during the very difficult years between 1937 and the present.
>
> When these changes were authorized, Innis raised a fuss behind the scenes. Falconer re-discussed the subject with Urwick and then gave Innis a further small increase which restored the former relation of seniority between the two men.
>
> In the result, when Urwick retired (and when I refused to come back from England and head the Department, as Cody pressed me to do) Harold Innis became Head.[16]

Jackson's remarks are interesting since they indicate that the university president was presumably originally in favour of this decision to block Innis's imminent appointment as department head. Further, they lead us to believe that, even in 1937, the new president, Henry John Cody, would have preferred appointing the British-trained business economist, Jackson, over the thoroughly established scholar, Innis.[17]

Bladen, while disagreeing with virtually all of Jackson's comments on the relative merits of Innis and Kemp, agrees with Jackson in seeing the political thrust of the event: 'The pecuniary aspect was negligible ... but the status problem, the apparent vote of non-confidence in [Innis] and of lack of interest in scholarship, the indication that the headship of the Department might go to Jackson or Kemp, these things hurt.'[18] Yet Bladen, too, confines the crisis to personal dimensions. It was not the exclusion of Innis from the key position of influence within the department that was germane, but the lack of recognition of Innis's scholarly accomplishments. Hurt feelings, not political dynamics, seem foremost in Bladen's view.

Bladen's version of his role in the affair as an ally of Innis is interesting, for he goes on to portray the close connection between personal background, influence in academic politics, and scholarly orientation within a colonial intellectual environment.

> Urwick became Head. Now Urwick was English. He had heavy pressures from Gilbert Jackson and Kennedy. I became a sort of administrative chore boy. I had for the next few years more influence in the Department than probably for years afterwards. Urwick understood me: I was English, I was Oxford. I was in his tradition and therefore I had a sort of unfair 'in.' I think I used it properly, but I did have a good deal of influence on him ... He could not understand this raw Ontario farmboy who was slugging away in the Archives and slugging away up in the Arctic, studying his country and talking such nonsense (as Urwick thought) about economic determinism.[19]

In other words, the intensely 'personal' reactions to the event stem from the frustrations that inevitably arise in a situation in which indigenous personnel are obliged to submit to the prejudices of expatriate direction and supervision. The colonial dynamics involved in this situation are much more subtle than usual. After all, neither differing languages nor skin colour acted as a clear demarcation between indigenous and expatriate positions in the Canadian university setting of the 1920s.

Although Urwick later became one of Innis's closest allies, from his point of view at the time – and that of Jackson, Falconer, and later Cody – Innis must have appeared unsuitable for academic administration because of an unpredictable and uncompromising strain in his character. It is unlikely that they took seriously Innis's wide-ranging passages outlining his hopes for the colonial university. In writing off his project as a quirk of personality, they were undoubtedly convinced that they were acting in Innis's own best interests, steering him away from administration and towards more research and teaching. Innis's violent reaction only reinforced their assessment of his overly sensitive and uncompromising nature.

Innis viewed the events from an altogether different perspective. The persistence and inflexibility of his actions stemmed directly from the tremendous sense of importance he attached to an overall vision that linked his research and the development of higher education in Canada with the maturing of Canada as a nation and with the continued vitality of Western civilization. All of his activities could in some way be related to this overall project. Accordingly, any specific opposition he met on any point in his program came to be viewed as a frontal assault on all aspects of it. Innis tendered his resignation for two reasons: because the salary decision represented a slight to his scholarly accomplishments, and because that decision meant, indirectly, that the policy decisions that he was vigorously championing in geography and in commerce and finance would not be taken seriously by the administration. To understand this political background to the resignation crisis, we must examine in some detail the structure of the department of political economy of the time.[20]

The Politics of Political Economy

The political economy department of the University of Toronto bore the stamp of the German academic tradition from the time of its founding under W.J. Ashley in 1888. It was an 'all-in' department whose inclusiveness was exaggerated further by the lack of structural elaboration of the university as a whole. The department not only included political science, economics, commerce and economic history but also served as the point of gestation for law, geography, social work, and sociology. Innis's lifelong diatribe against specialization in the social sciences, therefore, can be seen as an attempt to prevent the dissolution of the almost multidisciplinary structure of the department. His position was radical in its intensity, conservative in its thrust.

By the 1920s, the centrifugal tendencies in this small but sprawling structure were being felt seriously. Law had taken on a separate existence. Social work was also separate but was under the direction of Urwick, who, at the end of his career in 1928, was also acting as the caretaker head of the department. It was thought that he would effectively bridge the gap while the younger men in the department matured, in terms both of scholarship and of administrative responsibility. Not surprisingly, 'maturing' involved an inevitable jockeying for power in anticipation of the time when Urwick stepped down. This process was exacerbated by centrifugal forces imposed on the department. Calls for specialization and compartmentalization came from the increasingly elaborate curriculum of the university as a whole, from the example of the trends in the organization of metropolitan universities (particularly in the United States), and from the demands placed on the university by business and government for a more practical type of training to meet specific employment needs. In this atmosphere, departmental discussions were charged with personal empire-building efforts that, in turn, were directly related to personality clashes, as well as to more fundamental differences over educational policy and the role of the university in society.

Innis, with his war-heightened sense of mission, had entered these discussions with utmost seriousness. At first he had found little support for his opinions. Coming from a relatively poor rural background, he had no material resources, powerful contacts, or savoir faire in dealing with those in authority. We have seen how he attempted to change this by establishing a scholarly reputation and by corresponding directly with people in business and government who had an interest in his areas of work. At the same time, he was establishing his leadership among the younger scholars as an animateur of new journals and academic associations. Until he had built his reputation, however, he was in an exposed position, and his designation as 'economic geographer' must be interpreted in light of this. It seems to have been an ingenious move by MacIver, the department head before Urwik, to protect his young colleague. This allowed Innis to counter attacks centring on the 'excessive' amount of time he was spending on discussions concerning the publication of new maps. It also may have put others at ease by somewhat setting off a brash and opinionated young man from the more crucial areas of departmental policy related to political economy and commerce.

The Geography Crusade

Despite this, Innis remained by far the key disturbing element in the department. With his tremendous capacity for work, he made the starting up of a geography section his special project without in any way lessening his involvement in other areas such as commerce and finance. By 1930, when the university had the means to move him out of political economy departmental affairs by making him responsible for a nascent and separate geography department, Innis's scholarly reputation was such that he was able to resist. He wrote to President Falconer, not without a touch of pride: 'While I appreciate very much the honour which you have done me in suggesting that I should take the appointment [as the first professor in the new geography department] I have gone over the ground carefully, have consulted my friends in this and other universities, and have decided that I must keep my hands to the plough and remain in economics. If further arguments were necessary they were sufficed during the past week by the honour which J.M. Keynes has done me in asking me to write an article for the Economics Journal on the economic situation in Canada in 1931–32.'[21]

By recognizing Innis's concern for geography, the university authorities opened a legitimate path for him to correspond directly with the president without reference to his department head. After the university had sent him as its delegate to the 1928 Geographical Congress in London, and after he had given some serious consideration to the candidates available to fill the potential geography post, Innis became extremely insistent in his correspondence with the university administration. The tone of his correspondence was one of high-quality pestering. Not surprisingly, it elicited banal, off-putting replies from the president. Yet it tells us a great deal about Innis's perspective on the world and about the opposition he faced within the university.

From the beginning, Innis viewed personnel issues through a geopolitical perspective applied to academic matters.

> There can be no doubt that the social sciences, particularly geography, in the United States are becoming particularly concerned with Canada chiefly because we have done so little in the universities in the subject ... Foundations will almost certainly be turning to Canada to an increasing extent with a view to the development of research in geography. Under these circumstances it is extremely important that Toronto should take a prominent role – indeed should assume the leadership in order that foundations will turn to our men in the expenditure of special grants. In turn such leadership will attract graduate students and will enable the university to play an outstanding role in the development of research particularly in the north and the Arctic regions to which interest is rapidly turning.[22]

His was a highly nationalist approach going beyond concern with hiring Canadian personnel to the issue of building up institutions with an authentically indigenous perspective. In the case of the first geography appointment, the mediocrity or unsuitable perspective of available Canadian personnel led Innis to recommend the appointment of a non-

Canadian, albeit one who came from another European-settler society, Australia: 'Griffith Taylor ... commands the respect of geographers in the United States and throughout the world by virtue of his work on the Antarctic and in the field of geography generally. He would be a centre for the direction of research and from the standpoint of finance would attract funds for the development of any project ... His appointment would establish our supremacy in the field immediately and for a long period.'[23] Innis also argued that the perspective Taylor developed in work on another 'new' country, Australia, was more appropriate than the mainstream British or American point of view. Finally, he agreed with Taylor's assertion of the close connection between indigenous geographical research and successful national development.

The replies to these rather impassioned petitions indicate that the president simply could not fathom Innis's wide-ranging plea. The accepted view in Canadian academia pronounced British good and Australian bad. And, in any case, the administration's concern was simply getting a few new courses taught at the least possible cost. As Falconer put it: 'It is altogether probable that [Taylor] would be quite beyond what we would give for our inaugural position in geography. I have heard of one or two in England who are young men of great promise, and I am not without hope that we can secure one of them ... The fact that Professor Taylor is no longer a young man and that his life's interests have always been in Australia so far would have to be very carefully considered ...'[24] Innis, refusing to be put off, simply changed tack. A draft of his reply to Falconer reads:

> We are forced to leave out of consideration the appointment of an outstanding geographer and to confine ourselves to consideration of the possibilities of young promising men ... The difficulties of such an appointment have proved to be numerous ... Men trained under outstanding human geographers of other countries are almost certain to have the same slant of those men ... I am more and more convinced that the problems of a new country cannot be interpreted in terms of the problems of an old country and that the problems of a new country must be faced and interpreted from the standpoint of a thorough knowledge of conditions of the new country, and I am beginning to doubt whether we can safely appoint a young man who has not a Canadian background.[25]

Innis went on to argue that, if such an appointment is made, it should not have as a structural corollary the establishment of a new department. Rather, he saw 'the necessity of building up a department with its roots in existing departments thereby allowing it to draw strength from the old departments.'[26] In essence, the appropriate indigenous perspective on a discipline had to be ensured before that discipline gained a measure of independence within the university structure. Paradoxically, Innis felt that this measure of independence would be justified by the extent to which geography recognized its role as an integrating discipline, serving as a bridge for discourse between the natural and social sciences: 'For example, in transportation I doubt whether an economist appreciates the considerations which really influence the engineer in his final choice of a location. The serious gap between the engineer and the social scientist is a serious deterrent to the development of economics, of human geography, or of any of the social sciences. For this reason ... we need definitely to take steps in a geography department to bring together those sciences.'[27]

Innis's persistence in the face of the repeated rejection of his initiatives is amazing. He continually skirted the limits of what was civil for a young academic dealing with senior authorities. His resolute refusal to be put off by his academic superiors was rooted in the difference between his and their motives. While they were caught up in the day-to-day running of an institution with a minimum of controversy and cost, he was concerned with laying the basis for an independent country whose intellectuals would make a contribution to universal knowledge.

His universalist conception of matters often led him to define legitimate elements of the disciplines in which he was involved, be it geography or sociology, in terms of their relevance to his own project. For instance, 'geography' for Innis was synonymous with the study of economic development. When he referred to the subject, he more often than not qualified it as 'human geography,' meaning natural resources and those geographical features that have a bearing on their exploitation (through determining transportation infrastructure, for example).

Innis's tenacity of purpose had the effect of allowing him to achieve his ends by attrition. His nominee, Griffith Taylor, was eventually appointed to head the new department of geography in 1935. Yet Innis seemed unable to let go of a sense of responsibility in any area in which he became involved in the course of his career. In the geography department, he continued to intervene in personnel hiring, in research-fund allocation, and through the mechanism of book reviews. For example, long after he stopped pursuing research on the Arctic, he continued to provide an annual critical commentary on new books and scholarly articles on the topic in the *Canadian Historical Review*. As well as keeping his own reputation current in areas that he was no longer actively researching, this practice allowed him to update his existing publications with revised notes. The tremendous influence of the man in the last years of his life was largely related not to his research of that time but to his habit of never letting go of areas in which he had been working in the 1920s and 1930s. One might have expected, for instance, that Innis, having seen to the establishment of the new department of geography under the headship of Taylor, would have given up his machinations in this field. Instead we find him writing a personal note to the president in 1938: 'Professor Taylor tells me confidentially that he plans to appoint a junior man in his department. In recommending the appointment of Prof. Taylor I had in mind the appointment of a young Canadian who might be chosen with a view to succeeding Prof. Taylor after some time under his direction. The whole matter is of course none of my business but the future of geography will be immensely strengthened if Prof. Taylor's ultimate successor is a Canadian familiar with our problems and with the Canadian terrain.'[28]

In the late 1940s Innis intervened again, this time with the new university president, Sidney Smith. Smith had offered the position of department head to a renowned British geographer who was at McGill. When Innis learned that this offer had been made without what he considered to be an adequate consideration of Canadian candidates, he threatened his own resignation unless the offer was rescinded. Innis's reputation was then such that Smith, highly embarrassed by the affair, was obliged to, as he put it, 'roll with the punches' and retract the offer.[29]

The Clash over Commerce

Commerce and finance had originally been an honour course leading to a BA honours degree, but in 1930 it became a separate course delivered by Gilbert Jackson and led to a Bachelor of Commerce degree. The honour course in political science and economics was concurrently retooled, presumably to be less oriented to business needs. As he had in the case of geography, Innis entered the fray on the question of commerce and finance. We can gain greater insight into Gilbert Jackson's efforts to block Innis's promotion by reviewing the latter's role in this area of departmental activities.

When MacIver moved on in 1927, the university decided to appoint Urwick as head of the department of political economy. They had in mind the role of a caretaker head, and this led to increased efforts by some senior members of the department to strengthen their prospects – especially by consolidating those areas of the department in which they were particularly involved. In the case of Jackson, this meant the commerce and finance course. As a new initiative with implications beyond the bounds of economics, the course was under the supervision of the commerce and finance committee, with Jackson as its chairman. The committee included members from outside the department and was charged with setting the direction this rapidly expanding course would take. It made recommendations that were passed on to Urwick and C.R. Fay, the two senior political economy members. As one would expect, Urwick supported these recommendations and forwarded them to the university administration for action.

Jackson's key recommendation concerned the establishment of a course leading to a separate Bachelor of Commerce degree. He backed the recommendation with arguments which appeared designed to meet specific criticisms from other staff members. He was particularly strong on two points:

> Since the course was incorporated ... a high standard of scholarship has been maintained with the result that large numbers of students ... have failed and been compelled to leave it.
>
> ... Almost inevitably there arises among some of the parents of students who fail a feeling of resentment, which may become sufficiently widespread in time either to provoke an agitation for lowering of the standard ... or in some other way seriously to embarrass the University.[30]

The almost threatening tone of his argument brings the context into focus. The students entering the course were those set on a career in business, and presumably many came from families with significant business contacts – hence connections with the controlling circle of the university. It was the sort of clientele that would not put up with a high failure rate.

Jackson, unlike Innis, seems to have shared the same perspective on the commerce course as that held by this clientele. In short, he believed that whatever training the university might give to make these students more fit for the immediate day-to-day demands of the business world was all to the good. His second main reason for the establishment of a separate course was that 'the demand for Commerce and Finance graduates in Canada is *always* in excess of the supply.' It was the type of argument that would appeal to a Board of Governors eager to see the university pursuing work of practical significance to the world outside academia. Jackson, as a well-connected business economist little interested in research, was a force to be reckoned with.

Despite the backing of Jackson's proposal by Urwick and Fay, Innis took the extraordinary action of petitioning the president directly and repeatedly on the issue. He seems to have run a continuous and obnoxious opposition to Jackson's efforts within the department. He could do so with effect because of the support he had within the faculty, especially its younger members. This was based partly on his academic reputation but also on his key role as an organizer who, during the last four years of the 1920s, produced and distributed the annual bibliography listing new research on the Canadian economy. Innis was himself becoming known to business and government officials through his own research.

His method of argument was to describe the broader context in which Jackson's proposals were being considered. Once again he returned to his vision of a coherent project in which individual research, possible only within the protective institutional structure of the university, was essential for the effective pursuit of national-development plans:

> The whole question must be considered in relation to the problems of an economics department in an Arts Faculty. The problems of such a department are always serious, in as much as it is a frontier department from the standpoint of the public. In no other department is it necessary to consider with such care the probable effects of any policy from the standpoint of the University and from the standpoint of the public; and in no other department is it more difficult to obtain thorough consideration because of the possibilities of public pressure. This problem has become especially serious in courses training men for business. We cannot determine a policy for an economics department in the Arts Faculty on the basis of business demands. The business man is in very much the same position as an infantry man in the trenches. He is of vital importance, but one would never get anywhere in the trenches by asking his direction. He knows a small sector, and he thinks only in terms of this sector. The economics department must be a board of strategy, and it must determine lines of policy. No army could afford to rely only on the infantryman's advice, and our department cannot afford to rely only on the business man's advice. The policy would be fatal.
>
> An analysis of the position of the economics department in relation to

> the development of Canada will clarify this point. It is only from the economics departments of our Universities that men competent to do research on Canadian economic problems can come. These problems are of vital importance to Canada's national life. They involve the larger problems, with which the business man is concerned in only an indirect way. We cannot afford to sell our national soul to the immediate demands of business. I believe in the fundamental necessity of a strong economics department in the Arts Faculty from the standpoint of the relation of the University to Canadian development, and this involves a far wider point of view than the business man can possibly admit. The business man's importance cannot be neglected, but his place must not supersede that of the interests of the country.[31]

Once again, in the symbolism Innis uses, the sense of responsibility occasioned by the war comes to the fore.

The perspective implied by his project led Innis to see problems where Jackson saw only success. For Jackson, the fact that jobs were awaiting graduates was of positive significance. For Innis, the opposite was true: 'Commerce and Finance ... has proceeded in relation to business demands. It has grown rapidly and has tended to swing the energies of the Department of Economics into the training of men for business. On the whole, students tend to be less interested in the scientific point of view, and come into the course with the definite idea of obtaining a business training. It is apt to attract weaker students ... Only by concentrating our best energies on our best students can we hope to meet the situation ... The Course strikes directly at the heart of our economics work.'[32]

Jackson was preoccupied not so much with graduate studies and research as with the fact that the course graduates would be seen to acquit themselves well should they go on to metropolitan institutions. Innis argued that this lack of concern for building up a graduate-studies capacity meant that 'we are literally driving Canadian students to American Universities ... The loss of Canadian students to American Universities means a loss of the most precious asset we have, and even when these students do graduate work on Canadian subjects, it is done from the American point of view. Our problems must be approached from the Canadian point of view, which, as I see it, differs radically from the American.'[33]

Jackson proposed measures for increasing access to the course and for making it more relevant to business demands (for example, making mathematics compulsory). Innis suggested methods that would not only restrict access but would clearly indicate that the course went far beyond a concern with business training. 'The immediate remedy appears to lie along the lines of raising the standards and adding Latin to the requirements.'[34]

Finally, in his petition to the president, Innis attacks the structural basis of Jackson's influence in the matter: 'The present organization, in which the increasing majority of our students is under the jurisdiction of a separate Commerce and Finance Committee, is a serious danger to economics ... The only satisfactory solution ... is the abolition of the Commerce and Finance Committee.'[35] This document was circulated a few weeks before the salary recommendations discriminating against Innis were passed. At this same time, Innis was at the height of his initial and abortive lobbying campaign for the appointment of a geographer. Given these facts, it is likely that the decision on salary and promotion was a conscious act designed, on the part not only of Jackson but also of Urwick and Falconer, to send Innis a warning.

It is not difficult to understand why certain of his senior colleagues found Innis off-putting. His vision of the university was so intense that it often had the same effect on others as did the overly righteous attitude of the excessively religious. In other respects, he had a quite congenial personality. However, it would brook no infringement on the principles of his project. The nature of the project itself meant that he was constantly skating on the edge of impoliteness in making representations to those in authority. In the context of the post–First World War years, his stance could easily be viewed not only as anti-authoritarian but also, in his comments on business, as anti-capitalist. Yet, while he was suspected of being a radical, he continually annoyed those of progressive inclinations with his incessant critique of their pretensions to social planning.

Innis's impudence and the strength of reaction it called forth are well summarized in the following complaint of the mild-mannered Urwick to Falconer:

> I understand that Professor Innis has been to see you with reference to the position in this Department. In the particular circumstances I deeply regret that he has taken this step. It is not likely that he has explained the circumstances to you, and I, therefore, feel it to be my duty to do so.
>
> Professor Innis has for some time adopted an attitude of criticism of most propositions which have been brought forward at our staff meet-

> ings, especially when such propositions have had the support of the Supervisor of the Course in Commerce and Finance. This criticism culminated, a fortnight ago, in an attack upon the propositions of a committee appointed by the staff. The attack was made in a way calculated to give special offence to the chairman of this committee, Professor Jackson. The latter subsequently submitted to me a letter which he proposed, subject to my approval, to send to Professor Innis, asking him to give an undertaking not to continue these attacks. I sent this letter to Professor Innis, with a note to the effect that I considered the request reasonable. Professor Innis has refused to make any reply, beyond sending me a statement of various criticisms of the Department.
>
> With regard to the latter, I can only say that his statement has no bearing whatever upon the present difficulty, which he seems unable to appreciate. In his criticism of the Department, moreover, he is wrong in his facts, mistaken in his inferences, and stupid in his estimate of the position. That is not a matter of any great importance. But the fact that he has chosen this occasion to place his views before you is important. At any other time I should, of course, be glad to hear that he had done so. But in the present circumstances, I cannot but regard his action as an impertinence.[36]

It is a measure of Innis's growing importance that Urwick does not tell Falconer to ignore Innis's letter, but only requests that Urwick be given an opportunity to express his side of the dispute.

The Politics of Scholarly Reputation

Innis's tactic of forcing decisions within the university by threatening resignation became more effective as his scholarly reputation grew. Oddly, though, when Innis did attain positions of authority within the academic establishment, he did not in any sense regard himself as part of that establishment. His sense of detachment was not a trick played on him by his mind. On the contrary, it reflected the reality of his situation. Promotions within the university did not come naturally to Innis as they would have to a man who exhibited a similar level of scholarly production but without the encumbrance of a political vision concerning the role of the 'colonial' university. The greater the position of authority within the university, the more he felt the active opposition to his conception of the university. It seemed to him that he was surrounded by men who were all too ready to pick up the current dominant paradigms of the British or American institutions, with regard both to academic structure and to approaches to research. He was exasperated by the wholesale borrowing of such prepared systems as Marxism, Fabianism, and Keynesianism, viewing this borrowing as an intellectually lazy alternative to what he insisted on: the working through of an indigenous Canadian paradigm. Borrowing a paradigm or a bias in terms of one of the metropolitan centres automatically brought with it a set of allies. Innis, in contrast, was left in an exposed position by his insistence on the strategy of developing a new philosophical approach. As he put it in the wake of the resignation crisis of 1929, 'a new field has the advantage of being interesting and the disadvantage of being disheartening and at times lonesome.'[37]

All his major appointments within the university – as associate professor (1929), professor (1936), head of department (1937), and dean of the School of Graduate Studies (1947) – were carried through in the face of resistance from the university administration and on threat of resignation. Innis's project had a self-referring symmetry that reinforced his role as odd man out even as he advanced through the academic establishment. For Innis, the only way to prove that a new perspective was needed for a new land was to produce new scholarly research that pushed forward an understanding of that context. In accomplishing this, he became personally recognized within established paradigms as a scholar who specialized in colonial economic history. It was the most limited recognition possible, ignoring both the structural implications and the aggressive universality of the philosophical approach he was building. The university administration was put out by his shrill memoranda on the nationality of new staff and his related opinion on the appropriateness of their perspective to the Canadian context. At the same time, his colleagues viewed his already wide-ranging conclusions as eccentric or, more often than not, pretentious.

The comprehensive nature of his project imposed contradictions on Innis that would become increasingly severe with time. The resignation crisis of 1929 marks a significant stage in his development since it demonstrates the difficulties he experienced in attempting to link the scholarly and institutional levels of his project. Rather than developing these two levels in unison, his scholarly research became the platform from which he demanded an increasing share of political authority within the department. Over the years, the efforts necessary to extend his original research became immeasurably greater just as his level of authority grew. Yet the time available to be devoted to those efforts diminished drastically given the demands imposed on him by the exercise of increased authority.

A Premature Paradigm

[Innis's] reputation was made by a single book ... *The Fur Trade in Canada* ... It is doubtful that any other Canadian monograph, except perhaps F.X. Garneau's *Histoire du Canada*, has had an equal impact on Canadian intellectual life. After it, I suspect, few scholars wrote about Canada without wondering what Innis would think of their work. I know that I never did. The book's sweep was so thorough that it substituted an economic, geographical theme for the previous political, personal thesis of Canadian development.

– J.B. Brebner, 1953[38]

The significance of *The Fur Trade in Canada* lay not just in its illumination of new content from the Canadian past. At least as important was its demonstration of a new perspective: an innovative methodology for pursuing the past. Over the years, this methodology has come to be typified as the 'staples approach' or the 'staples theory' of economic growth. Several elements are usually included in the orthodox description of this theory. First, emphasis is placed on its limitations as a methodology of social analysis: it is a theory applicable only to cases of 'new' or 'empty' lands peopled by European settlers. Second, it is claimed, the staples approach uses two determining factors to explain the historic development of these regions; the first is the set of characteristics of the chief commodity produced by the region, and the second is the geography of the hinterland through which the commodity must pass. In fact, it is a case-study approach to specific commodities that takes on the aura of general theory through the mechanism of the particular writ large as 'staple.' It is often interpreted as an extremely deterministic theory – a kind of hinterland commodity fetishism raised to the level of an intellectual ideology.

This premature elevation of certain elements of Innis's early work to the level of a new paradigm suitable for probing hinterland contexts has had the effect of limiting our critical understanding of Innis.[39] The staples theory is characterized by a seductive symmetry of thought that has allowed it to become the orthodox interpretation of Innis. However, it can be a misleading interpretation, even of his early work. When we examine his writings closely, we find that they neither can be reduced to a few neat rules of analysis nor confined to the case of European-settler colonies.

There are good reasons why the staples theory and the essential kernel of Innis's thought have come to be improperly equated. For one thing, it is undeniable that Innis was concerned, especially during the

1920s, with the elaboration of highly deterministic explanations for Canadian economic development. His clashes with more senior colleagues in the department were reinforced at the level of intellectual work by his strident championing of the 'materialist' side of the idealist/materialist debate. Innis always found time away from his regular research and academic activities for wide-ranging, miscellaneous reading. During this period, the evidence we have of this type of reading indicates his fascination with monocausal explanations for a wide variety of social and historical phenomena. Innis's weakness for extreme deterministic presentations is demonstrated in his reaction to Ellsworth Huntington's *World Power and Evolution*. He was greatly taken by this work and seems to have incorporated it into his 1922–3 economics lectures. Huntington reduces everything from business cycles, death and sickness rates, human evolution, including the dominant position of the white race, and general trends in world history, such as the fall of the Roman Empire, to the effects of climate![40]

The rapid change in beliefs that Innis had undergone as a result of the war accounts for this bias towards deterministic explanations. That experience had hammered home the lesson that the highest personal ideals and the noblest sentiments concerning the sanctity of individual decision making were irrelevant when confronted with the brute power of technology. It was not a sufficient guarantee of nationhood to write about the sentiments of patriotism, evident in the act of Confederation or in Canada's participation in the First World War. The existence and continued development of the hinterland society that he strongly recognized as his own depended on the demonstration that this society was founded on objective factors, rather than in opposition to them. The war made evident the fact that human will was *not* enough. In this sense, Innis's famous aphorism that 'the present Dominion emerged not in spite of geography, but because of it'[41] had a philosophical significance extending far beyond Canadian economic history.

Innis organizes the presentation of his fur-trade study in a deterministic manner. He begins with a presentation of the characteristics of the staple, in this case, the beaver. The sedentary nature of the animal, the significant length of time it takes to reach maturity, and the inadequacy of its defensive techniques against human predators determined its quick depletion. This in turn led to a continent-wide chase to tap the diminishing stocks of beaver. The nature of this chase was itself determined by the two other key factors in Innis's equation: the structure of the river systems up which the beaver was pursued and the institutional structure of the commercial organizations through which the great chase was organized. Canada's boundaries were determined by the former and its political and economic structure by the latter. At first glance it appears to be a simple but enormously forceful vision of Canada's past.

If we were to trace the direct line of descent of Innis's approach to economic history, we would find that one of its main points of origin was a course on marketing given by C.S. Duncan at the University of Chicago. When Innis applies a commodity case-study technique developed in a mature economy, characterized by a wide mix of products, to a hinterland economy dominated by a few key exports, he moves from market research to imperial analysis almost by hazard. It is one of the earliest examples of Innis's Midas touch of intellectual eclecticism by which he transformed relatively mediocre source material into profound new insights. In the case of Duncan, Innis picked up a number of important themes. It was Duncan who first stressed the primacy of geography: 'It is necessary to have a panoramic view of the economic situation as a preliminary to a detailed discussion of individual problems. Such a bird's-eye view can best be had through a study of geography.'[42] Duncan also emphasized the importance of a thorough understanding of the characteristics of the commodity under discussion: 'The commodity [should be] considered more or less philosophically, by abstracting it and thinking about it in its various commercial aspects ... Just as the Interstate Commerce Commission has endeavoured to examine all classes of commodities that enter into traffic on the basis of their characteristics that affect transportation, so commodities for the market must be analyzed for all of their characteristics that affect their marketing.'[43]

In a sense *The Fur Trade in Canada* is a superior example of the type of commodity-analysis approach taught by Duncan. Had Duncan rather than Wright been Innis's supervisor, perhaps an attempt at the fur-trade study would have been made ten years earlier. Innis brought back Duncan's influence initially in the form of his teaching methods. Innis's campaign for new maps was a direct outcome of the realization that Duncan's type of approach was impossible in Canada because of a lack of prerequisite source materials.[44] In addition, the difficulties in directly applying Duncan's approach to Canada may have reinforced Innis's conviction that a new perspective was needed to understand the Canadian economy. Duncan's book had been aimed mainly at the training of business students entering a mature economy. As such, the method was meant to 'give a specialized knowledge that was invaluable so far as the specific commodity is concerned. Such an analysis will not, however, demonstrate broad principles ... One may know thoroughly that one subject, but he cannot generalize from it.'[45] It was a practical method that sought to enable individuals to master and profit from their own particular knowledge of a niche in the national economy. As such, it emphasized two types of commodities and commercial organizations: raw materials sold to the manufacturer (or to the consumer directly), and manufactured goods – transformed material passed on to the final consumer. Innis's application of this manner of thinking about the economy to the peripheral case of Canada automatically introduced an element absent from Duncan's treatment: the fact of empire. The 'marketing centre' as regards 'Canadian' raw materials was regularly found outside the country, in the United States and Britain. The study of the commercial organization would therefore involve a study of imperialism based on the exploitation of a relatively small number of resource commodities. Duncan's case studies were focused on a mature economy. Innis, in contrast, found that he could generalize over wide areas and long periods of history beginning with a single commodity analysis such as fur. What was a 'novelty' good for Duncan became a 'staple' for Innis simply because of their fundamentally differing vantage points on the economy. In this way, the 'staples approach' to economic development was conceived.

The influence of Duncan's work on Innis's treatment of the fur trade and on the staples theory of economic growth explains only a small part of what Innis was trying to accomplish in *The Fur Trade in Canada*. It is clear that Innis did not feel at this point in his career that he had successfully elaborated a new paradigm. His early work is filled with half-constructed concepts whose importance has been minimized by their not being included in the orthodox staples approach. These themes are not dominant during this period of his work, but they are important both because of their frequent occurrence and because they came to fruition in his communications studies. These themes include a tendency to jump back in analysis to the metropolis of the empire; a concern with the centrality of cultural factors; an awareness of prototype cases of cultural collapse and cultural balance; and a structuring of analysis of these cultural constellations through an examination of the principal communications (transport) systems of the time, taking into consideration the effect of the bias of these systems on both historical development and research efforts.

When we take these themes into account, *The Fur Trade in Canada* is seen as more complex, more universal, and less rigidly deterministic than commonly accepted. Innis never uses the staple as anything more than a focusing point around which to examine the interplay of cultures and empires. The staples theory is neither a theory in itself nor a good way to describe the methodological richness of his approach. By 1929, Innis had worked out a coherent sketch of Canadian economic development that revolved more around transportation systems than around staples. In the first period, water transport mediated the trade of fish, furs, and lumber; in the second, land transport (primarily railways) provided the infrastructure for trade in wheat, minerals, lumber, and pulp and paper. These two constellations of trade formed two phases of economic development in Canada. The period of transition from one to another was a time of crisis of cyclonic intensity: '[This] difficult period ... which dates roughly from the canals of the 1840's to the completion of the Canadian Pacific in 1885 is marked by the struggle for responsible government, the decline of the mercantile system, and Confederation, and these developments were more than a coincidence.'[46] Here Innis, positing the changeover from one transportation system to another with a different bias, through the introduction of a new staple and with great disrupting effects on all levels of society, is close to the approach found in his communications writings. At a later date we will find him analysing the replacement of one communication system by another following the introduction of a new technical innovation (paper, moveable type, application of power). In these writings, the results described will be the even more cataclysmic events of cultural collapse. Yet, even in the early writings, when Innis deals with key shifts in development of the marginal economy, his approach tends to lead back to analysis of dynamics at the *centre* of empire: 'Furs were essentially a product in demand as luxuries and adapted to mercantile policy, whereas lumber was required to bridge the gap in the shift from a relatively non-industrial to an industrial community, and from mercantilism to free trade. The rapid growth of towns and ships depended on lumber. This commodity furnished the scaffolding and moulding on which the present industrial equipment of Great Britain and the United States was built up. It served until industry could shift from organic to inorganic materials.'[47]

In 1929 Innis had already worked out the basic cyclonic tendencies in staples production. This went far beyond an analysis limited merely to hinterland growth and to the realm of the economy – he had already laid a basis for his later communications studies: 'The overwhelming efficiency of the new industrialism has necessitated the development of effective methods of disposing of the products to the unappreciative consumer. The rapid rise of advertising has been largely responsible for development of the pulp and paper industry.'[48] This tendency of Innis's research to lead into an analysis of metropolitan development was not inadvertent. Innis, unlike later commentators on the staples thesis, always believed that it would have important application at the centre of empire: 'Canada ... offers vast fields for research of the most significant character, *not only for her own needs, but also for those of highly industrialized countries*.'[49] Innis's intellectual project, therefore, did not limit itself to the development of a theory of economic development more appropriate to the study of hinterland economies than those currently in vogue in the metropolis. More radically, he was suggesting the need for a *global* theory of imperialism to be built around painstakingly concrete analyses of the many levels of interaction of the centre/margin of empires. He was using the staple to focus attention on the *cultural interaction* of different peoples at the edge of an expanding empire. The centrality of cultural factors in economic development was the main innovation made by Innis when he turned from his examination of the CPR to the fur trade.

With *The Fur Trade in Canada*, the staple becomes in earnest the focus of understanding the dynamics of economic development. The description of the beaver with which the book begins is a necessary prelude to the comprehension of how the cultural interaction of peoples took place. Innis points out that the spread of the influence of Western civilization across Canada occurred when 'in the language of the economists, the heavy fixed capital of the beaver became a serious handicap with the improved technique of Indian hunting methods, incidental to the borrowing of iron from Europeans. Depreciation through obsolescence of the beaver's defence equipment was so rapid as to involve the immediate and complete destruction of the animal ... with [its] destruction in the easterly part of North America came the necessity of pushing westward and north-westward to tap new areas of valuable furs.'[50]

Before the coming of the Europeans, the Indians of the Shield had been organized around a hunting economy based on the white-tailed deer. This economy had been balanced by the limitations of the existing technology of cooking (large stationary bark pots) and of hunting (bows and arrows). With the advent the Europeans, the Indians experienced a cultural revolution in these fields through the introduction of the portable metal cooking pot and the gun as well as such other readily accepted goods as knifes, hatchets, and cloth. Over the period of a few generations, the arrival of these goods made the Indians culturally dependent on the Europeans, and their old cultural traits and skills were forgotten. The new technology vastly increased the hunting efficiency and range of the Indians while their dependency on foreign trade goods allowed the Europeans to mobilize the skills of the indigenous population in the exploitation of the beaver. It is impossible to imagine the fur trade without the collaboration of the Indians and, in particular, without their contribution of culturally specific skills and artefacts. 'Such cultural traits as the canoe, the pack strap, the knowledge of animal habits enabling them to hunt for food and for fur, acquaintance with plants as food and medicine, their agricultural development, and the knowledge of the country, were stressed because of their importance in enabling the Indians to cover a wide territory and to get more furs.'[51]

But this is only one example of cultural analysis in Innis's treatment of the furtrade. In his conclusion, under the subsection 'The Importance of Staple Products,' Innis begins by saying:

> Fundamentally, the civilization of North America is the civilization of Europe and [my] interest ... is primarily in the effects of a vast new land area on European civilization ... People who have become accustomed to the cultural traits of their civilization ... on which they subsist, find it difficult to work out new cultural traits suitable to a new environment. The high death rate ... is evidence to that effect.
>
> The survivors live by borrowing cultural traits of peoples who have already worked out a civilization suitable to a new environment ... [But] the methods by which the cultural traits of a civilization may persist with the least possible depreciation involve an appreciable dependence on the peoples of the homeland ...
>
> Goods were obtained from the homeland by direct transportation as in the movement of settler's effects and household goods ... or through gifts and missionary supplies, but the most important device was trade. Goods were produced as rapidly as possible to be sold at the most advantageous price in the home market in order to purchase other goods essential to the maintenance and improvement of the current standard of living ... The migrant was consequently in search of goods which could be carried over long distances by small ... sailboats and which were in such demand in the home country as to yield the largest profit.[52]

Table 4: Social Sectors Involved in the Fur Trade

	European-based dominant class	Settlers	Indigenous people
Cultural strategy	fashion – fur felt hat	survival and recouping of European standard of living	solution of cultural impasses and revolution in potential living standard
Staple demanded	beaver	clothing, tools, and other articles related to metropolitan life	pots, guns, blankets, tools, and liquor
Consequence	strengthening of merchants, relative impoverishment of aristocracy	survival but stagnation owing to conflict between mode of production (fur trade) necessary to obtain the material basis for establishing another mode of production (agriculture) and that other mode	cultural transformation and collapse. Advent of the Metis

Now we begin to see that, in *The Fur Trade in Canada*, Innis has placed cultural analysis at the centre of imperial analysis. Cultural collaboration of indigenous and settler elements leads to the exploitation of the hinterland by fostering the dependence of its peoples on metropolitan inputs. What first appeared to be a simple matter of extrapolating the characteristics of the hinterland's geography and staple commodity has become a complex analysis of three different peoples – the metropolitan people, the settlers, the indigenous people – with three different cultural models and three different sets of demands.

The dynamic of empire is embedded in an interaction at various levels of distinct cultural constellations. The pattern presented in the book is distilled in table 4. We are not dealing here with a simplistic staples approach but with a complex model in which at least two sets of 'shadow staples' (iron trade goods for the Indians and a variety of goods demanded by the settlers) accompany the recognized staples trade.

Coming to terms with the extent to which an analysis of cultural models underlies Innis's treatment of the fur trade helps us to see far more clearly the relationship between this work and the communications essays. The apocalypse of culture disorientation that Innis would come to believe followed the introduction of writing and printing, and still later followed the application of power to print reproduction, has its prototype in the collapse of the traditional Indian civilization after the introduction of iron trade goods.[53] The revolution in lifestyle that accompanied increasing dependence on the Europeans made for the destruction of a relatively balanced natural habitat, unprecedented migration, increased and increasingly destructive warfare, and decimation by disease. In his later work, Innis would use this particular study of the cultural collapse of a marginal people faced with revolutionary technology as a pattern to fit the general cultural collapse of Western civilization that followed the application of mechanization to the vernacular.

Even in 1929, Innis seems to be conscious of this connection. The Innisian treatment of empire, in stressing the cultural complexity and dependency involved in imperial expansion, incorporates a theory of unequal development. It is not a modernization or diffusionist theory of imperialism.[54] Immediately after dealing with the cultural collapse of Indian civilization as a result of the spread of the fur trade, Innis notes: 'The unique nature of [Canadian] development has been largely a result of the sudden transfer of large areas tributary to the fur trade to the new industrialism.'[55] He then goes on to deal with the development of one form of this new industrialism in the pulp and paper field. The examination of the fur trade has added cultural complexity to his treatment of imperialism and has provided him both with a model of cultural collapse and with the staple with which to embark into the realm of communications theory.

However, although it sets the stage for his later work, the fur-trade study does not furnish him with the crucial link that will allow his consideration to turn primarily to the dynamics at the centre of empires. The fur trade is seen only as an indicator, not a dynamic catalyst, of metropolitan growth. 'Control of the fur trade was an index of world importance from the standpoint of efficient manufactures, control of markets, and consumption of luxuries.'[56]

The Importance of Veblen

Innis's treatment goes so far beyond the commodity-analysis approach stressed by Duncan that we must look for the influence of another main source in his work; it is unlikely that he would have developed these elements of his argument in such a short space of time *ex nihilo*. This other prime source is University of Chicago professor Thorstein Veblen.

Innis's 1929 article on Veblen clearly indicates the seminal effect of this unorthodox thinker on the young Canadian at the beginning of his academic career. Some of the concepts appeared immediately, in reworked form, in Innis's early writing. Others, such as Veblen's attitude towards the role and structure of the university, would guide Innis's activities within academic politics and only later appear in his publications concerning the aims of higher education. Still others, such as the history of human consciousness, the centrality and dynamics of advertising, and the theory of the leisure class (to which Innis's conception of monopolies of force and knowledge bears a striking relationship), would percolate in his mind for years, eventually coming to light in the communications studies. Echoes of Veblen emanate from Innis's writings on every major subject. However, we are mainly concerned here with the manner in which Veblen influenced the 'early' Innis.

It may be that Innis was drawn to Veblen as an intellectual model paralleling Innis's own experience as a precocious youngster from a relatively impoverished rural background. Both came from poor farm backgrounds. While Veblen's father was in character similar to Innis's semi-literate father, he was 'notably progressive in his use of farm machinery and interest in science [but] never learned English.'[57] Like Innis, 'Veblen emancipated himself from [the] piety [of his background] in certain respects ... [yet] he remained in many ways a ... Puritan protesting against the more commercially minded and urbanized.' The poor farm background meant that, like Innis, 'Veblen, ever fearful of scarcity, remained an ascetic and aesthetic enemy of consumption, above an efficient subsistence minimum.'[58]

Much of what is said about Veblen's attitudes and life, as, for instance, in this passage by David Riesman, could be applied verbatim to Innis:

> While a good deal of his economics may be interpreted as a farm boy's stubborn empiricism – a show-me attitude towards theory and refinement ... Veblen ... did not remain a populist ... Though proud all his life long of his ... ancestry, and happier with farmers ... than with professors who could not chop a tree ... he nevertheless identified himself with the international world of scholarship and only on occasion with the more folksy ...

> vulgarities of populism ... He came late to economics, as he came late to ... the genteel tradition and the cities ... [His] heritage made him a marginal man, linguistically and intellectually cosmopolitan, socially awkward, and emotionally expatriate.[59]

Veblen, like Innis, was obliged to tap the rare case of the Darwinian free-thinker located in a conservative, rural childhood milieu. Like Innis, he would pass through a small college with a theological atmosphere specializing in dispensing preparatory training to rural youth who would go on to higher education in the established institutions of the metropolis.[60] The result was an uncommon perspective held in common by both men. As Riesman says, 'during the same era that Durkheim and Freud, Brooks Adams and Sorel, Max Weber and Pareto were in one or another way preoccupied with the centrifugal tendencies in modern society, Veblen saw the role of leaders and elites from a village atheist's perspectives – as sheer unnecessary swindle and expense.'[61] But, above and beyond biographical similarities,[62] Veblen appealed to Innis because he raised this type of background to the level of a project. Veblen substantiated Innis's belief that the peripheral intellectual was central to the advancement of the social sciences and culture in general. As Veblen put it: 'Anyone who is required to change his habits of life and his habitual relations to his fellow men will feel the discrepancy between the method of life required of him by the newly arisen exigencies, and the traditional scheme of life to which he is accustomed. It is the individuals placed in this position that have the liveliest incentive to reconstruct the received scheme of life and are most readily persuaded to accept new standards.'[63]

Even basic biographical differences between the two men tended to reinforce Veblen's influence on Innis. Veblen had what Innis lacked: a solid background in philosophy. The strength of this background gave weight to Veblen's attack on classical economic theory. As a result, at the beginning of his career, Innis was able to borrow a critique (which he himself would not have been able to formulate) that allowed him to avoid the intellectual hegemony of metropolitan economic paradigms.

The manner in which Veblen criticizes the dominant paradigms is of great significance not only for our understanding of the staples works but also in accounting for Innis's shift to communications studies. Veblen's point of attack is not economics per se but the primitive conception of *psychology* that permeated the discipline. He takes great pains to demonstrate that the unchanging and universal image of man as a calculating hedonist was the foundation on which the entire superstructure of modern economic paradigms had been built.

> Since the principles of human nature that give the outcome in men's economic conduct ... are but simple and constant sequence of hedonistic cause and effect, the element of human nature may fairly be eliminated from the problem, with great gain in simplicity and expedition. Human nature being eliminated, as being a constant intermediate term, and all institutional features of the situation being also eliminated, the laws of the phenomena of wealth may be formulated in terms of the remaining factors ... Economic laws come, therefore, to be expressions of the algebraic relations subsisting between the various elements of wealth and investment capital, labor, land, supply and demand of one and the other, profits, interest, wages.[64]

As opposed to this hedonistic model of the general scheme of things, Veblen's work is based on a 'later psychology [that] gives a different conception of human nature.' It is an existential perspective according to which 'it is the characteristic of man to do something, not simply to suffer pleasures and pains through the impact of suitable forces ... According to this view ... economic activity ... is not apprehended as something incidental to the process of saturating desires. The activity is itself the substantial fact of the process.'[65] Many of the themes in this alternative conception of human psychology – themes that form the basis of Veblen's critique of classical economic theory – are rearticulated in Innis's work: the central importance of cultural factors in economics; the dialectic of individual consciousness and the cultural growth of the community as the central dynamic in human history; and a deep-rooted scepticism concerning the prescriptive claims made by science based on its presumed objectivity.

Veblen posits that intellectual paradigms changed following changes in the everyday manner in which people looked upon their particular time and place. Historically, he views the overall drift of this process to be one of 'disintegrating animism.' From this point of view, paradigms that have claimed to be scientific have exhibited unexpected and largely unnoticed recurrences of animism: 'For the earlier natural scientists, as for the classical economists, this ground of cause and effect is not definitive. Their sense of truth ... is not satisfied with a formulation of mechanical sequence. The ultimate term in their systematization of knowledge is a "natural law" ... The objective point of the efforts of the scientists working under the guidance of this classical tradition, is to formulate knowledge in terms of absolute truth; and this absolute truth is a spiritual fact. It means a coincidence of facts with the deliverances of an enlightened and deliberate common sense.'[66] Traditional science is seen as a new religion, the central credo of which is the ultimate unity of fact and common sense. The key mechanism involved is the retention of a conception of the normal as natural. The result is an extremely conservative bent to paradigms based on this traditional pattern of science. The common sense of the day is projected as the ideal for all time. Against this, Veblen assumes that his efforts have as their substratum

> the practical exigencies of modern industrial life [that] ... enforce the impersonal method of knowledge ... It is only a matter of time when that [substantially animistic] habit of mind which proceeds on the notion of a definitive normality shall be displaced in the field of economic inquiry by that [substantially materialistic] habit of mind which sees a comprehension of facts in terms of a cumulative sequence ... Under the stress of modern technological exigencies, men's everyday habits of thought are falling into the lines that in the sciences constitute the evolutionary method; and knowledge which proceeds on a higher, more archaic plane is becoming alien and meaningless to them.[67]

Veblen's approach, like Innis's, is on first reading a straightforward technologically determinist one. Yet both men, in raising to a conscious level the preponderant effect of industrial technology on the cultural environment of modern man, go beyond a rigidly deterministic position. By pointing out the effect of new technologies, they also indicate the possibility of intervention in the development of these technologies and, therefore, in the structuring of human nature.

Innis picks up both Veblen's emphasis on culture in general and his preoccupation with the special case of the effect of the spread of industrialism on peripheral cultures. Here Veblen is at his least deterministic. If he had developed a general theory in which industrialism inevitably caused certain changes in human society, we would expect to find a diffusionist theory of economic development. The new technology would gradually spread outward from the old metropolis, transforming the peripheral societies in essentially the same way in which it transformed the 'old' countries. In his first paper given to an international scholarly meeting, Innis, following Veblen, chooses to emphasize that this is not so. He stresses that the effect of industrialism on rural settlements in

Europe was quite distinct from that in western Canada (and new countries in general), where settlements 'had their *raison d'être* in modern industrialism.' He goes on to make two main points. First, citing Veblen, he indicates that industrialism undergoes a significant change as it spreads to peripheral societies: 'Important, as [the] sudden spurts of industrialism have been to the new countries, they must not be permitted to obscure the significance of steady and persistent experimentation essential to the evolution in technique of machine industry. The technique involved in the countries which have had the longest experience, as in England, has been modified and improved and borrowed wholesale by the new countries ... The more recently the country has been industrialized, the more rapid tends to become its industrialization.'[68] Second, in a corollary point, Innis says that social-science work done on the effects of industrialization in Europe is of limited relevance to new countries: 'It is doubtful if [their] conclusions can be applied satisfactorily to Western Canada. Certainly settlement in Western Canada differs fundamentally from settlement in Eastern Canada and in the old world.'[69] Innis's paper was, in fact, a case study of a Veblenian theory of unequal development in which new technology interacts with local cultural institutions in a radical, but also radically indeterminate, way. One point is clear in Veblen's schema. It is in the peripheral societies that we can expect to find the most creative and progressive use of the new technologies. To understand why this is so, we must understand a key concept of Veblen's that was taken over and used by Innis in *The Fur Trade in Canada*: the concept of 'borrowing' technology. Veblen's theory of technology and culture is best summarized by the phrase 'the merits of borrowing and the penalty of taking the lead.' The paradox of this phrase arises from the recognition that any technology, to become operative, must be implanted in a cultural milieu that has somehow recognized the technology as legitimate. Veblen writes:

> It is a virtual matter-of-course that any material innovation ... will be attended with a fringe of magical or superstitious conceits and observances ... [These] embody the putatively efficacious immaterial constituent of all technological procedure ... These magically efficacious devices have grown into the prevalent habits of thought of the population and have become an integral part of the common-sense notion of how these technological elements are and are to be turned to account ... [These practices eventually become] standing conventions out of the past [that] unavoidably act to retard, deflect or defeat adaptation to new exigencies

> ... All this apparatus of conventions and standard usage ... acts in some degree to lower the net efficiency of the industrial system.[70]

Maintenance of outmoded equipment, wasteful consumption, and active limitation of innovation and production are the result.

However, borrowing involves only one side of the culture and technology dichotomy. Borrowing automatically implies the introduction of a multiplier factor in the productivity of the technique concerned.

> The borrowed elements of industrial efficiency would be stripped of their fringe of conventional inhibitions and waste, and the borrowing community would be in a position to use them with a freer hand and with a better chance of utilizing them to their full capacity, and also with a better chance of improving on their use, turning them to new uses, and carrying the principles (habits of thought) involved in the borrowed items out, with unhampered insight, into farther ramifications of technological proficiency. [There] is a secondary effect of borrowing ... The borrowed elements are drawn into a cultural scheme in which they are aliens and into the texture of which they can be wrought only at the cost of some, more or less serious, derangement of the accustomed scheme of life and the accepted system of knowledge and belief ... They are vehicles of cultural discrepancy, conducive to a bias of scepticism, and act ... to loosen the bonds of authenticity.[71]

Innis, in *The Fur Trade in Canada*, is in effect producing a practical study of this process of borrowed technology transforming cultural patterns. In many ways, the book is a direct expansion and elaboration of passages found in Veblen. For instance, in the latter's *Imperial Germany and the Industrial Revolution*, we find a significant observation: 'In some instances of such communication of alien technological and other cultural elements the terms on which a settlement has been effected have been harsh enough, as, e.g., on the introduction of iron tools and fire-arms among the American Indians, or the similar introduction of distilled spirits, of the horse, and of trade – especially in furs – among the same general group of peoples ... In these cases the intrusion of alien, but technologically indefeasible, elements of culture has been too large to allow the old order to change; so it has gone to pieces.'[72]

Veblen's concept of borrowing technology entails a concern for cyclonics especially in the application of new techniques to virgin resources in new countries. The tremendous release of productive capacity involved in transplanting the technology, coupled with the unprecedented level of available resources, meant that overexploitation, waste, exhaustion of resources, and economic collapse (rather that the initial fostering of resource development) were the key problems of new countries. Innis, too, was concerned with these cyclonic tendencies in peripheral economies.

The conclusion of *The Fur Trade in Canada* traces in Veblenian fashion 'the superimposition of machine industry on an institutional background characteristic of the fur trade.'[73] This 1929 version of the economic history of Canada has already identified pulp and paper as one of the key hopes for the future of the economy because it is renewable; it is associated with positive linkages such as the development of surplus power; and, as a Canadian Shield industry, it 'represents a direct contribution to the reduction of overhead costs.' Veblen's stress on the dialectic of technology and culture implies a view of politics as a symptom of more fundamental social dynamics. It is ironic that this perspective, which offered little hope for practical politics, led Innis into early editorializing concerning the key commodity of pulp and paper. In the March 1929 issue of the *Canadian Forum*, under the title 'The Newsprint Dilemma,' Innis discusses the difficulties entailed by the control of the pulp and paper industry by American newspaper chains. The quandary is a Veblenian one of cultural stagnation versus economic exhaustion on the periphery.

> The interests of Canada are at stake, with the result that we lose when the mills are operating at capacity, from the exhaustion of our resources; and when the inefficient mills are closed, with the burdens imposed on the settler. No definite progress can be made until the provinces decide on a policy which will enable them to control a situation which is now obviously beyond control. Confederation and the placing of the mines and forests in the hands of the provinces may be the occasion for the loss of our resources. Some united and definite policy on the part of the provinces, with conservation and settlement as essentials, is imperative.[74]

It is also ironic that the entire editorial is based on a traditional economic assumption that 'a higher price [for newsprint] will probably be responsible for a falling off in consumption.' During the changed conditions of the Depression years, Innis's observation that this assumption had no necessary validity in the case of a product that created its own demand is a key jumping-off point for the new perspectives of the communications works.

Innis began his career in the odd-man-out role of social critic in an optimistic age. The tone of his early works went against the spirit of the times in which they were composed. As A.F.W. Plumptre, who studied as an undergraduate during these years, recalls:

> Innis had an undertone, if not of pessimism, certainly of deep misgivings at a time when we undergraduates really thought the world was going to be all right ... The League of Nations was getting going, etc.
>
> Innis at that time was very much concerned with where industrialism was taking the world. I can remember him talking about something which in those days we'd never heard of – 'Frankenstein' – which later became famous, but at the time was still a novel by Mrs. Shelley ... The mechanical monster took over and ran its creator. [Innis] kept coming back to the worry that technology was going to do this to civilized man.[75]

This concern with the problems brought on by the spread of industrialism into new areas was passed on directly from Veblen to Innis. Both men attempted to articulate a system of cyclonics. Both wrote about a dialectic of overproduction and resource exhaustion, on the one hand, and of the checking of production and innovation and the advent of conspicuous consumption, on the other. And both presented this system not in abstract theoretical terms but in the course of practical investigations of a particular social structure.

Yet Innis comes to fundamentally different conclusions in developing this theme. These differences reflect underlying ones in the cultural milieu that gave rise to the two thinkers. Despite many similarities between their backgrounds, the two men are separated by their nationality. Veblen's iconoclastic comments on American society are biting because, at a fundamental level, he remained an American. As one scholar puts it:

> Veblen was a profoundly conservative critic of America: he wholeheartedly accepted one of the few unambiguous all-American values: the value of efficiency, of utility, of pragmatic simplicity. His criticism of institutions and the personnel of American society was based without exception upon his belief that they did not adequately fulfill this American value ...
>
> As a critic, Veblen was effective precisely because he used the American value of efficiency to criticize American reality.[76]

For Veblen, it is not the effect of new industrial technology per se that is suspect but the truncation of the natural development of that technology by the cash nexus permeating the institutions in which the technology took root. The subordination of the 'engineers' to the 'captains of industry' within these institutions is Veblen's great complaint.

Innis is at once more pessimistic and more profound as a social critic. In the bright days before the arrival of the Depression, he had already recognized that the spread of new technology in Canada not only had led to a new set of problems but had left the population myopic and intellectually exhausted. The efforts of the engineer were not adequate. It was in this context that his academic work became intensely political in nature. In 1929 he wrote about the advent of a new scholarly series, 'Studies in History and Economics,' which was the forerunner to the *Canadian Journal of Economics and Political Science*.

> Heretofore Canadians have exhausted their energies in opening up the West, in developing mines, hydro-electric power, and pulp and paper mills of the Canadian Shield, in building transcontinental railways, grain elevators, and cities, and in taking a share in the war. The rapidity and energy with which these gigantic tasks have been accomplished have only been possible with the technical advantages of modern industrialism and the concentrated efforts of a small population. The engineer and the technician have been to the front. But the very rapidity with which the task has been accomplished, and the efficiency with which it has been accomplished, have given rise to new problems. Foremost among these problems is that of providing an adequate market for the raw materials which Canada has succeeded in developing so rapidly. Scarcely less important is that of protecting natural resources from rapid exploitation ... overproduction ... and exhaustion ... which follow from dependence on older, more highly industrialized areas. The development of ways and means by which this protection may be obtained is an important task of the economist ... The welfare of the nation for the future must depend to an increasing extent on the economist. The promotion of this publication is only a minor and very humble attempt to provide a media for a sane, unbiased, and scientific discussion of economic phenomena peculiar to Canada and to new countries and so contribute to the solution of their problems.[77]

Already in this passage we find Innis's perception of the Promethean task imposed by technological innovation on man. The application of the new technology requires a tremendous exertion of social intelligence. Success in economic development calls forth a new range of problems demanding attention. However, the very technology responsible for the advent of these new problems has also caused a dislocation and exhaustion of the intellectual capacity of the society to such an extent that the challenge of the new problems can be met only with the greatest of intellectual efforts. From the beginning, Innis perceived that the new technology had a tendency to get out of control because of its ability to intervene in the mental as well as material condition of humanity's existence. Innis differed from Veblen in believing that this process would continue regardless of whether the 'captains of industry' or the 'engineers' were ostensibly running the productive apparatus of society. This realization would soon leave him in a lonely position, sceptical of both right-wing and left-wing panaceas in a Depression world that was longing for panaceas.

Innis was well aware that he owed more to Veblen than to perhaps any other single intellectual. Writing on Veblen's contribution to the social sciences, Innis showed an early appreciation of a number of themes that would continue to occupy his attention in the years ahead:

> He was among the first to detect the relation between the industrial revolution and the Darwinian theory and the later theories of Physics and Chemistry ... His approach was from the inductive side in keeping with his philosophy ... [His] articles were models of analysis of economic facts ... Veblen and his followers who have protested most against the inclusiveness of price economics have done most in the study of price phenomena ... [He made] a devastating attack on the marginal utility theory ... His first important volume was designed to show the weakness of economic theory on the consumption side ... He attempted to destroy the hedonistic calculus [of] Jeremy Bentham ... His main argument was ... that machine industry was overwhelmingly and increasingly productive, and that the problems of machine industry were incidental to the disposal of the product ...Veblen's interest was in the state of the industrial arts which had got out of hand ... He was also interested in the effect of the industrial revolution on economic theory ... He insisted upon the existence of laws of growth and decay of institutions and organizations.
>
> His life work has been primarily the study of the processes of growth and decay ... His interest in anthropology, his terrific irony ... protect[ed] him from absorption into the partialities of modern movements. His anxiety has always been to detect trends and to escape their effects ... He [was] in revolt against mass education and standardization.[78]

In these lines, written in 1929, Innis could have been writing his own epitaph.

CHAPTER FIVE

The Great Betrayal, 1930–1940

If we ask ourselves what will happen to a humanity where every group is striving more eagerly than ever to feel conscious of its own particular interests, and makes its moralists tell it that it is sublime to the extent that it knows no law but this interest – a child can give the answer. This humanity is heading for the greatest and most perfect war ever seen in the word, whether it is a war of nations, or a war of classes.

Julian Benda, 1927[1]

The 1930s marked a new phase in Innis's career. The previous decade had required him to work in relative isolation as he built up his knowledge of Canada through 'dirt' research. In the 1930s, by contrast, he was recognized as an authority on the Canadian economy, and his expertise was called on by a wide range of individuals and institutions; this led him to play much more of a role as a leader and shaper.

This recognition was partly the result of the growth and increase in stature of the political economy department in the decade of the Great Depression, when people turned to economists in an unprecedented fashion to find a way out. The department expanded considerably. When McMaster University moved to Hamilton in the early 1930s, the University of Toronto's political economy department moved into the old McMaster building at 273 Bloor St West. Twenty years after entering the building for his first year at McMaster, Innis returned as a senior faculty member of the University of Toronto.

A raft of promising intellectuals were hired as lecturers, including a talented young woman, Irene Biss, and the first of Innis's own students to reach faculty status, D.C. MacGregor and A.F.W. Plumptre. Innis played a major role in shaping the careers of young intellectuals like these through his influence on appointments and research resources. This was particularly the case with two major collective-research projects. The *Canadian Frontiers of Settlement* series was edited by Innis's friend and colleague, W.A. Mackintosh of Queen's University. He often sought Innis's advice on topics and personnel to pursue the series, and Innis himself contributed 'Settlement and the Mining Frontier' to the series in 1936. From the beginning, too, Innis was centrally involved in a larger undertaking, the Carnegie Endowment–funded project *The Relations of Canada and the United States*. He fought for and won a great deal of autonomy for the Canadian contribution. While the American James T. Shotwell remained the overall editor of the series, Innis was given editorial control over a block of twelve volumes to be written by Canadian scholars, and he wrote the prefaces for four of these volumes. Especially in the midst of the Depression, his control of the resources of the Canadian volumes in the series increased his influence enormously.

Through the decade his own career advanced steadily. In 1933 he was the university's delegate to the Sixth Annual Conference for the Scientific Study of International Relations. In 1934 he became a fellow of the Royal Society of Canada. In 1936 he was made a full professor at University of Toronto. In 1937 he became president of the Canadian Political Science Association. Finally, in July 1937, he succeeded E.J. Urwick as the head of the department of political economy.

While his professional workload increased, he maintained a commitment to public education – lecturing at the Liberal-Conservative Summer School (1933), giving a public lecture at the University of British Columbia Summer School (1935), and organizing a series of public lectures on the occasion of the fiftieth anniversary of the department of political economy (1938).

The focus of his personal interest turned from the fur trade to the cod fisheries during this time; one half of *Selected Documents in Canadian Economic History 1497–1783* (1929) dealt with this latter subject. While much of the 'dirt' research on the fisheries had been undertaken in the 1920s through his archival forays and field trips, he completed this research by travelling extensively in the Maritimes and Newfoundland in 1930, 1932, and 1936. In 1930 he presented his first essay for the Royal Society of Canada, 'The Rise and Fall of the Spanish Fishery.' He spent the rest of the decade completing his magnum opus for this period and wrestling it into publishable form. *The Cod Fisheries: The History of an International Economy* was published in 1940.

While he travelled less to remote parts of Canada to undertake 'dirt' research than he had in the 1920s, he directed the travel of younger researchers who were completing the review of Canada's staples industries. He continued to travel extensively, to be sure, but mainly to attend conferences, summer schools, and international meetings. In the summers of 1932 and 1933 he travelled to see the devastating effect of the Depression on the west, and in 1934 he experienced his first whirlwind of government travel in Nova Scotia as a member of a royal commission charged with studying the provincial economy. All the while, he exercised tremendous influence through a ceaseless stream of lectures, journal articles, and reviews on a wide range of topical subjects, and his reputation grew. In 1935 he helped to launch the second scholarly journal in his field of interest, the *Canadian Journal of Economics and Political Science*.

With his growing stature, one would expect that he would be a contented man by the late 1930s. This was not the case. His contention was that a monk-like devotion to 'dirt' research would lead to a new worldview and a better understanding of Canada. He also believed that this painstaking development of an indigenous paradigm was a prerequisite to looking at subjects far beyond Canada, and to making a genuinely Canadian scholarly contribution to universal knowledge. He was, above all, convinced that scholars should work from the specifics to the universal, from the concrete to the abstract.

He feared that scholars who borrowed pre-existing paradigms from Keynes, Marx, the Fabians, and others had betrayed their scholarly calling. In working from the universal to the particular they undercut the value of 'dirt' research. A 'monopoly of knowledge' was accepted as a starting point, and facts were interpreted to fit within it. Intellectuals became caught in a self-referring realm of their own making. The possibility of a new paradigm being developed by recognizing the patterns inherent in vast quantities of new facts was abandoned.

The Depression increased the dangers of this happening as politicians and public opinion cast around for solutions to the slump. Existing intellectual paradigms that appeared to offer comprehensive solutions to social problems were appealing during such times. They appeared to offer ready-made solutions, whereas Innis's approach insisted on the primacy of more 'dirt' research and only the tentative conclusions of an indigenous paradigm in the making.

That is why Innis in the 1930s engaged in an extended and at times vitriolic debate over 'preaching' in the social sciences. By this he meant using an existing paradigm with the certainty that it provided correct answers to social problems in the same sense that a religious creed provided guidance in life. By the end of the decade, Innis realized that he had largely lost the debate just when he felt he was about to prove his point by applying his methods to topics far beyond Canada. In 'For the People,' for instance, he rejected the conclusions of *Social Planning for Canada*, a program written for the new socialist party in Canada, the Co-operative Commonwealth Federation (CCF), by left-wing intellectuals heavily influenced by British socialism. His vast knowledge of Canadian facts made his arguments quite compelling, but he had little to offer in terms of an alternative comprehensive prescription for society other than more painstaking research detached from political and commercial interests.

A more serious conundrum for Innis was posed by the humanist Urwick (his department head) in his essay 'The Role of Intelligence in the Social Process' in the first issue of the *Canadian Journal of Economics and Political Science* (1935). Urwick argued that objectivity in the social sciences was unattainable since it was impossible to separate the social scientist from the society which he was studying. Innis did not effectively counter Urwick's argument. He could only contend that a central part of the social scientist's work should be the examination of the biases of individuals and institutions. In many respects, Innis, in returning repeatedly to the issue of bias, would struggle with the conundrum posed by Urwick for the rest of his life.

When Innis realized that he was losing the public debates on these central issues in social science, he adopted an unpopular stance. In an era when the violent anti-intellectualism of the right was at its height, he began to view these debates as a menace to the advancement of social science, characterizing them as a tyranny of talk. As a result, at the very time when his scholarly accomplishments were winning wide recognition, he became an increasingly lonely and intellectually isolated figure.

At first this was not readily apparent. Largely through his own work on the fur trade in Canada, he had won over his old colleagues to the validity of looking at Canada through an exhaustive examination of staples industries. This had been a collective endeavour that was nearing completion by the end of the 1930s. None of his old colleagues, however, viewed this work as anything more than an end in itself. It provided a better understanding of Canada (and perhaps other new lands). But they did not share Innis's belief that it was a point of departure for other, more wide-ranging investigations.

Nor did Innis win over a new generation of scholars to this endeavour. Many younger scholars were involved in the completion of staples-industry studies. They, too, were unconvinced that it provided a methodology that had wider applicability. More often than not, they became an intellectual comprador class which brought back more developed paradigms from graduate studies in the United Kingdom and the United States and simply applied these to the Canadian context.

Nowhere was Innis's struggle to win over bright young intellectuals to his vision of scholarship more evident than in his relationship with Irene Biss. In 1929 she had been recruited as a lecturer with the department of political economy after a stint at the London School of Economics, Cambridge, and Bryn Maur. She worked closely with Innis during the 1930s and admired the man immensely. She advanced steadily in the department and Innis exerted his utmost influence to keep her engaged in the type of scholarly career that he exemplified. Despite his blandishments, she chose marriage (to a politically engaged socialist) and a family instead. Her loss was a great blow to Innis.

Not one member of the new generation of scholars embraced Innis's version of his scholarly mission. Like his older colleagues, many remained focused on the economic history of Canada but did not view this as a starting point for more extensive investigations. Others were won over to alternative paradigms and associated themselves with politics. Still others went into government service. Aside from those pursuing Canadian economic history, Innis viewed them all as sell-outs and cut his relations with many of them. He always asserted that the scholarly calling was a demanding one, and he was as hard on his colleagues as he was on himself when he judged that these demands had not been met.

By the end of the 1930s, his reputation as a scholar was higher than ever. At the same time, however, he recognized that no one was following him into what he regarded as the next logical phase in his scholarly project: to apply the methodology and perspective of his detailed study of Canada to more wide-ranging topics. He made the decision that he would soldier on alone in the next phase of his research. His lonely scholarly work of the 1920s had set the agenda for the great collective endeavour to understand the history of Canada from an economic and geographical point of view. Perhaps if he undertook an equally lonely research tour de force, he could make his point and win back a future generation of young scholars to his approach.

The natural evolution of his own research facilitated this decision. Unlike the fur-trade work, the cod-fisheries study had naturally led him away from Canada to show how the fishery had played a role in the international economy and in the ebb and flow of empires. He had been attempting to interest Irene Biss in pursuing a joint book with him on hydroelectricity and the pulp and paper industry. When she chose marriage instead, Innis opted to pursue the topic on his own. It would lead him on a journey never to be completed.

The Depression

Given Innis's deterministic and pessimistic tone during the 1920s, one might have expected that he would have been more thoroughly prepared than his optimistic colleagues for the Great Depression that was to follow. Such was not the case.

The Depression pushed Innis to the emotional breaking point. The atmosphere of the era was one of perpetual and often acrimonious discussion. At no time in his life were personalities so directly linked to those general social trends that fascinated and frequently repelled him. I have attempted to preserve this aspect of Innis's Depression experience by dealing with three major themes in conjunction with three personalities – E.J. Urwick, Irene Biss, and James T. Shotwell – who were particularly responsible for bringing these issues to Innis's attention.

The Depression highlighted the extent to which social dynamics continued to unfold beyond the conscious intervention of man. One response to this observation was to denigrate the pretensions of the social sciences to scientific – that is, objective – validity. This intellectual challenge to Innis's determinism was most effectively put by Urwick. It set Innis to considering the question of bias in social-scientific research in far greater depth than he had in the past.

It is curious that the Depression, in demonstrating the bankruptcy of previous schemes of social engineering, elevated the image of the social scientist as social engineer. For the first time, the intellectual replaced, in the public mind, the General, the Politician, and the Priest as the source of a way out of the crisis.[2] If objectivity was not possible, some argued that the only moral stance on the part of intellectuals was to elaborate a position that would support reformist or revolutionary schemes designed to benefit the dominated and exploited majority of society. Particularly because of its moral overtones, this question of the social commitment of intellectual work greatly troubled Innis. For his part, he struggled to win over Irene Biss, the young British immigrant, to his conception of an authentically indigenous approach to scholarship that would remain free of both metropolitan paradigms and party politics.

A corollary of Innis's 'political' defence of 'apolitical scholarship' was that such a position could not prevail exclusively through polemics and the publication of counter-strategies for public policy. The validity of his position could be clearly demonstrated only by the continued publication of new scholarly work on the Canadian economy. Innis's influence would triumph not by winning this or that point in contemporary discussions but by asserting itself as an accurate and indigenous perspective that would form a more solid basis for public-policy discussions. In a sense, Innis aimed at making a contribution to the cultural rather than the political life of his country. His participation in political debates was therefore an annoying diversion from his main priority during the Depression: the completion of the core elements of the collective project of writing Canadian economic history that had advanced so rapidly in the 1920s. His status as an intellectual and his influence within academic, government, and business circles depended primarily on these efforts. His editorial work with James T. Shotwell on the *The Relations of Canada and the United States* series provides us with a point of departure for investigating this area of the Depression experience.

Urwick and Objectivity

Urwick's British background and lack of sympathy with Innis's determinism explain the antipathy that characterized their initial relationship. I have cited in the previous chapter Urwick's letter to President Falconer of the University of Toronto complaining about Innis's 'impudent' attitude. Yet, in the space of a few years, this relationship was transformed. On 29 May 1936 Urwick wrote to the new president, Henry John Cody, about Innis:

> With regard to your suggestion about bringing in some outstanding Canadian Economist such as McGibbon or MacIntosh. The latter, I am quite sure, would not come if his coming in any way interfered with Innis' prospect of being appointed Head of the Department. McGibbon ... would disrupt the Department ... If he was appointed Head of the Department I fear that Innis would leave. In any case his appointment would cause some astonishment among Canadian Economists – unless Innis had previously left. For I think there is no doubt that Innis now has (and

deserves) a reputation, here and abroad, second to none in this country. And few Economists would understand his being passed over in favour of anyone else.

Another matter troubles me seriously. When you discussed the possible promotion of Innis to a full Professorship, you wished to postpone this step, chiefly, I believe, because the governors did not wish to make any more such promotions until the proper salary could also be paid ... Auld has been promoted to a Professorship. Without questioning his desserts I feel compelled to suggest that his claims are much inferior to those of Innis ... Innis cannot but feel slighted, – and I am afraid the members of the staff in Economics will also feel that our Department is now considered of less account than the Division of Law ... Innis ... is deeply hurt, and I venture to suggest that the governors should be asked to put him on an equality with Auld at the earliest possible opportunity.

If I may assume that the governors do not wish to lose Innis, I would also give it as my strong conviction that he is likely to leave if he is now passed over in the matter of the headship of the Department. I have, in fact, assiduously 'groomed' him for the task of managing the Department. Would it not now imply a definite lack of confidence if he was passed over?

There is one other matter. You suggested the possibility of bringing over some outstanding British Economist, for a short period ... I doubt whether anyone would come for more than one session without a prospect of a permanent appointment, which would raise difficulties of which you are of course aware. It is, as I know, extraordinarily difficult for any stranger to understand the Canadian setting especially in the Economic field – in less than three or four years.

... If it is now the intention of the governors to pass over Innis for the Headship of the Department, I should feel that I had failed in my first duty of keeping them fully acquainted with the needs and difficulties of the situation.[3]

The letter is interesting in three respects. First, it indicates a continued resistance on the part of those in authority towards Innis's career advancement. The university apparently approved his appointment to full professorship and to head of political economy only under duress. Reading between the lines, one suspects that stalling on the question of a full professorship for Innis may have been a tactic to allow the university authorities to deny him the departmental headship. Urwick's letter is only apparently an informal note. In fact, it is a closely argued account of why it is impolitic and unjust to consider any option that would not result in Innis's becoming department head immediately.

Second, these discussions were taking place during the period in which Innis suffered a collapse and was bedridden with 'nervous exhaustion.'[4] Undoubtedly, they helped to contribute to the gravity of that crisis. Also, the discretion with which his breakdown was handled (even close friends were unaware of it) may have been related to its timing. He was in bed from 10 to 24 March 1937, and his appointment as department head was announced in June 1937. As we will see, not only his professional status but also a number of other aspects of his life reached a crisis point around this time.

Third, the letter indicates an unusual about-face in Urwick's attitude. Not only has Urwick revised his opinion of Innis and his work, he has also adopted some of Innis's basic attitudes. In particular, he reiterates Innis's assertion of the importance of familiarity with the 'Canadian setting' as a prerequisite for scholarly work in Canada. Clearly, something has happened in the interim. In fact, Urwick appears to have experienced something akin to a personal revelation in the colonies. As he later wrote to Innis following his retirement to the west coast:

> I came here to escape, but I begin to see that there is no escape. Victoria is not just a single peculiar place. I am afraid the inhabitants of that otherwise pleasant city are only a type ... of a very common and world-wide order ... Their one obvious virtue is blind loyalty, or better, a swarm of blind loyalties instilled into them by school and college and club and social class ... They never think or have thought about the objects of their loyalties – Empire, Religion, the British way of life, the Upper Classes, the importance of being a Gentleman, and all the other idols of blind class interest ... These dear people are everywhere; and they don't all come from England!
>
> I have to confess that I have been associated with these people and almost one of them for much of my life. No foe as bitter as a renegade! I have been gradually awaking during the last dozen years ... I am [now] pretty fully awake, and for the moment rather hopeless.[5]

Urwick had come to understand Innis's position as a colonial intellectual. He had also begun to appreciate that the peripheral intellectual might have the unique ability to transcend more readily the traditional prejudices that inhibit a critical approach to understanding society. Furthermore, this realization produced the same sense of pessimism in Urwick as it did in Innis.

Two incidents throw further light on Urwick and Innis's changing relationship. In the late 1920s, Urwick became unwittingly involved in a controversy that may have led him to re-evaluate Innis's persistent and cantankerous attitude as an appropriate one given the exigencies of the colonial academic milieu. Through his son's travels, Urwick had learned of conditions in the mining towns of northern Ontario. Struck by the primitive facilities existing in these communities, Urwick commented during one of his lectures that increased social services would decrease the incidence of drinking and gambling among the miners. This innocuous comment led to '[the] Premier [of Ontario] boiling over with indignation, and telling the Professor, in effect, to stop meddling and go back to his place as a teacher.' One newspaper proclaimed, 'PREMIER SHOWS URWICK HIS PLACE AT TEACHING DESK.'[6]

Urwick was astounded at the reaction and complained, 'It never occurred to me that the drinking, gambling and immorality ... had any political significance.' As a new arrival, he was accused of not knowing what he was talking about since he had not had the 'opportunity to learn the true state of affairs in the north.' It was the sort of attack against which Urwick had no defence. At this point, Innis, who had extensive first-hand knowledge of the north, stepped in to confirm Urwick's report on social conditions with direct examples that could not be summarily dismissed. The episode gave Urwick (and should give us in the altogether different academic milieu of today) an idea of how sensitive the politicians of the day were to criticism (even if was more implicit than explicit) emanating from such a position of intellectual authority as the head of the University of Toronto's political economy department. This was not only because of the sorry record of these politicians in dealing with social issues, but also because Canada remained in many respects a 'colonial context.' Urwick must also have realized that he was not under as much pressure as Innis would have been in the headship. A controversy involving Urwick could always be diffused by his being written off as a foreigner. This would not have been possible with Innis, a Canadian. By the same token, Urwick must have realized that an outsider such as he could never expect to have the same impact on public-policy discussions as a native Canadian intellectual. This realization seems to have convinced Urwick that he should, indeed must, work actively for Innis's appointment. The incident reminds us in a graphic manner that Innis's reference to his academic work as 'politics' was not merely a figure of speech.

The second event that led to a greater understanding between the two men was, ironically, one that underlined the fundamental differences in their points of view. In his essay referred to above, 'The Role of Intelligence in the Social Process,' Urwick argued, in an almost flippant manner, against the materialist and determinist tone that pervaded the social sciences, especially in North America. In its 'philosophical approach,' the article was a direct attack on the direction Innis was taking in his research. Urwick argued that 'social science' was a contradiction in terms – that scientific 'objectivity' was impossible given that the social 'scientist' was caught up in the very life processes that he was purporting to study. He concluded that social-scientific pretensions were really a complex form of game playing and should be recognized as such. For Urwick, what counted in these fields were faith and feeling, not the illusion of objectivity.

Urwick's paper implicitly criticized Innis's project at the levels of politics and epistemology. If human intelligence cannot produce an accurate vision when applied to its sphere of practical operations, society, then the application of intelligence through public discussion is even more severely limited. Yet Urwick concluded by praising discussion as an essential consensus-building mechanism that by its nature is 'not really a *logical* process ... [but] a display of feeling and very little else.'[7]

This argument troubled Innis greatly. On the one hand, it denied the possibility of objectivity. On the other hand, contrary to Urwick's personal conclusions and albeit in a roundabout manner, it upheld the position of academics who were an anathema to Innis: politically involved intellectuals who claimed prescriptive powers for the social sciences. If objectivity was impossible owing to the subtle influence of social sectors on the intellectual, then the important thing was to be quite clear and open about which social sector one's intellectual output favoured. 'Objectivity' as a term became linked to a class position rather than being a characteristic of the disinterested individual qua intellectual. If one could not be sure of anything with scientific certainty, one could at least champion a particular blueprint for society with the fervour of such certainty provided that one saw this action was benefiting the oppressed majority.

Innis's reply to Urwick's article, in the same journal later that year, is intense. From the outset, he separates the general from the particular problem. He makes clear his strong conviction that employment of the social scientist by governments, political parties, ecclesiastical organizations, and specialized research institutes perverts the possibility of a social science. He then goes on to echo a theme that will surface later in his communications studies: 'The adjustment of social science to current demands has been facilitated by improved methods of communication, such as the radio, which strengthen the influence of lower levels of intelligence. An attempt to escape the difficulties by resort to history has led to the simplification of thinking in the social sciences in terms of the Marxian analysis and the class struggle.'[8] While agreeing that the bias imposed by the demands of various social agencies seems to make objectivity impossible, Innis does not sink into relativism or cynicism. He stresses that the social scientist 'may take comfort in the argument that thought in the social sciences grows by the development and correction of bias ... Since the social scientist cannot be "scientific" or "objective" because of the contradiction in terms, he can learn of his numerous limitations. The "sediment of experience" provides the basis for scientific investigation. The habits or biases of individuals which permit prediction are reinforced in the cumulative bias of institutions and constitute the chief interest of the social scientist.'[9]

Innis continues his argument by saying that, once the nature of individual or institutional biases has been determined, then the task of the social scientist becomes not the provision of immediate answers to questions of public policy but repeated 'attempts to predict general trends.' He makes it clear that 'periods and regions with sufficient stability to support freedom of inquiry' are rare and that it is this fact rather than the approach of any particular social science that limits genuine contributions. Thus, the effective sheltering of intelligence by force – what he will later describe through the symbolism of Minerva's Owl – is a central factor in determining new contributions to knowledge. As examples of significant contributions to social science, he mentions Marx, Veblen, and Adam Smith. We see the generalist drift of his thought at this time when he closes the section of the article dealing with objectivity in the field of economics with the remark: 'Incidentally, the significance of a classical training to the social scientist should rest in the breadth of the approach and its emphasis on the range of human activity.'[10]

The article is followed by a long postscript in which the university is presented as a social-science tool par excellence in that it plays the essential role of providing the milieu that allows the intellectual to grapple more readily with the problem of bias: 'The University can protect the social scientists in part. Even in the countries which have witnessed the most serious disturbances apparently work in the social sciences can continue without serious interruption, chiefly because the complexity of the subject renders conclusions of little value to those in control of policy, and terminology becomes a defence against the inquisitive. Moreover, Universities with centuries of tradition serve as a defence.'[11] Since the university is *the* essential collective tool that provides a calm point of reference for social investigations, 'the problem of the social scientist is the problem of the University.'

Whereas Urwick's article is primarily philosophical in approach, Innis's reply is intensely political. Although he has been forced to put his own basic premises in question, he does so in a curiously restrained manner. He seems uneasy dealing with 'objectivity' in a general or abstract manner. Instead, he concentrates on the perversion of objectivity through the influence of various 'vested interests' in the social sciences. He never insists that objectivity *is* possible. He limits himself to maintaining the central importance of the *attempt* to attain it through avoiding fanaticisms supported by vested interests. In this regard, he is in profound agreement with Urwick on the danger posed by the increased pretensions that accompanied the trend towards specialization in the social sciences. 'Specialization runs mad, and when it does so, never leads to understanding. Its natural result is strife and violent dogmatism.'[12] In these words of Urwick's, which Innis recalled in writing Urwick's obituary in 1945, the positions of the two men were finally reconciled.

'The Role of Intelligence in the Social Process' had a seminal influence on Innis out of all proportion to the profundity of the article. First, it forced him to go public with the sort of political commentary that heretofore had characterized his confidential university memoranda. Second, it was his first sustained treatment of the question of bias. Finally, the treatment of bias in the debate with Urwick was fundamentally similar to his treatment of the subject in the later communications works. Bias was viewed as being embedded in individual consciousness but always as a result of a process of imprinting determined by social and technological circumstances.

The article was written during a time in which the various aspects of Innis's project – his individual research, the building up of peripheral academic institutions and a scholarly perspective, the development of an effective national policy and Canadian contribution to universal knowledge – were all under attack by the forces of the Depression. The chief dynamic arose out of an intellectual cultural imperialism that, ironically, accompanied a period of renewed economic nationalism. Innis became virulent in denouncing imperialism in the intellectual realm, particularly when it manifested itself in the guise of progressive social reform.

> In countries in which traditions are less conspicuous, in which an old generation of University Presidents is being replaced by energetic young men guaranteed to do things, and where the development of the social sciences is so weak that conclusions are understood by those in control of policy, difficulties become serious even though the cultural background has been less seriously disturbed ... Because conclusions are reached by scholars in Universities with long-established traditions, they are held to be valid and applicable universally. The more subtle of the social scientists will proceed by careful study to discover the weak points in the intellectual armour of the community and to utilize the conclusions with effect, and consequently they say that because Great Britain has a certain type of machinery therefore Canada should have it, that capitalism is a good thing, that we should produce for use and not for profit, and that we should read our Bibles. There are apparently advantages and profits to be gained by advocating any one or more of these 'remedies', including production for use and not for profit.[13]

But Innis had another concern, too: the intellectual seduction of promising young academics by such groups as the League for Social Reconstruction (LSR). In fact, this worried him even more than the effects of Urwick's ideas on social debate.

The Scholar and Social Commitment

The Depression-era debate on the role of the intellectual in the social process was the most personal and intensely emotional one of Innis's career. It was also one that, in the end, he would lose. This loss would launch him into the last phase of his career as a lonely scholar probing the great themes of empire and communications. But, at the beginning of the 1930s, this outcome was not a foregone conclusion.[14]

Innis fought the battle in two stages. At the beginning of the 1930s, he hoped to convince a group of Canadian intellectuals of the validity of his brand of 'apolitical' scholarship grounded in intensive 'dirt' research. When this failed, he concentrated on the conversion of one young, socially engaged, and highly intelligent economist, Irene Biss, to his brand of scholarship.

What is often not recognized is the extent to which Innis entered the Depression years as a progressive. He came from a rural community and knew first hand what it was like for ordinary people to struggle to make a living in a peripheral resource-based economy battered by a Depression. He instinctively sympathized and stayed in touch with farmers and workers who were the principal victims of the economic downturn. Indeed, he felt the effects of the Depression personally as he was forced to sort out the mortgage problems of the family farm brought on by the collapsing prices of agricultural commodities.

At the beginning of the 1930s, he was naturally inclined to take a progressive public stance. Indeed, I believe that he was much closer to the activist intellectual left than is usually credited. We have already dealt with his highlighting of negative social conditions in the north in defence of Urwik. A.E. Havelock, a professor of classics at Victoria College in the University of Toronto, provides us with another report of Innis's populist inclinations at the time.[15] He recalls that a group of progressive intellectuals concerned about the effects of the Depression assembled in Toronto on 17 November 1930 and resolved 'that those here assembled undertake the organisation of a group which, dissatisfied with the policies of the two major parties in Canada, wish to investigate public affairs with a view to forming a definite body of progressive opinion.'[16] Innis became a dues-paying member of the group and delivered a talk to his colleagues on 12 January 1932 on 'Economic Conditions in Canada.' His presentation was bracketed by those of Frank Underhill on 'Canadian Nationalism' and J.S. Wordsworth on 'The Canadian Political Situation.' On 23 January 1932, the more militant intellectuals among the group went on to meet, in the former offices of the political economy department (Baldwin House), with like-minded colleagues from Montreal to form the League for Social Reconstruction. Then, on 26 May of that year, these intellectuals met with the 'Ginger Group' of independent left-wing members of Parliament in Ottawa to form a new socialist party, the Co-operative Commonwealth Federation, with Wordsworth as its first leader. The LSR continued as the 'brain trust' of the new party, with Underhill writing the first platform of the party: the Regina Manifesto. With the forming of the LSR, the old University of Toronto group had gone out of existence, but Innis's influence on the new organization was confirmed when its first publication, *What to Read*, included three of his publications.

As we have seen in his decision not to join the clergy in 1916, Innis was a person who did not like to be 'pushed' into things. I suspect that the very passion with which his left-wing colleagues urged him to support activist causes made him suspicious rather than engaged. When Innis declined to join friends and colleagues who were establishing the LSR and CCF, it was not because he was against scholarly work influencing public policy. For Innis, that was one of the main points of independent scholarship, and it formed a central part of his project. Rather, it was that he was against intellectuals participating directly in party politics. He believed that doing so would corrupt the critical independence of scholarly work.

In fact, Innis himself was significantly engaged in trying to influence public policy during the period. The paper he gave to the Toronto group did not end up appearing as an LSR publication but as 'Transportation as a Factor in Canadian Economic History' (May 1931), which he read before the Canadian Political Science Association. This was not a one-off effort. In the summer of 1932, he wrote an article on the Depression for the *Financial Post*. In October of that year, he was corresponding with Prime Minister R.B. Bennett, arguing that 'the machinery for hammering down fines ... in ... the Combines Act should be made as efficient as possible. Our natural tendency in Canada is towards a concentration and it's important to check the disadvantages of that tendency ...'[17] Innis's preface to *Problems of Staple Production in Canada* (1933) dealt with the Depression, as did the book he co-edited with A.F.W. Plumptre, *The Canadian Economy and Its Problems*. For the *Economic Journal* (under the editorship of J.M. Keynes), he wrote 'Economic Conditions in Canada 1931–32.' He also participated in attempts to raise the level of public debate by facilitating the publication of popular essays, such as 'Smoothing the Bumps in Business' (Ryerson Essay no. 55), for which he wrote the preface.

Another incident took place at this time which may have made Innis wary of being more closely associated with the left. On 15 January 1931 a front-page headline appeared in the *Toronto Telegram*: 'Free Speech Urged by Varsity Faculty in Protest to Police Board.' Sixty-eight members of the University of Toronto faculty, including Innis, had written to complain that the zealous actions of the police were infringing on the right of free speech in the name of fighting communism. While the letter was signed by professors with a wide range of political views, it was initiated by Underhill and Havelock and the same 'radically-minded' small group of academics that had begun meeting in the fall of 1930.[18] They were probably able to obtain the signatures of a wide range of faculty with the understanding that those signing were doing so as individuals without reference to the university. Predictably, however, the *Telegram* had obligingly added the department and college affiliations of the signatures (obtained from the university calendar) before publication. The issue that triggered the letter was the cancellation, under police pressure, of a meeting organized by a liberal organization, the Fellowship of Reconciliation, to discuss police brutality in suppressing communist gatherings. Some of the individuals who signed must have been mortified when the letter appeared as front-page news and included their university affiliations. Embarrassingly, the entire staff of the department of political economy had signed.[19] It must have been obvious to even the more conservative members of the department that they were writing in support of 'one of the proudest heritage of the British peoples' – free speech. But they were probably surprised to find that, in the polarized atmosphere of the era, support from the public and politicians favoured the police and not the professors. The *Globe* opined: 'The fact that communism has become less militant in Toronto ... is sufficient evidence that the police of the city took the correct course in dealing with it.'[20] In the end, the signing faculty members were not disciplined, but the university's Board of Governors passed a resolution dissociating the university from their stand.[21] The controversy surrounding the incident undoubtedly reinforced Innis's belief that academics should not become directly involved in politics.

The incident of this letter and the earlier founding of the Toronto group of progressive academics indicate the extent to which the Depression had the effect of polarizing members of Canadian intellectual circles. Many individuals influenced by Fabianism or Marxism moved left and, with the zeal of the newly converted, denounced all non-self-proclaimed leftist intellectuals. Underhill, for instance, called them 'the garage mechanics of capitalism.' Others drifted off into fascist sympathies or right-wing paranoia. H.J. Cody, chairman of the university's Board of Governors, visited Italy during the period, became infatuated with fascism, and spoke positively of it in his speeches. He seemed oblivious to the contradiction that he was doing exactly what his board had warned the faculty not to do.[22] Another example was Gilbert Jackson, the business economist who was Innis's nemesis in the commerce debate, who wrote to the president of University of Toronto: 'Ever since the rapid expansion of numbers in this University began, there has always been the danger of developing a discontented intellectual proletariat, such as is found in many European countries, and there are signs of its appearance here already. It seems to me that this should be the care of Canadian business as well as of Canadian University authorities.'[23]

Polarization in the social sciences was particularly strong because of the widespread belief that the new specialist knowledge of these disciplines might produce a solution to the Depression. Everyone longed for an end to the misery they were experiencing. In these circumstances, what was wrong with an intellectual (especially an economist) attempting to bring his knowledge to bear on the problems more quickly by associating himself with a political formation?

In Innis's opinion, everything was wrong with this approach. Unlike Urwick's attack, it was a position that appropriated for itself a monopoly of moral rectitude. In so doing, it drew Innis's concentrated fire. Urwick had maintained only that faith and feeling remained important because objectivity was an impossible goal in the social sciences. The 'socially committed' intellectuals went further and presumed that objectivity was undesirable. They viewed the idea of objectivity as an ideological myth that was the natural by-product of bourgeois liberal democracy.

Innis's response to this challenge was complex. On the one hand, he was obliged to maintain that politics and advances in the social sciences did not mix. On the other hand, he put forward this position 'politically' at every public and academic forum to which he was invited to speak. Innis's political articles during this period reached a greater quantity and intensity than at any other time during his career with the exception of the final years of his life. The advent of the Depression forced him to defend his position prematurely, which accounts for both the ambiguity and the extreme bitterness that characterized his replies to socially committed academics. The work of the 1920s was just beginning to lay the groundwork for both Canadian economic history and an indigenous critical methodology. Absent the Depression, Innis would have been fully preoccupied carrying forward his research and writing work during the 1930s. Because of the Depression, he was forced to fight continual defensive actions to sustain the legitimacy of this collective research. Nor did the tremendous detail of the staples studies afford security for his intellectual position. 'Social commitment' meant borrowing fully developed metropolitan paradigms through which to view not only Canada but also the world. To Innis's consternation, the staples approach had not yet given rise to an indigenous paradigm that could hold its own against more sophisticated and politicized world-views.

Innis adopted four tactics against the socially committed intellectuals. First, he invariably took a negative tack, criticizing the oversimplifications and myopic tendencies of the imported paradigms, especially when these were applied to the Canadian case. Second, since the simple coherence of his opponents' fully developed paradigms placed him at a disadvantage in public debate, he began to damn discussion per se as an impediment to knowledge. His concern with the fate of the oral tradition expressed itself here for the first time as a fear that 'talk' was destroying the possibility of intellectual contemplation. From the start, Innis ironically associated the silence of reading, writing, and thought, far more than the sound of public debates and pronouncements, with the oral tradition. Third, Innis called for outside intellectual support from individuals who shared his political orientation and interests.[24] Finally, he was obliged to accelerate his elaboration of his own positive 'macro' vision of those dynamics in history and human thought that had led to the impasse of the Depression. In these political articles of the 1930s, we find the beginning of his long-term, large-scale vision of the world, Canada, the social sciences, and the role of communications – including many of the themes that would come to dominate the works of the communications period. This is how Innis viewed the Depression in 1934: 'We are faced with the far-reaching results of the technological drift of modern industrialism. The success of measures designed to solve the problems of the depression is necessarily determined by their relation to problems of the secular trend ... which have accentuated the decline characteristic of the business cycle ... An analysis of the factors peculiar to a long run development is essential to an understanding of immediate difficulties.'[25] The 'secular trend' to which Innis refers is at best a vague notion. Nevertheless, he fixes on this term rather than some more immediate and banal mechanism as the central determining factor of the period. Chief among the elements of this notion of secularism is its tendency to exaggerate natural fluctuations in the economy. Innis posits a not-at-all self-evident connection between the lessening influence of ecclesiastical institutions and their thought, on the one hand, and the loss of elements of continuity and moderation in modern society, on the other. Again there is much of the autobiographical in this view. In a sense, Innis is projecting his own loss of faith writ large as the essence of the world trend towards modernization. In so doing, he is not proposing a return to faith in the manner of Urwick. Rather, he is suggesting that attention be paid to fostering those elements of long-term perspective that have traditionally held a place in society outside organized religion. Only when an increasingly secular society can understand its own biases and, therefore, its own limitations, will its radical fluctuations be brought under control. In typical fashion, Innis, having rejected religion, also rejects the pretensions of uncritical secularization.

Innis goes on to associate the unfolding of this secular trend with the history of technological innovation. His schema is basically that of Patrick Geddes, the Scottish social evolutionist, town planner, and mentor of Lewis Mumford, in which the paleotechnology of coal, steam, iron, and water transport leads into an era of neotechnology dominated by oil, hydroelectricity, steel, and rail transport. The key difference in Innis's treatment is his profound sense of the fragility of progress. Others might expect a cleaner, more labour-free Utopia to be brought about as the new technology replaced the old. Innis expected neither that the old technology would be superseded peacefully and disappear nor that the new would necessarily usher in a better world. He believed that he was living through a massive changeover to the new technology. The old techniques had formed the basis of the free-trade era and had continued to dominate the world economy down to the First World War. They had been characterized by the concentration of production facilities around steam-powered installations near to coal-producing areas with ocean access, and had been predicated on the international availability of a vast resource base mobilized through water transport. A corollary of this process was the development of new metropolitan centres that served as the termini for ocean and land routes shipping raw materials to the centre, and manufactured goods to the periphery, of empire.

Since electricity did not dissipate as rapidly as steam over short distances, neotechnology permitted the decentralization of production facilities. Oil reinforced this tendency since it was more easily transported and distributed than coal. Furthermore, hydroelectricity demanded the relocation of industry to areas previously viewed as relatively inaccessible. Hydroelectric power implied heavy overhead costs that necessitated increased efforts at resource exploitation in the interior. Rail links enabled large areas of the interior to be effectively penetrated for the first time.

Innis believed the political consequences of the changeover to the new technology to be profound. Metropolitan centres that had served as commercial centres under the old technique now took on greater independence and developed their own industrial base. Innis viewed this process as largely unconscious and therefore beyond control. Its limits were those of the resources on which it fed. A cumulative dynamic was set up in which the new technology exposed vast new vistas of virgin resources to exploitation by the industrial system, thus increasing the pace of industrialism and augmenting the proportions of the crisis that was bound to take place when new virgin resources were exhausted.

The linchpin of these virgin resources was land itself. Technological innovation, the catalyst of wars, and the discovery of new energy sources sent out an army of European settlers to serve on the front line of the new technique. However, the exhaustion of productive 'vacant' lands changed the international economy irrevocably. As Innis put it:

> The rapid strides of technological improvement which accompanied the war and the use of new sources of power, have involved relative exhaustion of the last virgin natural resources. The disappearance of free land has coincided with the rise of nationalism ... The peace consolidated industrialism's gains from the war. New nations were carved out and industries which had grown up during the war were supported by nationalism and tariff barriers ... Nationalist sentiment and organization, developed during the war, provided support for tariffs to protect new industries. In turn competition from more recently industrialized areas based on hydro-electric power and the drive of overhead costs, and on oil, on improved technique, and on less exhausted natural resources was followed by tariffs imposed by highly industrialized regions to protect older industries based on coal and iron, more exhausted natural resources, less tractable labour, and plant affected by depreciation through obsolescence ... Nationalism fostered by the war and the boom period became more intense as a result of depression.[26]

Innis's schema for the history of the world economy places economic nationalism at the centre of the problems that occasioned the Depression. This does not mean that he longed for a return to the days of free trade. He was not tempted by the myth of a golden age of the free market because he recognized that such a time had never really existed, and that there was no going back in history. He viewed these types of sentiments, particularly with regard to the situation in the Maritime provinces, as a form of pernicious nostalgia that drew attention away from action on current problems. Nor was Innis's a diffusionist view of economic development. He did not expect a wave of higher living standards occasioned by some general phenomenon termed 'modernization' to spread evenly and benignly throughout the world. His recognition of the significance of overhead costs served as an antidote to a belief that the unfettered operation of free-market forces would lead to the best of all possible worlds. His vision was closer to one of unequal development: economic advances moving rapidly forward in certain regions or sectors and at certain periods while at the same time blocking development in closely associated areas.

In all of this he recognized a general movement away from older economic centres towards peripheral regions. This was followed by the attempts of traditional centres to resist the transfer of economic weight to the margins. The adoption of protectionist strategies simply served to exacerbate the process.

> Metropolitan centres in recently industrialized regions which are still important producers of raw materials are strengthened in part by new sources of power and lower prices of raw materials, in contrast with metropolitan centres in highly industrialized regions which have been weakened by higher prices of raw materials incidental to protection and economic self-sufficiency, and particularly in relation to wheat. The trend of modern industrialism has been toward the more recently industrialized regions with new sources of power, and less exhausted natural resources, but this trend has been strengthened by cumulative forces in economic nationalism.[27]

Just as in an earlier free-trade era the disparity between the mother country and its colonies tended to become exaggerated, so in the era of economic nationalism did the disparity between the industrial heartland and the resource periphery, or between city and country of each nation state. The ideological corollary of this changed dynamic is a switch in emphasis from a rhetoric of imperialism/colonial independence on a world scale to one of political centralization/regionalism on the level of each nation-state.

Innis himself was in the process of applying a specific subsection of a metropolitan paradigm – the economics of overhead costs – on a grand scale to the political economy of Canada. He believed that this question of overhead costs was central to the dynamics of the Depression. His Canadian case, therefore, went far beyond the bounds of a study of a regional economy and became a fundamental critique of price theory per se. With the Depression, he said, 'we are faced with the problem of overhead cost on a vast scale, prices have become less satisfactory as indicators, the solution depends on the introduction of economic intelligence which avoids monopoly and perfect competition – nationalism with intelligence ...'[28]

Innis argues that a thorough understanding of the overhead-cost structure of the Canadian economy is a prerequisite for rational public-policy proposals. In like manner, coming to grips with this problem at the international level is a necessary prelude to adopting measures to end the Depression. Yet, at this point, from his contemporaries' point of view, he is at his least constructive, most critical, and most exasperating. As he attempts to sketch out the economic history of Canada leading up to the Depression, he is always careful to stress the tentative nature of his work and the contrary interests imbedded in the economy.[29]

His sympathy with the marginal regions of the Canadian economy is a recurring theme: 'An important phase of the Canadian problem is the marked decline in standards of living of the wheat-growers of the Prairie Provinces and of producers for export in other areas such as the Maritime Provinces: a country cannot endure half slave and half free.'[30] But even when this sentiment leads to the recommendation of an income tax to 'tap the wealth of Central Canada obtained as a result of protection and expenditures in subsidies to assist the development of regions compelled to bear an undue share of the burden,' this is motivated by a concern for 'fairness' rather than economic efficiency. It is also qualified by the stipulation that far more research into the long-term effects of such a step is needed before it is put into practice.

Innis returns again and again to two main positions. First, he condemns the centralist bias of the period and its heavy dependence on imported paradigms, whether they be couched in the form of the Rowell-Sirois report or that of the LSR manifesto. Second, he stresses the need for more 'dirt' economics to determine more fully the nature of the *Canadian* problem. His suggestions of a constructive nature are limited to the naive belief that, once the indigenous scholarly research has been produced on Canada and personnel trained for the civil service have been imbued with this new perspective, appropriate public-policy consequences will follow automatically.

Innis's great fear was that the social sciences in Canada would never get around to doing the necessary 'dirt' research to influence public policy positively. He became alarmed that forces were at work within the various disciplines that undermined the application of indigenous intelligence to Canadian problems. He viewed this as a universal malaise linked to increased specialization in the social sciences, a theme he dealt with directly in articles such as 'The Passing of Political Economy,' 'Discussion in the Social Sciences,' and 'Economics for Demos.' In these essays, Innis almost inadvertently began his work in the field of the sociology of knowledge. They were crude and unfocused attempts to account for the intervention of critical intelligence in human history, an endeavour that would take on greater coherence when he used the influence of various media to focus his analysis during the post-1940 period.

'The Passing of Political Economy' traces the development and disappearance of the conditions that led to the holistic approach of Adam Smith in particular and critical thought in general. Innis partly attributes Smith's insight to his position as a peripheral intellectual, a Scot whose work was pursued outside Oxford and London. Such a social outlook allowed Smith to escape the stultifying biases of metropolitan scholarship. It also permitted him to experience an enriched environment characterized by widespread cultural interaction and economic growth. Innis fixes on the strength of the classics and theology as one indicator of the intellectual fertility of Smith's milieu.[31]

Innis felt that, with the end of the free-trade era, the circumstances conducive to holistic critical thought had vanished. With the extension of industrialism, 'thought' itself has become a product, bought to fulfil an immediate purpose at a fixed price: 'The end of the nineteenth century and the twentieth century were marked by the extension of industrialism ... the decline in freedom of trade and the hardening of political entities in the intensity of nationalism. With these has come the end of political economy, the emergence of specialization in the social sciences, and its subordination to nationalism. Social scientists are in great demand at the prevailing rates among business firms, including publishers, governments, and political parties.'[32]

Innis's comments on his colleagues' activities during the Depression years were extremely bitter. For Innis, it was an era of wholesale intellectual prostitution. '[Some] scorn cash, and take promissory notes on the revolution, or take both notes and cash; and still others, having found truth, have their own rewards.'[33]

In the midst of this bleak vision, the media of communications first make their appearance as a grim motive force for cultural collapse:

> The circumstances are not propitious for another great epoch of thought. The rise of literacy and improved communication promoted the rapid growth of groups, associations and nations and reduced social scientists to a position as defenders with the zeal of proselytes of this and that particular cause. Under the influence of modern industrialism in the printing press and cheap paper, universities have become increasingly specialized, and increasing demands for space in the curriculum have enhanced the activity of administration and promoted the growth of vested interests.[34]

While Innis was respected as a defender of academic freedom, his personal opinions went far beyond this principle. His alienation was reflected in the increasingly extreme and unpopular nature of his opinions. Not having a fully developed indigenous paradigm from which to launch a counter-attack, and confronted with the increasingly aggressive stance of Canadians who had found solutions *à la* Keynes, Marx or Sidney Webb, Innis began to damn 'discussion' and 'freedom of speech' per se. 'It is impossible to discuss the increasingly complex problems of economics with the untrained, as it is impossible to discuss the [actually less difficult] subjects of mathematics or physics.'[35]

One of Innis's most heatedly political articles, 'Discussion in the Social Sciences,' appeared in 1936. It was a vitriolic attack on 'freedom of discussion' itself. In the midst of the Nazi era, Innis adopted the unpopular stance of arguing against the assumption that more open public discussion of various policy options would facilitate a solution of the problems of the Depression. As previously mentioned, Innis viewed the increased level of discussion as partially the cause of, rather than a solution to, the Depression. Innis thought that conferences, royal commissions, and other public forums should be assessed in terms of the long-run secular trend brought about by industrialism. He argued that the myopia imposed by this secular tendency confused the role of discussion and the role of contemplation in the advancement of knowledge. He saw this as particularly dangerous because increased discussion inevitably lessened the possibility of getting on with the low-key contemplative style of scholarship. He began to conceive of the university as the secularized heir to the monastic tradition that had kept Western culture alive during the Dark Ages. Following his ideal of the scholar monk to its logical conclusion, Innis developed a lifelong antipathy to marriage as an institution. He recognized that such an ideal was out of tune with the times and with his personal circumstances. Yet he continued to hold to it. His own marriage had provided him with the psychological and material support to get on with his work, but he reciprocated by devoting little time to his four children. He recognized the contradiction in this and complained to S.D. Clark that he 'should have been a monk rather than subject his family to his life style.' Nevertheless, he continued to apply this unrealistic ideal in judging the career prospects of others.[36]

This image of the scholar as monk against which he consciously or unconsciously assessed fellow intellectuals was made all the more alienating by the extent to which Innis himself lived up to it. Throughout the period, he forged ahead with his own research while at the same time taking an increasingly active part in controlling and assessing the work of others. His 'political' interventions were, in his eyes, a distracting nuisance – a necessary defensive action so that the public field would not be left uncontested to the 'travelling comedians who masquerade as economists.' He developed a distaste for intellectuals in general. As a group, their wholesale seduction by the prospect of recognition through participation in public discussion underlined their failure to live up to his image of the solitary scholar. Furthermore, and paradoxically, their very participation in this discussion obliged Innis to waste time engaging in the same sort of activity in order to point out that this was *not* what scholarship was all about.

The extent of Innis's alienation from trends in the intellectual world can be gauged from a curious anomaly that repeatedly arises in his work from this period. Innis began fixing his hopes for the application of intelligence in the social process to the sector that was least vocal in public discussions and therefore presumably most monk-like in character: the civil service! His praise for this group during the Depression is all the more striking given his utter distaste for the same sector during the post–First World War and Second World War periods. In 1935 he writes: 'It is probable that no federal administration has ever been manned by an abler group of civil servants.'[37] He went so far as to suggest that the best possible hope for a solution to the Depression would be 'an intelligent dictator (e.g., civil service).'[38]

Innis viewed the problem of applying intelligence to the social process as one of the key problems of democracy. On the one hand, the circle of adequately qualified technical experts capable of discussing rational strategies for tackling increasingly complex problems was small. And it was progressively becoming smaller, or rather more fractious, with increased specialization. On the other hand, mass literacy and universal suffrage had resulted in these problems being discussed in much wider circles than had previously been the case, among people who did not understand them. At this point, Innis introduced changes in communications technology as a way of understanding these overall social dynamics:

> The character of discussion ... has been tremendously influenced by recent industrialism and inventions. In the nineteenth century, with the development of the printing press, economic expansion and the growth of literacy, discussion from the standpoint of the press was concerned with an attack on abuses which concerned those capable of reading or those capable of subscribing to the papers. It paid in the newspaper business to attack abuses, or to conjure up abuses and attack them. The reforms of Great Britain and the struggles for responsible government in the Dominion coin-

> cided with the emergence of the industrial press ... The struggle for cheaper rates on newspapers and a free press and for public education and public schools may have been a struggle for democracy, but incidentally it was a struggle for bonuses to literacy and indirectly to newspapers. No other industry has been so lavishly supported from public funds.[39]

It is anything but a romantic vision and seeks to deflate the pretensions surrounding the early crusading press. But it is a mild treatment compared with what follows on the contemporary situation:

> Changes have [already] taken place within the industry ... News collection has been tremendously improved as a result of the telegraph and its successors, the telephone and wireless; distribution has been widened with improved transportation facilities; the press has been expanded in size and improved in accuracy and speed, and the raw material has been increased almost inconceivably by the shift from rags to wood. The pulp and paper industry is a fundamental development. With these epoch-making changes, the business of newspapers has changed as well. Circulation has become even more important with the increase in size and efficiency of the newspaper plant. The spread of literacy has provided a market ... Advertising has become a basic source of revenues.
>
> The contrast between literacy and intelligence implies a shift of emphasis from an attack on abuses to devices which will attract the interest of the largest possible number of readers ... Politics have dwindled to a position in which circulation is of first importance ... Defence of freedom of speech has become an attractive means of attracting public attention ... Recent improvements in facilities for discussion, particularly the radio, have tended to displace the newspaper ... [but] even government regulation and government ownership have failed to improve materially or to check the character of the discussion which dominates the air. Whether under government control or under private control, the appeal is to the largest number of possible listeners, and there are even more listeners than readers, or more people capable of listening than reading. The radio, like the newspaper, is concerned with marketing and distribution, and its discussion is probably on a lower level than that of newspapers.[40]

Given the dominance of the communications technology, the free-thinking intellectual becomes increasingly more anachronistic:

> He remains as a vestige of an era of discussion which has passed. He is valued by universities as a means of displaying to the public their contin-

> ued belief in academic freedom – the steeplejack who dances about on the upper structures of the framework to demonstrate its soundness. No self-respecting university can afford to be without at least one ... The intellectual writes informatively for a respectable group of people who still believe they discuss the complex problems of society intelligently, and is employed by the paper accordingly, or failing the paper where his efforts are narrowed perhaps to a small column imprisoned as a memorial to freedom of the press, he writes for subsidized journals dedicated to the maintenance of the belief in the importance of freedom of discussion.[41]

Innis goes so far as to conclude that 'discussion has become a menace rather than a solvent to the problems of a complex society ... The tyranny of talk has ominous possibilities.'[42]

This florid language is not characteristic of Innis and reflects the tremendous emotional pressure he felt as a result of the political debate of the 1930s. The position of those intellectuals who opted for socially committed research represented a direct challenge to his own project for the slow strengthening of an indigenous paradigm. Not having such a paradigm at hand he was obliged to fight back in primarily negative terms. He went so far as to vilify the position of his opponents within the secure atmosphere of his own undergraduate lectures. His bitterness is linked to his recognition that, while he might counteract some of the more extreme pretensions of social commitment, he could never win the debate by employing a primarily negative strategy. In this sense, his concern with the effects of communications technology originated as a positive attempt to meet the challenge posed by the Depression debate on the role of the intellectual.

'I shall continue to struggle as long as you do!'[43]: Irene Biss and Innis

Innis' arguments regarding the role of the intellectual reached an emotional height with Irene Biss. From the start, Biss represented a challenge to Innis because her background seemed so uncongenial to the approach that he was championing. In 1929, at the age of twenty-two, she came to the university fresh from J.M. Keynes's lectures. As she put it herself, 'I arrived from Cambridge thinking I knew all about everything.'[44] She was confronted by Innis, who quickly pointed out to her the limits of metropolitan economic theory. Innis made no secret of his belief that several years of Canadian experience were necessary before a scholar could begin to understand the peculiar problems of the Canadian economy. He actively helped her to organize the field trips that would turn her into a 'dirt' economist, advising her, for instance, on how to put questions devoid of economists' jargon to ordinary people. He did the final vetting of some of her academic publications and was undoubtedly responsible for placing some of them in academic as opposed to political journals. Biss, for her part, assisted Innis with his fisheries research, travelling to Nova Scotia to help him gather material for *The Cod Fisheries*.[45] She was one of the few people who could decipher Innis's handwriting. For two years, she worked with him on the cod-fisheries manuscript, trying to wrestle the unwieldy mass into publishable form.

Biss was young, English, and idealistic and had been exposed to the English Fabians. When she was confronted with the misery of the Depression, it is not surprising that she became attracted to the group of intellectuals who were opting for an expression of their social commitment through active participation in such organizations as the LSR, the CCF, and the Radio League of Canada. It was the sort of involvement that Innis fundamentally opposed. His relationship with Biss seems to have evolved into a struggle against these other quarters for her intellectual soul. How she reacted was an indicator of how Innis was faring in the general debate of the period. Innis quite blatantly applied a traditional carrot-and-stick technique to try to win her over to his position. If she were to forget outside issues and concentrate on long-term research, she could expect his full cooperation as a senior faculty member. If she insisted on sympathizing with Underhill and on becoming active in the CCF, however, she could expect him to stop viewing her as a serious scholar.

Innis opens one letter to Biss by asking if he can dedicate to her a prospective book that would include his articles on discussion in the social sciences, approaches to Canadian economic history, the role of intelligence, and the Depression. He then goes on to discuss the possibility of teaching a joint fourth-year course with her. She is to teach a section on 'the immediate problems of advance in the social sciences.' Innis continues:

> At this point one becomes disturbed. As you know I have insisted that an approach which may be expected to make any contribution to advancement of the subject must have depth and involve one's whole outlook ... When you think, however, of having the subject in hand by that date and of completing it next summer and when you spoke of doing research work for the party it is apparent that we are thinking along entirely and fundamentally different lines. You tend to emphasize the superficial that

they call research which is done by graduates and post-graduates at so much per year or per term. I am not particularly interested in that sort of work nor do I think that the university should be interested in it certainly so far as its staff is concerned ... I do think that ... the approach is fundamental and that it is the long run and not the short run and that the student should be left with a philosophy of attack and not a superficial solution. One should have as a first essential a bias toward truth and not toward this or that solution.

... I have more faith in your ability than the above would indicate or than my general feelings toward your whole work would suggest. Your value to the university or your contributions to the social sciences may be of immeasurable importance or of serious damage. You said that your beliefs were beginning to break up – which makes one hopeful or rather which makes me hopeful. If you feel that they have broken up to the point that you are willing to face the issues I should be only too happy to help you in every possible way with your graduate work ... I shall not give up my confidence in you until the last word has been said – when you finally hand yourself over to politics. Until then the door is wide open if you want to use it. Politics imply compromise in a way which is impossible to the social sciences.

The fourth year course is a test case ... My position with regard to your work and the course may be regarded as the strongest possible protest I can make against your position in regard to politics.

I wish I could conscientiously give you all the best wishes for a new life but I believe in your abilities too much for that. I have too keen an appreciation of my own limitations and of what needs to be done urgently in the future.[46]

It is a remarkable document, highlighting Innis's sense of his personal limitations and his emotional intensity in communicating with Biss.

Biss objected to Innis's position because she sensed that it contained an element of defeatism, leaving the field clear, in the area of public policy, to contemporary vested interests. Innis brushes aside this argument: 'I have no fears on that score. This country is full of so called leaders always anxious to seize upon anything which has the ring of truth in it and a thoroughly sound and scholarly piece of work will be taken up by all parties.'[47] Furthermore, Innis argues, while 'England can afford the luxury of [intellectuals acting as] preachers ... the terrific handicaps of a young country in the social sciences demands the greatest possible economy of intellectual effort.'[48] Political involvement by intellectuals on the periphery was not just detrimental to their scholarly efforts. Innis believed that it turned scholars into 'traitors' by taking them away from the long-term work needed to develop the critical perspective that would best serve the national interest.

Innis was evidently becoming overwrought in this debate. His sense of exhaustion and under-accomplishment was similar to his physical and psychological state following the Great War. He wrote that 1935 was a wasted year, despite the fact that it had been his most productive up to that time as far as published articles and reviews were concerned. One letter to Biss opens: 'I am not certain that you will ever get this and you may never know that I was glad to get your letter but it did serve to stimulate renewed interest in my own struggle which I suspect is even more acute than yours and to prevent relapse into a final coma.'[49] In comments such as these, it is clear that Innis treated Biss as a confidant much more than he did any of his other colleagues. In the mid-1930s, his dialogue with her provided him with an avenue to work out his emotional stress, reminiscent of the role Mary Quayle had played in the immediate aftermath of the First World War. The result was the same. I believe that Innis fell in love with Irene Biss.

But the context was different. He was now a married man with a family, and a scholar with an established reputation. I do not know if they had an affair, but I suspect that the pitch of emotion evident in these letters was the tip of the iceberg of Innis's attachment to Biss. From his side, taking up formally with her would have resulted in professional and personal disaster for both of them. The most Innis could offer her was the role of intellectual nun playing opposite his scholar monk. But he had a family already. Irene Biss did not, and she badly wanted one. It was the one thing Innis could not give her. Worse, he knew that the rule of the times was that, if she married, she would be obliged to resign her academic position.

Biss knew very well what was bothering Innis. She wrote to him, clearly frustrated at the impossible position in which he was putting her:

> I have said that I am interested in trying to see how the thing works, clearly and squarely, and that if any engagement or marriage begins to interfere I will stop pretending to be an economist.
>
> I am not quite clear as to what you mean by being 'willing to face the issues.' If you mean break my engagement then I am not.
>
> ... If by my 'position in regard to politics' you mean my engagement – I can only say I am sorry.[50]

At other times she strikes a more plaintive note:

> The incompatibility of [scholarship] with matrimonial commitments, so plain to you, does not seem to me to exist.
>
> ... Finally I know that it is no use planning a life of pure intellectual activity for myself. I am not made that way. Whatever I can do must be done within the boundaries of an ordinary human life.[51]

This was written in the summer of 1935 and, for three more years, Innis and Biss struggled with this frustrating state of affairs. The situation was made worse by the fact that the man she would marry, Graham Spry, had become unemployable as a result of his political activities in the polarized environment of Depression-era Canada. The harsh reality was that they needed her income and therefore delayed their marriage.[52]

Why has not this important relationship of Innis and Biss been identified before? In fact, there has always been significant gossip about the relationship. Neither one, for obvious reasons, would confirm that they were romantically involved and no documentary evidence to such a relationship seems to exist. We have discussed the probability that Innis's correspondence was subjected to a selective procedure in which anything of an emotional or political nature has been destroyed. I believe this is also the case with the Biss material. Despite the fact that the two worked closely for three more years before she married, there is no documented communication between them during this period that I have been able to find in the mainstream Innis holdings at the University of Toronto Archives. Yet their 'struggle' must have continued. It was during this same time-frame that Innis experienced his undisclosed emotional breakdown.

Copies of Innis's side of the correspondence first came to my attention in the secondary holdings of the Innis College Archives. (This collection has since been dispersed.) The copies had originally been deposited at the Innis farm (which was owned for a period by the college) by Irene Biss Spry herself.

Previous to this, Irene had recognized the significance of the letters when she forwarded the originals to Donald Creighton, who was then writing his biography of Innis. Creighton was close to Innis during the period in question and would have been aware of Innis's emotional attachment to Biss. For this reason, he promises to use the letters with discretion, and indeed he does. He excerpts a non-personal selection from the letters and, without noting the actual source, cites it as Innis writing to 'one correspondent.'[53] When Creighton returns the originals to Spry, she writes back to thank him for his discretion.[54] When I found the copies of Innis's letters in the late 1970s, I met with Irene Spry to see if she had any other Innis correspondence. She seemed not at all surprised when I told her that I believed Harold Innis had been deeply in love with her and that she had had a profound effect on the course of his studies. She smiled, but did not comment. I took this as confirmation that I was correct in my observation, since I believed that she would have dismissed it immediately were it not true. She clearly did not want to discuss this further, and I respected her reticence. She said that she did have other correspondence with Innis but, in preparing her papers for deposit in the archives, she had destroyed the Innis ones since they were personal notes rather than letters with historical interest. When I asked why she had not simply put an embargo on access to them for a period, she said, rather wistfully I thought, that perhaps she had made a mistake, but that what was done was done.

The survival of the Biss side of the correspondence was more serendipitous. '[Innis's] administrative files contain a great deal of personal correspondence, which he kept at his office, indicating the dividing line between personal papers and administrative records was not as sharply defined as it is now.'[55] Fortuitously, the Biss letters are among these departmental papers deposited at the university archives and were not subjected to the selection process that I believe the mainstream Innis holdings were.

There is a story often told of Innis: a young colleague approaches him with the happy news of her upcoming marriage. His reaction is not to congratulate her, but to upbraid her on the promising scholarly career she will be abandoning. He is so severe that she leaves his office in tears. The story is told as a general illustration of the austerity of Innis's belief in scholarship. What is usually omitted is the fact that this young colleague was Irene Biss. After her departure from the university, Innis terminated all contact with Irene Biss – he would never see, write, or speak with her again for the rest of his life. The next book that he had wanted to dedicate to Biss was instead dedicated to his children.[56]

From Biss's perspective, one would have expected that she would bear a life-long anger for being so shabbily treated. Not so. She followed Innis's career and after his death was a faithful participant at virtually every event organized to revisit and celebrate his work. In fact, Biss blamed *herself* for the rupture: 'I certainly did let Innis down.

He was training me to carry on the work that he thought should be done by a social scientist, and instead I went and got married and had children and did war work, and never produced the book I was supposed to write about the energy sources in Canada. I perfectly understand why Innis wrote me off; he had every right to do so.'[57] It is difficult to believe that she would be so magnanimous towards Innis had she not recognized the extent to which he was smitten by her. Innis lost the debate over scholarship in the most personal and hurtful manner possible – Biss's marriage to Graham Spry, the epitome of the type of engaged intellectual that Innis despised.[58]

And yet it is difficult not to posit that the debate was merely continued at another level. Curiously, at precisely the time at which Biss abandoned her career in favour of a man who was having such a direct influence on public policy in the communications field, Innis's research veered into an extended preoccupation with communications theory. Brebner would recognize this link by writing to Innis in 1950 concerning *Empire and Communications*: 'It's quite a jump you've taken from the book that Irene Biss was to have written for Shotwell.'[59] We can still legitimately debate which of the two men had the greatest influence on the field of communications: the practical politician or the dedicated scholar.[60]

Innis's project for the development of a critical social-science indigenous to the Canadian context was political in the sense that it required a collective intellectual endeavour. Innis had to convince other scholars of the validity of his approach in order to realize his project. His articles of the 1930s on social commitment and the social sciences are passionate because they touch on this key attempt to win over his colleagues to his vision. In many respects he already had. In 1934 C.R. Fay, one of his colleagues at the University of Toronto, was already announcing in the *Economic Journal*, edited by J.M. Keynes, the advent of 'The Toronto School of Economic History.' But this was recognition of the work that was being done on the staples industries. It was not at all clear how such work offered a prescription for getting out of the Depression. Innis's articles during the period are strident in tone because he knew that he was losing the debate. Whether of right- or left-wing political persuasion, people could respect Innis's championing of academic freedom and monk-like scholarship at the same time that they themselves were becoming more and more involved with work outside the university.[61] This tendency grew to become an exodus of scholars from the universities during the Second World War. By that time, Innis was forced to recognize that he had lost the debate. No one believed in a Canadian paradigm of universal significance. Even those economic historians who had produced critical studies on the basis of the approach that had been developed by Innis and his colleague, W.A. Mackintosh at Queen's University, used this approach only in connection with Canadian or, at most, the 'vacant land' contexts of European-settler colonies.

Innis's lonely effort to forge ahead and produce significant new perspectives on universal history in the communications work should be viewed as a strategy developed in response to losing the political debate of the 1930s. If he could not directly convince other scholars of the necessity of avoiding getting caught up in contemporary discussions, he would do so indirectly by forging ahead with his own research using an indigenous perspective on a hitherto unprecedented scale. More than any other single event, Irene Biss's decision to marry represented the tipping point for launching Innis's solitary excursion into the communications works.

Shotwell and Professional Influence

As already noted, Innis's insistence that scholarship remain apolitical was, ironically, intensely political. Innis was maintaining that scholars would have a greater long-term influence on society through the generation of new ideas and perspectives than through the proselytization of existing points of view. His own personal experience bore this out. The increased weight of authority his pronouncements took on during the Depression corresponded directly to the new research and publications he produced in the same period. Innis had a personal relish for politics of a particular type: politics in which monk-like scholars influence practical politicians through their quiet research and through their role in training future civil servants. The reverse side of his distaste for intellectuals aligning themselves in public debate was his passion for an autonomous, self-governing university. He not only believed that scholars should run their own affairs, but, unlike the stereotypical scholar, he also developed a certain flair for directing intellectual endeavours from within academic institutions.

In many ways, this is the murkiest of Innis's realms of activity. Commentators are virtually unanimous in asserting that Innis maintained a paramount political as well as intellectual position within the social sciences in Canada from the late 1930s until his death in 1952. It would be difficult to exaggerate how powerful he became in the social sciences. At the same time, however, he continued to play the odd-man-out role of feuding with the university administration. Moreover, he would consistently deny the extent of his influence. When young scholars wrote to thank him for supporting them in the adjudication process for various awards, he would profess ignorance of the matter.

Surprisingly little material exists in Innis's papers dealing on a personal level with academic politics. There are several reasons for this. Letters of recommendation, except of the most trivial kind, have not been included in the papers. Apparently, Innis carried out much of this type of activity orally, either at conferences or through personal visits during his extensive travels. Finally, his correspondence was almost always handwritten and sent out with no copy being kept for his files.

Despite the difficulties introduced by these variables of documentation, certain facts about Innis's strategy for building influence are quite clear. From the start, it is evident that he centred his activities on two key avenues to academic power: publications and funding. He ruthlessly forced his own work into published form often when it was in a state that clearly needed further editorial attention. Nothing interested Innis less than polishing his work for publication. He never viewed social-science publications as an individual effort in the sense that a novel was. As we have seen, he began his career with the compilation of bibliographies of unpublished work and the distribution of mimeographed essays in the twenties. He moved on to initiate the establishment of occasional papers and then full-blown journals when he thought that the depth of Canadian scholarship was sufficient to ensure that these publications would continue to be supplied with first-rate material. Finally, he remained pre-eminent in certain fields of scholarly publication such as the Arctic, the fur trade, economic geography, and railway studies, long after he had abandoned active research in them, by producing comprehensive reviews of new books in these fields. He had an unerring ability to place himself close to the centre of major research and publications projects from their inception. In 1933 he worked with the Canadian Institute of International Affairs to bring out the series of essays *The Canadian Economy and Its Problems*. In addition, he played an important part in the two main social-science publishing efforts of the 1930s: *The Relations of Canada and the United States* (Carnegie Endowment) and *Canadian Frontiers of Settlement* (Social Science Research Council) series.

His practical nationalism soon led him to recognize the potential of

American funding sources for Canadian academic endeavours. As Carl Berger points out, the Rockefeller Foundation and the Carnegie Corporation were jointly responsible for 30 per cent of the total endowment to Canadian universities during the years from 1911 to 1935.[62] Innis, during the 1930s, became the pre-eminent adviser to both these organizations on Canadian grants policy. From a foundation administrator's point of view, he was ideally suited to such a role. He was renowned as a scholar. He knew personally virtually all the institutions and individuals engaged in social-science work in Canada. And he worked diligently to induce scholars to translate their research into print. On the basis of this influence in the Carnegie and Rockefeller organizations, he went on to wield similar power in the Canadian programs of the Guggenheim and Nuffield foundations and of other foundations as well. Furthermore, after having been largely responsible for the initiative of setting up the Canadian Social Science Research Council (CSSRC), he served as the chairman of its grants-in-aid committee from 1940 to 1948. It would be difficult to exaggerate the influence that Innis built up in the academic world in Canada primarily on the basis of his work during the Depression. It should also be noted that he was able to do so largely because of his championing of apolitical scholarship. In short, his career benefited from the position he maintained against his more left-wing colleagues.

The new documentation on economic development produced by Innis during the 1920s and 1930s only partly accounts for the tremendous increase in his influence on the social sciences. New content or data on the periphery economy could readily be absorbed within the existing orientations of established academic institutions. Innis's great achievement was to produce this new material in a form that embedded it in a new methodology or perspective and to insist that it be disseminated in conjunction with this new manner of looking at economic development. His influence grew as he succeeded, in defiance of established schools of thought, in imprinting this non-metropolitan perspective on such collective scholarly endeavours as *The Relations of Canada and the United States* series. The process was far from automatic. It depended on Innis's ability to complement his flair for scholarly research with his skills of political manipulation.

Both the relative strength of his scholarly work and his pre-eminence in the social sciences during the last part of his career are now so obvious that they tend to obscure the extent to which Innis was obliged to push against the grain of the academic milieu in order to establish his influence in the Depression. Commentary tends to jump from the content of his publications to his appointments to various positions in recognition of his growing reputation without dealing with the dynamic area of academic politics that linked his research and his recognition. We tend, for example, to ignore or underestimate his very early role in planning some major academic projects. Fortunately, Innis's correspondence with James T. Shotwell, director of the Carnegie Endowment, has been preserved *in toto*, making their collaboration on *The Relations of Canada and the United States* series a detailed example of Innis's growing influence during this era.[63]

This series was initiated within metropolitan academic institutions as a vehicle for propagating a continentalist perspective in the social sciences. Its immediate aim was to provide an important sectoral investigation fitting in with the general review of American foreign policy that was being undertaken at the time. Beyond this, Shotwell had a personal motive in backing the project: his long-time interest in international peace led him to believe that a study of Canadian-American relations would demonstrate an exemplary case of pacific interaction between nation states.[64] The series produced twenty-five volumes during the period from 1936 to 1945 under the general editorship of Shotwell, with Innis maintaining editorial control over the Canadian-authored volumes. These were the first studies of foreign affairs written by Canadian scholars to be published.[65]

Innis became involved in the project at the earliest stages but seemed reluctant to devote much of his time to it at the outset. Ostensibly, this was the result of pressures of work, but it is more likely that he was prickly about the domination of the series by American-based scholars. Whatever the case, Innis became more and more deeply involved in the series as Shotwell exhibited a flexible spirit and as the project offered one of the few sources of research funds in rather lean times. Shotwell was in many ways the reverse image of Innis, which made them ideal collaborators. Coming from a rural background, he left Canada to study in the United States before the First World War. Whereas Innis would pen his more philosophical communications work at the end of his career, Shotwell did his most philosophical work at the beginning of his. The First World War weakened Shotwell's ties to Canada as much as it strengthened Innis's. In contrast to Innis the signaller, Shotwell experienced the war as a diplomat, which led him away from philosophical concerns and into the practicalities of U.S. foreign policy. In stark contrast to Innis, Shotwell proudly proclaimed that he had never noticed significant cultural differences in moving from Canada to the United States.

Innis submitted research plans of a wide-ranging nature to his American colleague.[66] It was on the basis of these and plans from a few other scholars that Shotwell sought funding for the project from the Carnegie Endowment board. Innis called for industry-by-industry studies that more often than not adopted a primarily Canadian focus (or a more general international approach) as opposed to concentration on cross-border trade. This approach fundamentally subverted the continentalist thrust of the series.[67] Innis's continued participation is a testament to Shotwell's genuinely liberal spirit, for Innis was always blunt about his orientation. In 1932 he wrote: 'I have argued at considerable length ... that the boundary line is not accidental and that the economic background is fundamentally different in the two countries and that one might expect the attitudes to differ as a result.'[68]

Even scholars who had been allies of Innis in the past, and some who would become his allies in the future, opposed his early influence on the design of the series. R.M. MacIver, who had been head of the department of political economy at Toronto from 1922 to 1927 and who had then joined Shotwell at Columbia University, rejected the staples approach as parochial: 'I very much doubt the desirability of a specific history of the Maritimes or such of any other area. I also doubt whether we should include specific histories or monographs on industries as such. The way they enter into Canadian-American relations should appear in the treatment of capital movements, tariffs, technology, etc.'[69] Brebner, who would become one of Innis's greatest admirers, warned Shotwell: 'Innis is likely to think in terms of Canadian economic history and temporarily to lose sight of our study of relations.'[70]

All in all, Innis's approach stressed conflict and competition more than it did cooperation between the two countries.

> My general scheme ... might be outlined as follows. Beginning with competition of two water routes, St. Lawrence versus New York tracing the effects of our efforts to make the St. Lawrence a competitive factor which involved the expenditure of millions of dollars ... we have built up in Canada a planned economy with most extensive government intervention to compete with American transportation ... Within this large scale the emergence of later staples, mining and pulp and paper must be pitted as American contributions which steady at one point and disturb at others the Canadian economy. My own feeling has been that we cannot go on

> being exposed to these disturbances and that more adequate machinery must be designed in Canada to meet the situation.[71]

Innis went on to propose a program for the Canadian side of the series that included twelve separate volumes organized on an industry-by-industry basis, with separate studies on commercial organization (Creighton) and transportation (Glazebrook). As early as 1933, he had assigned himself another study included in his overall scheme, that of the pulp and paper industry. He also undertook to do the fisheries volume and the final synthesis volume. The Carnegie advisers did not appreciate his ambitious plan and seem to have suspected that he was bending the project to his own ends. The following personal memorandum by one of them to Shotwell is typical of their response to Innis's approach to the series.

> There is in my opinion a serious difficulty in connection with the original Innis program of studies ... This program seems to be designed to produce materials of a research character for chapters in a book on the basic industries of Canada. These chapters might be very useful in an economic history of Canada without reference to its relations to the United States. A pertinent question arises in my mind of what interest to a relational study is the subject of specialized agriculture ... of interest for a relational study, lumber, coal, fish, potatoes, and apples might be useful, hardly anything else, and ... all of these ... could be easily put into a single data paper of four or five thousand words.
>
> This would eliminate, for the time being at any rate, any consideration of a synthesis volume which Innis proposes. If it is difficult for you or embarrassing to communicate this to him, why not adopt this procedure, namely, inform him that in due time all the materials of an economic sort which have been gathered by his group of associates will be released to him ... and out of these he may write his own economic history of Canada and publish it under some other auspices.
>
> We have talked considerably about the desirability of having Innis do any of the final writing. In my estimation it would be a mistake because of his inability to write readable material and his unwillingness to accept an Olympian point of view ... I ... feel that the disqualifying characteristics of his writing are such as to cause you in the final judgment to eliminate him.[72]

In practical terms the writer of this memo proposed emasculating

Innis's proposal by cutting its budget down to 25 per cent of the original projection. As if this was not bad enough, Shotwell was obliged to announce: 'You certainly have been doing a fine piece of work in your planning, but it is on that very ground that I am most troubled ... The situation here, in all confidence, is that one of the trustees of the Corporation raised objection to the voting of any further funds, because he was afraid that we might be stirring up Canadian antipathy to the United States instead of helping matters.'[73] Shotwell asked Innis to modify his plan to correspond to the sharply reduced budgetary allocations. Ruthlessly fixed on pursuing his project, however, Innis wrote back suggesting that *all* the studies be pursued as planned, with their authors receiving proportionately less funds (and Innis none at all) in order to bring the program to completion as originally conceived. In the face of such zeal, and the continued backing of the plan by Shotwell, the opposition to Innis's outline seems to have been overcome.

By 1934, Innis had already embarked on the long process of expediting his twelve-volume section of the project through to publication stage. He exhibited a significant strain of nationalism throughout this process, although he often raised these nationalist sensitivities as if they were only an objective component of the milieu in which he worked and had little to do with his own opinions. These concerns extended to the level of the smallest details. In 1935 he complained: 'I doubt whether the adoption of American spelling is a happy arrangement. If it is at all possible volumes written by Canadians should follow the English spelling and by Americans the American. Considerable controversy has raged about this point and I feel it would be wise to avoid stirring up issues over minor points which might be of serious damage to the major project. This is an important consideration.'[74] Innis's insistence on this sort of detail led Shotwell to propose that a Canadian publisher bring out the Canadian-authored studies.[75]

Innis continued to be prickly on the matter of Canadian particularities. Furthermore, he saw these issues as fundamentally political and of immediate importance. In the spring of 1935, he commented to Shotwell: 'I am sure our correspondence would make an interesting study in Canadian-American relations. I think you fail to appreciate the difference between the Canadian and other publics. In spite of your arguments I still insist it makes considerable difference as to how the series gets started, particularly in Canada. You will probably reply that I have been living too closely with politicians in Canada and am too sensitive as to public interest. In the long run I would grant your case, but in the long run we are all dead and an excellent start appeals greatly to a small population.'[76]

The proposed 'start' to the series to which Innis objected was the publication of a plan of the research work to be undertaken stressing the similarities between the two countries and the appropriateness of pairing them as a unit of investigation. To the continentalist scholars who initiated the series, these observations were its aims or principles. To Innis, however, they were conclusions that might or might not be found to be valid only *after* the various studies had been completed and reviewed: '[The] publication [of such an outline] would be premature ... The material has been collected but the task of interpretation has been scarcely begun.'[77] Having secured an editorial position on the project, Innis thus went on the offensive, using his position to question the original underlying purpose of the series. He was particularly annoyed by the tendency of the American scholars to overlook certain fundamental differences between Canada and the United States: 'There is much loose writing. Perhaps I can illustrate the problems [I have] in mind. On p. 204 "the same language, racial and cultural characteristics" neglects the French population and blurs a picture in which the differences [between Canada and the United States] are of significance. These differences can be brought out by a Canadian.'[78] Innis repeatedly returned to the significance of a French presence in Canada as one of the fundamental differences glossed over by the more continentalist scholars working on the series. In March 1938 he wrote in a confidential note to Shotwell: 'There is a point which is less apt to occur to you but which I try to keep in mind namely that this is a bilingual country. I have generally tried to resist translations in order to bring the fact home to Canadians but it is probably too much to expect of Americans.'[79] Shotwell replied in a jocular but essentially belittling way: 'I fully appreciate your reminding me that there are some French in Québec! But I don't believe that there will be many of the habitants who will be interested in deep-sea fishing. I think we will have to satisfy them some other way.'[80]

Through the correspondence one can glimpse how Innis became less and less reticent about imposing his perspective on the project as the years went on and he gained experience in exercising his influence. By 1937, he was commenting in connection with a particular study: 'It becomes fairly evident that I shall have to write an introduction or something of the sort to give the whole significance. I am afraid the same will have to be done with labor and mining. I am reluctant to suggest this, but the series should at least try to do more than collect facts

... However, let me know what you think about this as it involves a matter of policy.'[81] The 'matter of policy' Innis raised was in essence the changing of the original goal of the series, which, as mentioned above, was precisely the collection of 'facts' with minimal commentary. Yet Shotwell, probably impressed with Innis's prodigious working ability and now assured of funding for the project, quietly acquiesced to this relatively major change in direction.

As Innis grew accustomed to his new institutional influence, he began to use it to further personal initiatives that went far beyond the immediate aims of the Carnegie series. One such passion was to strengthen the University of Toronto Press, founded in 1901, to make it a prestigious scholarly press.

> I would like to see Canadian American relations improved by the series in every possible way and I think we have made more than our contribution to the Ryerson Press and that we should turn to the University of Toronto Press. Such a move could strengthen enormously the position of the University of Toronto Press and the position of the University of Toronto. From the standpoint of cultural development I can think of few more important steps than the building up of the University of Toronto Press as a national press for University work which is something that does not exist at the moment. That is a very large and a very important aspect.[82]

Because Ryerson Press had been recently designated as the Canadian publisher for the series largely because of Innis's insistence, one would expect a degree of awkwardness in his championing of the University of Toronto Press.[83] It seems, however, that once he had decided to adopt this cause, he felt no reticence in abandoning Ryerson. In fact, he became the most vociferous critic of Ryerson Press, if not openly, at least in his letters to Shotwell. He badgered Shotwell about changing publishers in the same way he had pestered the president of the University of Toronto about appointing a geographer. Ryerson Press, Innis reported, was at best incompetent, but, more likely, it was actively engaged in 'sabotaging' the series.[84] Yale, which was publishing the American contributions to the series, was too distant and too slow. But 'if *The Cod Fisheries* were printed at the University of Toronto Press it could be started immediately.'[85] Far from being a naively principled scholar, as he is often presented, Innis was a practical academic politician.

Berger's report of this event provides us with an example of how even the most careful of commentators can confuse Innis's motives. Berger states:

> On large matters and small Innis was a prickly nationalist who could become incensed at aspersions on Canadian scholarly effort. When the claim was made that Ryerson Press did not have sufficient expert resources to edit several volumes in the Carnegie series on Canadian American relations, Innis angrily told Lorne Pierce that the 'suggestion that the editorial work be handled by Yale is disquieting and should be resisted as a reflection on Canadian scholarship as well as on myself and the Ryerson Press.' As this remark suggests, Innis was not above occasionally identifying his personal interests with the cause of 'Canadian scholarship.'[86]

In fact, the actual situation was much more ironic. We know from the Shotwell papers that Innis himself, acting behind the scenes, was the main perpetrator of the claim that Ryerson Press was incompetent. The key to the apparent contradiction in Innis's attitude is found in Shotwell's reply to Innis's request that Ryerson be dropped in favour of the University of Toronto Press. Shotwell investigated this possibility and then reported to Innis:

> The Yale Press turned out to be very reluctant to cut loose from Ryerson, and as they had been good to me I said I would be willing to continue with Ryerson on one condition; that I dealt only with Yale Press and that they had all of the dealings with Ryerson. That means that I put the manuscripts henceforth in the hands of Yale. The New Haven people are anxious that it should be settled this way and as long as they assume the responsibility it is easier for me. If, however, as I rather suspect may be the case, they find out on their own account that it will not work, then they agree that it shall be the University of Toronto Press. But at the present there is to be no change in the eyes of the public.[87]

Since Shotwell's letter occasioned the Innis letter to Pierce that was subsequently cited by Berger, it is difficult to escape the conclusion that Innis was hoping to facilitate the disintegration of working relations between Ryerson and Yale to such an extent that the contract would pass to his own university press.

Innis thus honed his skill in academic politics in his work on the series. He also extended his influence by using his connection with the series to control other Carnegie Endowment interventions in the Canadian academic scene. In November 1938 he wrote a personal letter to Shotwell expressing his unsolicited views on the potential appointment of senior visiting faculty. He ended his comments on the matter with the following: 'I have not been able to send a memorandum to you about it as I promised because the President [of the University of Toronto] asked me to hold off. However you might explain the situation to the proper authorities.'[88] In other words, Innis continued to exercise his influence with official funding bodies even when he was obliged to do so covertly and in blatant insubordination to the authorities of his own institution. The series also allowed Innis to extend his academic power since it offered him opportunities to function as an international scholarly statesman.[89] Through direct contacts with men such as Shotwell at the Carnegie Corporation, Innis could bypass the formal authority of other academics.[90] This makes the extent of his power difficult to trace thoroughly because so much of it was based on informal but direct influence with people in authority.

Innis viewed the Canadian-American relations series as an exemplary case of academic cooperation precisely because it allowed him to manipulate the endeavour in such a way as to further his project for an independent intellectual base in Canada. In March 1939 he outlined the effects of the series to Shotwell:

> Practically all over courses on Canada have been revised in light of the work published in the series, and the students reading changed accordingly. [The series publications] provided a new basis of approach which is fundamental in the long run. Graduate work as well starts from a new premise.
>
> ... Brady was commenting the other day in connection with his annual review of the social sciences on the tremendous change which has come over the subject in Canada in the last decade as a result of our series ... Whether you appreciate it or not, your attitude and assistance have placed us in a position where we can begin to hold up our heads in the social sciences. I can think of no greater contribution that could have been made to the intellectual interest of Canada and to the rounding out of her position to the point where she can become a literate nation.[91]

Innis was not unaware of the irony that American personnel and funds were allowing him to pursue more effectively his project for the development of a specifically Canadian social-science perspective. Help had come from the United States at a time when many Canadian scholars were betraying Innis's vision by adopting metropolitan paradigms holus-bolus. These circumstances undoubtedly reinforced Innis's complex and non-parochial attitude to nationalism and national interest.

Innis's 1946 decision to stay in Canada rather than accept a lucrative offer at Chicago has often been mentioned as an indication of his nationalist sentiment, and indeed it was. Yet, before making his decision, he chose to write to Shotwell on the matter: 'I have wanted to write you a personal letter asking for advice ... I find it difficult to know how to answer [Chicago's offer] ... Would it be better to stay at Toronto and try to damp down the intense nationalism which is constantly rushing up or to go to Chicago as a protest against it. By staying in Toronto it may be that I am actually accentuating the nationalistic animus.'[92]

Shotwell provided Innis not only with the means by which a collective intellectual project could be mobilized but also with the incentive to pursue new personal research projects. The series gave Innis a vehicle for pursuing two areas that would ultimately surface in the subject matter of the communications works. The original plan, as already noted, called for him to produce a study on pulp and paper as well as a wide-ranging synthesis study, in addition to the fisheries volume he was already working on. The advent of the war, however, led to the abandonment of publication plans for the pulp and paper and synthesis volumes. In a very real sense, his communications works are a reformulated version of the studies he had planned to produce for Shotwell during the Depression.

Innis owed a great debt to Shotwell not only for his own increased influence within academic circles but also for Shotwell's editorial support, which came during a difficult period. *The Cod Fisheries* would not have been published in anything like as comprehensive and polished a scholarly version as did appear were it not for Shotwell's tremendous attention to detail on the book. Shotwell supplied the title, subtitle, and introduction to the study.[93] More important, in the year after Innis's crisis of nervous exhaustion in 1937, Shotwell diplomatically assigned an experienced editorial assistant, A.E. McFarlane, to Toronto to bring the unruly manuscript to press. Writing to thank Shotwell and McFarlane for their editorial assistance, Innis identified the basic difficulty that was holding up the study:

> The problem of the whole volume has been partly psychological since it meant a complete break from the analysis on which I had spent so much time in the fur trade. As an Upper Canadian I understood the St. Lawrence and have lived through and felt its powerful centralizing tendencies. It was consequently much more difficult to escape from the St. Lawrence and to appreciate the decentralizing tendencies of the Maritimes. The readjust-

> ment involved a terrific amount of difficulty on all sides. I venture this explanation by way of apology for the length of time which it took and for the burden which it has imposed on you and on McFarlane. I am afraid I should describe both of you as midwives but for whom the idea would never have been born ... I am tremendously grateful in that I have survived and in that I see the Canadian problem and the North American problem in a much clearer light than would otherwise have been possible.[94]

Innis felt that the chief intellectual accomplishment of *The Cod Fisheries* was the honing of a new, more critical perspective on economic development. This sort of perspective grew out of Innis's conception of a political project that stressed his own role as a peripheral intellectual. Shotwell's liberal beliefs were ideal for providing Innis with the flexibility and support he needed but were totally incapable of recognizing the 'political' side of Innis's endeavours as expressed in his obsession with a new perspective. For Shotwell, what was germane was the specific rather than the general, the content rather than the methodology. Shotwell wrote of Innis's breakthrough to a wide-ranging perspective on empire in the conclusion of *The Cod Fisheries*: 'The last seven pages leave the fish far behind and deal with a fundamental problem of North-American history in terms which reach far into the theory of government. In my opinion, you have struck gold in this section ... But no reader would expect to find gold off the Grand Banks and I think the place for this is in your real magnum opus which is your general survey of economic history.'[95] Even Shotwell, who was generally sympathetic to Innis, could not grasp that a new and universal critical perspective could or should have been the outcome of such a peripheral study. After an interval of six years, he repeated essentially the same strategy for manipulating Innis's endeavours as that proposed by the Carnegie board member vis-à-vis Innis's original plan for the Canadian-American relations series: confine him to statistical spade work and suggest that he construct his new paradigm on his own time and with his own resources.

A Leap of Research

The wide-ranging conclusion was retained against Shotwell's initial advice in the published version of the study, thus emphasizing Innis's contention that *The Cod Fisheries* had required him to adopt a perspective that was fundamentally different from that of *The Fur Trade in Can-*

ada. The logic of the fur-trade study was found in the structure of the inland waterways through which the fur trade was pursued. The root-like symmetry of the drainage basins imposed a relatively neat explanation of why political integration took place in these peripheral regions. Despite the inclusion of other elements in the analysis, Innis is chiefly concerned with the penetration of the hinterland of North America by European influence. It is an approach that focuses on effects on the hinterland. When it reaches the mouths of the rivers and gives way to the open ocean, Innis's analysis is forced to become more complex. The case of two-way trade dominated by a particular empire yields to that of competing empires juggling a balance of raw materials and finished products to come up with the most effective commercial project overall. *The Cod Fisheries* provided Innis with a point of entry to an assessment of those overall imperial projects. In this sense, the cod fisheries study triggered the leap in Innis's research to the centre of the empire. In the preface, Innis states: 'It is not too much to say that European civilization left its impress on North America through its demand for staple products and that these in turn *affected the success of empires projected from Europe*. The author's study of the fur trade is therefore followed by a study of the fishing industry in the hope that it will throw light on the significance of that industry for the economic, political and social organization of North America *and Europe*.'[96] Here Innis is stating a fact that has been ignored or underplayed by his commentators but that is essential to an understanding of the continuity between his staples and communications work. *The Cod Fisheries* is far from being an analysis of the economic development of an 'empty' land only. It is, on the contrary, 'The History of an International Economy,' as the subtitle puts it. As such, it tells us more about the Basque country or the impoverishment of the Spanish Empire or even about the slave trade and commerce with India than it does about the consolidation of a political entity in the northern half of North America. This wide range of analysis is evident in the conclusion to the study:

> Dried fish, as food for seamen, gave shipping the range of the tropics and played its part in the activities of Spain and Portugal in the New World. Specie was secured in return for the fish ... England, thus supplied with specie could trade more effectively with India ... The demands of the Catholic peoples for cod and of metropolitan areas for luxury goods gave greater meaning and value to the anchorage ... The effectiveness of specie in developing trade was evident in the production of commodities of high

> value and light bulk ... ranging from furs to tobacco and sugar. The former depended on the aborigines of North America, the latter on the aborigines of Africa. Slaves brought to the New World to produce tobacco and sugar meant a demand for the poorer grades of dried fish ... Encouragement given to the production of tobacco and sugar by England was the means of draining specie from Spain.[97]

To the extent that Innis is dealing with developments in northern North America, he is concerned with a more negative task than that of his fur-trade study. He seeks to delineate the reason for the blockage of development in maritime Canada at an economic level and the reluctant integration of this region with the rest of Canada politically. As is clear in the above quotation, his focus for exploration once again is not so much the staple and its characteristics as it is the cultural constellations of the peoples involved and the qualities of the communications systems through which their productive efforts were channelled.

However, if the basic structures of the communications networks essential to the production of staples determined the effect of this activity, it was cultural factors that determined its level and pace. Whereas *The Fur Trade in Canada* depended on the mercurial changes in upper-crust European fashion, the market for cod was linked to a much more substantial cultural base: 'In the tropical countries, demands for high protein content in the diet of the people, the difficulties of cold storage, the rigidity of customs and consumers' preferences, and the dominance of Catholicism were stabilizing factors of demand.'[98]

Another culture-bound factor was the recognition of nations, and particularly of England, that national development could be ensured only by actively fostering that subculture within its populace that was a necessary prerequisite to a strong navy before the advent of steam: the highly skilled proletariat of sail. Directly and indirectly through these cultural factors, 'cod from Newfoundland was the lever by which [England] wrested her share of the riches of the New World from Spain.'[99]

As in the fur trade, we find that the staple that is under consideration is coupled with a number of 'shadow staples' whose characteristics are in some cases nearly as important as that of the staple itself. Here I am thinking of gold, sugar, tobacco, and even more so of the staples of cod production: timber, fishing tackle, and salt. On this last ingredient to the staples equation, Innis built his analysis of the English system of 'dry' fishing. Having relatively small supplies of salt and being late arrivals, the English concentrated on the smaller fish found closer to shore and developed the technical innovation of shore curing. This led to a better trade product as well as eventual English colonization of the shoreline. In contrast, the southern Europeans and French, using their readily available supplies of sun-dried sea salt and rock salt, continued to concentrate on 'green' fishing (heavily salted, barrel-packed fresh fish). This resulted in a product less suitable for trade and negated the need for shore landings. In a sense, Innis is describing here the ingenious breaking of a monopoly of knowledge – in this case in the technique of staples production. The same process will later appear in a similar form in his communications works.[100]

Two other aspects of *The Cod Fisheries* illustrate this continuity when we look at the work in retrospect. Innis starts the fisheries study by clearly presenting the biases of the documents that served as a basis for his research. On the one hand, one finds a great deal of material surviving in the diplomatic archives at the centre of the empires engaged in the fishery. This material represents what Innis came later to call the written tradition. It is greatly deformed by the political exigencies of the time. (Domestic imperial production was minimized and the production of competing empires exaggerated.) Innis notes that the only way such information could be deciphered was by the lengthy comparison of claims of competing empires and by a careful notation of sources. On the other hand, in contrast to this mass of extant material originally gathered far from the fisheries, one finds a lack of sources on the fringe of the empire, in the fishing areas themselves. The reason for this was that the production of cod through small, independent, sea-bound units favoured individualism and illiteracy or, more accurately, an orally biased tradition. The precise records have died or are hidden in the memories of the staples producers, while the history of imperial politics has deformed surviving records: 'The enormous wealth of diplomatic material is in sharp contrast with the paucity of accurate detailed information upon technique and personnel. The illiteracy of the fisherman *is the reverse side* of the literacy of the diplomat.'[101]

The second minor aspect of the fisheries study deserving mention concerns the development of a conception of 'balance' upon which Innis's analysis repeatedly turns. One could point to the success of the British as a particularly ingenious case of product balance, but I am more interested in his treatment of Nova Scotia. Here he develops for the first time a conception of cultural balance that will later appear in a more extended form in his analysis of the Byzantine Empire. The balance results when a region that has been established in the age of wooden ships, mercantilism, and internationalism (or, rather, solid integration with European metropoli) makes the transformation necessary to enter a new stage of development based on the trawler, commercialism, and nationalism. The transition was possible for Nova Scotia because of its relatively rich resource base. At an economic level, it is exemplified in rail connections to the east and the development of iron and coal industry. Politically, the transition is made through responsible government to Confederation.

The most interesting aspect of Nova Scotia's balance, however, has to do with the cultural level. Nova Scotia's links to an international economy are cited as leading to the 'conspicuous advance in extraterritorial sovereignty [of Canada] after Confederation.'[102] The extension of the Canadian diplomatic and financial network and the advent of British Commonwealth are all traced back to this regional balance. Innis points out the contribution of the province in education – first, directly through legislation, and second, indirectly in the migration of highly trained personnel to central Canada: 'A renaissance has been apparent in the cooperative movement sponsored by St. Francis Xavier University, the activities of maritime universities, in the rejuvenation of the interest in cultural growth, in the development of museums, in the preservation of archives and in a revival of pride in a notable past.'[103] I am not concerned here so much with the accuracy of the statements as I am with the fact that they represent one of the first examples of what Innis would later symbolize with Minerva's Owl: the cultural flowering of a region at a time of declining empire, a region whose communications biases (land and water, rail and ship) offset each other.

Innis completed the groundwork for *The Cod Fisheries* early in the Depression. By 1933, he was informing Shotwell that he expected to complete his study within the next year.[104] This major miscalculation in scheduling indicates Innis's difficulty in developing the confidence necessary to put forward the panoramic view of the international economy summarized in the conclusion of the study. *The Cod Fisheries* was to provide a positive example of the type of non-partisan scholarly work that he had championed in a negative way in his diatribes with Underhill and Urwick. As a consequence, the study had to contain not just data but a new perspective that would counteract the force of analysis of the ready-made paradigms borrowed by the more politically engaged scholars. In short, *The Cod Fisheries* became the chief practical vehicle of the Depression through which Innis would define the critical and independent role that the peripheral intellectual had to play in articulating a new world-view.

During the period, Innis wrote a number of seminal essays that were his first attempts to outline something on the scale of such a new world-view. 'Unused Capacity as a Factor in Canadian Economic History' (1936) and 'The Penetrative Powers of the Price System' (1938) are the two most important. These essays are the literary thread that connects the negative polemics of the 1930s on 'objectivity' and 'party politics' in scholarship to the full-blown portrayal of the fortunes of empire found in the conclusion of *The Cod Fisheries* and later in the communications works. In these essays, Innis pushed to the breaking point some elements that appeared in his staples research to be hinting at a broader perspective. He felt impelled to do so prematurely in order to develop a general intellectual perspective that could resist the complete encroachment of metropolitan paradigms in Canadian institutions. Like the later communications works, these tentative essays can be described as highly interesting and yet, because of their lack of congruity, unsuccessful.

An Intellectual Leap

When Innis, in *The Fur Trade in Canada*, wrote that 'the economic history of Canada had been dominated by the discrepancy between the centre and margin of western civilization,' he had in mind principally the need of European settlers to reconstruct their European cultural framework in the New World. The 'demand pull' of these settlers (and the Indians) for European products was the central factor in the effective pursuit of the fur trade. Innis's study put forward what was basically a comprador theory of economic development arising from the adoption of an intellectual point of view situated in the colonies. It was a logical perspective for a farm boy from Otterville.

However, by 1936, when 'Unused Capacity' appeared, Innis had switched over to a consideration of economic development principally from the perspective of the metropolis. The centre/margin discrepancy in this essay is represented predominantly by the imbalance in cargoes going to and coming from the colonies in connection with a variety of economic endeavours. Innis uses the idea of balanced cargoes as the key indicator of imperial efficiency and the focus of his analysis. This permits him to review the ground he has already covered in *The Fur Trade in Canada* in a breathtakingly schematic style.

Much of the analysis later to appear in the fisheries study was introduced by Innis in this essay. However, we are concerned here chiefly with those elements that will eventually be used in the communications work.

Innis suggests that each staples trade has associated with it a degree of unused capacity in its related transportation system. The problem of developing and maintaining an effective empire from this perspective becomes the problem of reducing the overall unused capacity to a minimum by engaging in a number of complementary staples trades in various regions of the world. The successful existence of an empire through time is an indicator that this problem of balancing the unused capacity of various staples trades has been addressed in a relatively effective manner.

It is a short step from this to the communications approach. From the perspective of communications systems, each type of system contains a bias just as each staples-transportation system contains an unused capacity. In communications, an effective imperial project is measured by the extent to which these biases are offset one against the other by the inclusion of a number of complementary media in the administration of the empire. The existence of empire is an indicator that a relatively successful mix of media has been developed to allow for the maintenance of a balanced policy.

In his essay, Innis recognizes that the conscious steps taken by imperial planners to cut down on the problem of unused capacity might well have the opposite effect. For instance, the chronic imbalanced cargo via the St Lawrence in connection with the fur trade dictated a strategy of increased westward expansion first under the French and then under the English. Yet this expansion of the same staples trade inevitably led to an exaggeration rather than diminution of the problem of balance in transatlantic cargoes.

In his later application of this type of analysis to the communications field, Innis would posit that too extensive a reliance on a particular medium within an empire led to serious long-term problems. A medium that was characterized by rapid transmission but impermanence might lend itself effectively to the requirements of military operations. However, an attempt to solve a problem threatening to disrupt the continuity (that is, stability) of the empire by using such a medium would have a tendency to recommend further expansion of the empire through military action. This in turn would tend to exacerbate the imperial crisis, for the integration of more territory would simply add to the difficulty of continued domination of that territory through time. Just as in the schema developed in 'Unused Capacity,' imperial stability could be guaranteed only by using a different medium (or staples trade) to correct the bias of the original medium (or original staples trade).

In its theme, style, and obscure nature, 'The Penetrative Powers of the Price System' foreshadows even more closely than 'Unused Capacity' some of the elements that would appear in the communications work. Yet both essays grow out of the pushing of some traditional elements of the staples analysis to their limits. In 'Penetrative Powers,' Innis fuses the traditional staples approach and the sociology of knowledge by considering specie (coins or metal currency as opposed to paper money) as a staple. 'Under the stimulus of treasure from the New World the price system ate its way more rapidly into the economy of Europe and into economic thought.' He attempts to reach a level of understanding more profound than price-theory economics by coming to grips with the material conditions leading to the development of this paradigm. In a way, Innis is attempting to transcend the central paradigms of capitalism not only in their intellectual manifestations but also in their appearance as the everyday common sense of modern consumer society. To describe it using terms from the communications period, we could say that the essay attempts to trace the decline of the oral tradition in trade (barter) and its replacement by a more effective space-binding technology (specie).

The essay is not successful. Innis seems ill at ease with the development of a conceptual framework. The 'price system' is not defined. Instead, Innis slips repeatedly into a tortuously compressed presentation of universal economic history, beginning with the discovery of the New World. These passages are connected to vague references to the corrosive powers of the price system active in the supersession of feudalism by capitalism, commercialism by industrialism, palaeotechnic technology by 'Neotechnic,' and so on.[105]

The essay deals with the field of communications directly but only as a manifestation of neolithic technology perfecting the penetration of the price system. In particular, it stresses the role of the newspaper and radio in this process. Another theme of this essay that would become central to the communications work is the question of the social basis of knowledge.[106]

In 'Penetrative Powers,' Innis is on the brink of making the intellectual leap that will lead him to the communications work. He concerns himself above all with the long term, using changes in the general characteristics of trade goods as an indicator of long-term trends. He recognizes that the operation of capitalism over the centuries had resulted in 'the trend in the movement of goods from light and valuable raw materials to heavy and cheap raw materials and to light and valuable finished products.'[107] However, this type of observation, which later would prove so fruitful in tracing the effects of the supersession of different media technologies, leads nowhere when it is applied initially to commercial goods.

Innis comes closest to the language of the communications work in his treatment of the newspaper:

> The cheap newspaper is subordinated to the demands of modern industrialism and modern merchandising. Overhead costs have contributed to lack of precision in accounting, and the allocation of costs between the purchaser of goods from the department stores and the purchaser of the paper, or between the purchaser of paper and the purchaser of hydroelectric power from plants owned by paper companies, is extremely difficult to determine. In paying for electric light or for groceries one cannot be certain how much is paid for newspapers. The patterns of public opinion or the stereotypes have become blurred, and amalgamations of newspapers and the fusion of editorial policies lead to demands for general programmes which appeal to the business mind. Broad stereotypes are typical, such as the belief in the stability of governments or ... the dangers of civilization crashing, whatever that may mean.[108]

The Second World War provided Innis with a concrete indicator of what that might mean. He spent the remaining years of his life engaged in an effort to understand the determining powers of the media, powers that he knew were great enough to deform even the traditional mechanisms of the price system itself.

Part Two

To the Margin, 1940–1952

CHAPTER SIX

Hunting the Snark

In the midst of the word he was trying to say,
In the midst of his laughter and glee,
He had softly and suddenly vanished away –
For the Snark *was* a Boojum, you see.

Lewis Carroll, *The Hunting of the Snark*[1]

An extraordinary parody of Innis's scholarly project is found in his favourite Lewis Carroll book, *The Hunting of the Snark*. This hunt serves neatly as a metaphor for the efforts of Canadians to develop an indigenous approach to social science. The story begins with a bizarre group of characters setting off in search of a creature that is both rare and somehow profoundly significant. Their strength is the blank map that serves as their guide. A beaver ensures their arrival at the hunting ground, and once there it seals a friendship with a scribbling butcher who is keen on numbers. The most introverted member of the fantasy crew is a baker who has been warned by his uncle that not all snarks are benign. After a long quest, the impatient baker sprints ahead of the hunting party to find the snark and it is at this point that tragedy occurs. Unhappily, this particular snark is also a 'boojum,' a kind of black hole of the world of prey. And so the audacious hunter vanishes without a trace. He leaves his companions behind without even having given them a description of the nature of the beast. Indeed, his sad friends are unable even to locate the spot where he vanished.

Innis's communications works were his own intellectual boojum. During the last phase of his career, his reputation as a 'hunter' in the social sciences continued to grow. This reputation, founded on his earlier staples research, led to honours and increased influence. Yet, throughout this same period, the essential Innis – Innis the researcher – had switched the focus of his work from staples to communications. In this sense, he disappeared from the intellectual universe of his colleagues, for none of his peers followed him along the new research path. Like the vanishing baker, Innis found it impossible to communicate, in terms his colleagues could understand, the nature of the beast to which his research had led. Later, he would be remembered chiefly for his dedication to the hunt – the pursuit of the truth – rather than for the nature of his prey – communications theory.

The last decade of his life, then, was characterized by disappointment. Despite all the honours and influence he acquired the objectives he sought to achieve were more and more out of tune with the times. He became more influential in an environment that was progressively less amenable to the ends to which he employed his influence. In other words, Innis's personal power increased as his vision of an overall political project became less and less practicable. He died a desperately overworked man, embittered both by the realization of the severe limits of the institutional power he came to wield and by the negative reception of his communications works, which he regarded as the epitome of his intellectual contribution.

Various factors drove Innis to create the flawed but insightful intellectual synthesis laid out in the communications works. Although he was at the height of his influence and recognition during this period, he was pursuing his own research alone and in areas that left his former colleagues far behind. In addition, he was burdened with the emotional aftermath of the 1930s, largely caused by his bitter struggle in the debate over the role of the intellectual and politics; his lost battle to win Irene Biss over to his side of the debate; and – what he, as a veteran, feared most – another world war.

A Decade of Success and Recognition

The war brought to an end significant financial resources available for the *Canadian Frontiers of Settlement* and *The Relations of Canada and the United States* series. They were nearing completion in any case. His own great work of the 1930s, *The Cod Fisheries*, went to press in 1940. Work in these two areas – the series and the book – had demanded an enormous amount of time in the previous decade, and, as a consequence, their completion gave him some time to invest in new endeavours. The pressure of work and emotional intensity of the debates of the Depression era had taken a toll on his health. He had suffered a nervous breakdown and had been bedridden during the month of March 1937. He decided to slow his pace somewhat and rebuild his health. His schedule was not as intense and travel, except to North American-based conferences, was kept to a minimum. For the first time, the Innis family rented a cottage and took a vacation, a practice that continued throughout the war. Innis's vacations may have been shorter and more work laden than his colleagues', but they were vacations nonetheless.

If anything, Innis had a higher degree of engagement in professional associations during the period, reflecting his status as a senior scholar. In the early 1940s he helped plan the founding of the Economic History Association in the United States and its professional publication, the *Journal of Economic History*. He was to become the second president of the U.S.–based group of scholars. In June 1941 he taught summer school at Stanford University in California. In September 1940 the Canadian Social Science Research Council was established on the basis of a report co-authored by Innis, followed at the end of 1943 with the founding of the Humanities Research Council, again with Innis playing a key role as organizational midwife. These organizations reflected Innis's obsession with keeping the universities fully intact during the war in the face of mounting efforts to curtail courses viewed as non-essential to the wartime effort. In 1942 he co-authored a letter to Prime Minister Mackenzie King arguing for student exemptions from conscription schemes. Honours began to flow in. In 1944 the University of New Brunswick granted him an honorary doctorate. His old alma mater, McMaster, followed suit.

To some extent, the profundity of the changeover in his research subject matter during these years was masked by Innis himself. As we have already seen, for instance, long after his travels and active research on the Canadian north had ended, he maintained his reputation as a current authority in the field by regularly writing reviews of new literature appearing on the Arctic. Beyond this, Innis was instrumental in organizing the Arctic Survey, the first cooperative research project of the CSSRC.[2]

In 1945 he received an invitation (along with two other Canadians, Hans Selye of McGill University and A.E. Porsild of the Canadian Geographical Society) to attend the 220th anniversary celebrations of the founding of the Academy of Science of the Soviet Union. He spent

6 June to 4 July on his Russian trip; it was to be a milestone in his life, starting with his first experience in air travel via a Royal Canadian Air Force flight to Moscow.[3] For Innis, the Soviet Union was unsettling. Here was a whole society organized on the basis of the Marxian paradigm. Clearly, it demanded serious study. He published excerpts from his trip diary in the *Financial Post* immediately after his return.[4] A speech at Queen's University and a further article for the *Financial Post* provided material for the essay 'Reflections on Russia,' which eventually appeared in *Political Economy in the Modern State.*

In 1945 Sidney Smith was appointed president of the University of Toronto, and it was clear that under his leadership the university would undergo a serious phase of expansion and reorganization. Smith appointed Innis to chair a presidential committee to reorganize and expand the School of Graduate Studies during the 1946–7 period. In the spring of 1947, Innis was appointed dean of the newly organized School of Graduate Studies while retaining his role as head of the department of political economy.[5]

In the summer of 1946 he taught summer school at the University of Chicago. The university renewed its generous offer to him of a senior academic post with few teaching and administrative duties. It would have been an ideal situation for Innis to pursue his communications research, but he turned it down again. His priorities included not only his own research but also the strengthening of higher education in Canada and the providing of expert advice to the politicians when called on to do so. His role in the development of the University of Toronto's School of Graduate Studies and his service on the Manitoba Royal Commission on Adult Education (1946) represented these two different priorities in his project. In May 1946 he became president of the Royal Society of Canada and delivered his seminal essay 'Minerva's Owl' as his presidential address. That same year, he published *Political Economy in the Modern State,* which contained a selection of his essays and speeches from 1933 to the mid-1940s. It was a transitional collection conceived as a textbook for returning veterans that straddled both his staples studies and his new interest in communications.

He was granted honorary degrees by Laval University and the University of Manitoba in 1947. In 1948 he was invited to give lectures at the University of London and Nottingham University, and in the same year he was also invited to give the prestigious Beit Lectures at Oxford. During the long summer visit, he attended the Commonwealth Universities Conference in Oxford and was awarded an honorary degree by Glasgow University. It was his first trip to Europe in eleven years.

In 1950 Innis's Beit Lectures were published by Oxford University Press as *Empire and Communications*. In the spring of 1949, he took up a post as commissioner on the federal government's Royal Commission on Transportation. Even by Innis's standards, this involved an incredibly heavy workload, with a great deal of commuting between Ottawa and Toronto as well as travel across the country for hearings. He invested a large amount his time over the next two years in this commission work despite poor health and intensive work on his communications research; he accepted the appointment in his belief that it was his duty to Canada. When the commission's report was finally published in 1951, Innis had only twenty months left to live.

In June 1951 he returned to Europe, between bouts of prostate cancer, visiting the university community in Dublin, Glasgow, Paris, Oxford, Cambridge, and London one last time. In the fall he attended a series of learned societies meetings in Canada and in the United States, culminating in his election as the president of the American Economic Association in Boston in December of that year. The various essays and speeches he had prepared on communications topics during this busy period were collected and published as *The Bias of Communication* (1951) and *Changing Concepts of Time* (1952). After a long, painful battle with cancer, Harold Innis died, on 8 November 1952.

The University and the War

As Innis had feared, the advent of the war greatly disrupted the operations of the university. Enrolment went down only marginally – from 8,000 in 1938–9 to 7,000 in 1943–4 – but this masked proportionally large declines in the liberal arts courses.[6] The real crunch, however, came on the staffing side. By 1942, almost two hundred staff members of University of Toronto had joined the armed forces, and several from the department of political economy, including Wynne Plumptre, Joseph Parkinson, and C.B. Macpherson, had taken wartime positions in Ottawa.[7] At the national level, it is well to remember there were only sixty-five economists and political scientists in all the English-speaking universities in Canada. Of these, one-fifth left their teaching posts to take up wartime government work. Another one-fifth left the universities on shorter-term assignments with the government.[8]

With so many intellectuals leaving the university to take up war work, and because the mobilization regulations posed a severe threat to the continued functioning of the university, Innis came to the fore during the war years as the defender par excellence of the university tradition. The polemical essays he wrote during this period linking the continuity of the university tradition directly with the sacrifice of the veterans, approach, in their vituperative tone, the anti-American essays of his last years. However, as well as providing a theoretical or scholarly defence, Innis was the centre of practical resistance to the disruption of Canadian higher education. Innis had been the person responsible for submitting a memorandum to the prime minister on behalf of the CSSRC concerning the necessity of maintaining the liberal scholarly tradition even within a state at war.[9] This stance received considerable publicity. For example, when Douglas LePan, serving on the staff of Canada House in London, became concerned with wartime schemes for higher education for servicemen and with planning for post-war reintegration of veterans, it was to Innis that he wrote for information.[10] Innis's position was made all the more singular by the readiness of a significant number of Canadian university leaders to agree that those disciplines not directly related to the pursuit of the war effort should be deactivated.

For Innis, the situation must have provided an emotional sense of déjà vu, particularly given that he grappled with these problems from the same office in the old McMaster building as Chancellor A.L. McCrimmon had occupied twenty-five years before when he was dealing with similar problems in the First World War. Innis wrote: 'As one who went through the universities at [the] time [of the First World War], I can say that men in the armed forces were treated little short of scandalously. We were told about the great advantages of education we could have on our return and we were greeted with depleted staff and rush courses.'[11] It was in his role as a wounded veteran of the last great war as much as that of an established scholar that Innis led the defence of the liberal arts tradition. He pointed out that 'if certain courses are marked out as not essential to the national interest or the prosecution of the war they will never recover from the stigma involved and would receive a blow to their prestige that would weaken them in peace as well as in war. This means in turn a blow to the prestige of the universities ...'[12]

In the winter of 1942–3 the debate over liberal arts during wartime came to a head. A special meeting of the National Council of Canadian Universities was called for January 1943, and rumours flew that its purpose was to get the universities to agree with the proposal of Principal Robert C. Wallace of Queen's University and Principal Cyril James of McGill University that arts education be curtailed for the duration of the war. Innis co-authored a petition defending the liberal arts for the CSSRC, while also encouraging the humanities scholars to organize a similar petition. In the end, the Wallace/James proposals did not go forward. Arts faculties would not be curtailed. There would be no accelerated courses in the arts such as Innis had experienced during the First World War. Arts students would continue to be non-eligible for conscription as long as they maintained results that put them in the top half of their class.

It would be misleading to characterize Innis's intervention as the main factor leading to this outcome. The conscription debate had been divisive, both along language lines nationally and within the Liberal Party. As Michael Horn points out, 'Mackenzie King wanted to leave [the universities] alone ... The last thing he wanted was a resolution urging his government to meddle in higher education – political poison, especially in Quebec.'[13] Nor were the arts faculties left totally unaffected. Male students increasingly gravitated to the 'essential' science and engineering faculties, where the top-half-of-the-class rule did not apply. Moreover, these disciplines underwent a significant changeover in research from pure science open to peer assessment to military-related research cloaked in secrecy. Innis abhorred this trend towards specialization, secrecy, and external funding of scholarly efforts that could yield immediate practical applications.

Innis was under no illusion as to the outcome of the debate on the liberal arts tradition during wartime. He had made it his goal to maintain liberal arts functions at the university at pre-wartime levels, fearing that the veterans would return to find the universities knocked back to the situation he had experienced in 1918. And he knew that something of significance had been accomplished: 'an intact and integrated but badly weakened organization'[14] that could still be cranked up quickly to meet the needs of the returning veterans.

Aside from this accomplishment, he could also share credit for two other positive outcomes – one intended, the other circumstantial. The former was the wartime creation of the key institutions in the social sciences and humanities: the Canadian Social Science Research Council and the Humanities Research Council of Canada. The latter was the 'feminization' of liberal arts as women took advantage of openings in higher education that had not been readily available before the war and would not have been available during the war had liberal arts been curtailed.

Donald Fisher has written an excellent essay on Innis's role in the formation of the CSSRC,[15] in which he outlines in some detail the intense debates that took place during the war over the role, governance, and funding of the social sciences in pursuing research related to post-war reconstruction. He concludes:

> Innis' influence had been decisive. Not only did he have the confidence of the foundations, but he was also the only scholar who had a place in just about all the discipline associations and the interested organizations. He was president of the CPSA (1937–38), a member of CHA Council (1937–40), and a fellow of the Royal Society of Canada, and he was recognized as both a geographer and someone who was active in international relations. He was the trusted and acknowledged 'dean of the corps' both at home and abroad ...
>
> As a result of his involvement, academic social science was separated from the state and placed firmly in the hands of university researchers.[16]

Innis's vision for this key organization would dominate it even after his death.

In working against changing the liberal arts structure of the university, Innis was being consistent with his overall project: he was thinking to the future and the role of the social sciences in the post-war world. Some of his initiatives did not come to fruition as rapidly as the CSSRC. In the midst of the war, for instance, he noted the need for changes in the peculiar structure of University of Toronto, where courses in the social sciences, history, and philosophy were being delivered by the university while the colleges within the university retained courses in languages, the classics, and religion. In 1943 Innis wrote to University President Cody highlighting the importance of breaking down this division so that the social sciences could benefit from closer exposure to the work going on at the college level. (Innis's own experience of having his research revitalized by exposure to the college-based classicists may have led to him raising the issue.) While Cody did not act on Innis's suggestion, his successor, Sidney Smith, asked Innis to head a committee looking at this issue in 1951. Without Innis's leadership as a result of his death, the committee became deadlocked, and the necessary changes that he had recommended in 1943 were not implemented for another thirty years.[17]

Academic Freedom in Wartime: Underhill Again

Frank Underhill of the University of Toronto's history department had argued throughout the Depression for a more independent Canadian position on foreign affairs that would be more isolationist and increasingly oriented towards the United States. He drew fire more readily than his colleagues, however, not so much for what he was saying but for how he was saying it: he had the knack of coining upsetting figures of speech.[18]

Innis was drawn to his defence at least twice during the early stages of the war. Underhill was attacked in the Ontario legislature (during the debate on the educational budget) over his famous comments that 'we must ... make it clear to the world, and especially to Great Britain, that the poppies blooming in Flanders fields had no further interest for us.' In fact, the comments had been made years before and been republished in a survey book on Canadian foreign policy,[19] but this did not mitigate the seriousness of the attack. Both the provincial premier and the leader of the opposition called for the university to fire Underhill, and they were supported in this by two of the main Toronto papers, the *Globe and Mail* and the *Telegram*.

Underhill was called before the Board of Governors to explain himself. Innis spoke out in his defence, recalling Underhill's record of service in the Great War and indicating his intention of resigning should Underhill be dismissed. The impact of Innis's intervention on Cody and the board must have been substantial, coming as it did from such a senior scholar, especially one who had had bitter disagreements with Underhill over his political involvement. After much controversy, Underhill survived the episode at the price of a letter of contrition to Cody and the board.

In the summer of 1940, Underhill came under attack again. France had fallen, the Soviet Union had signed a non-aggression pact with Germany, and the United States had not yet entered the war. During this bleak time for Britain, Underhill was quoted in the press as saying: 'We now have two loyalties – one to Britain and the other to North America. I venture to say it is the second, North America, that is going to be supreme now. The relative significance of Britain is going to sink, no matter what happens.'[20] Yet again a number of faculty rushed to Underhill's defence on the principle of academic freedom, with Innis playing a leading role. He made an emotional plea when the faculty members met with Cody on the issue; this must have been influential, for he later had his 'statement typed for the President's personal use.'[21] It is a remarkable document. Innis quickly focuses on the nature of the bad publicity that would surround the dismissal of Underhill. He then engages in what appears to be an extended non sequitur:

> It is possibly necessary to remember that any returned man who has faced the continual dangers of modern warfare has a point of view fundamentally different from anyone who has not. Again and again have we told each other or repeated to ourselves, nothing can hurt us after this. Courage in the face of criticism of friend or foe means nothing to anyone requiring the courage to face imminent physical danger and death. All of you will have seen evidences of this fearlessness on the part of men who have seen active service and my own feeling is that it arises from that fact. If my resignation would save the President's position, the unity of the University, and Professor Underhill I would give it gladly. You may say that would take courage – I assure you it is nothing compared to the courage necessary to save a man's life under fire. The Board of Governors or the legislature cannot really hurt those of us who have seen modern warfare. If a man's academic life is endangered because of reckless fearlessness resulting from participation in war I should be glad to run the risk of losing my own academic position to save him. So much for the fraternalism born of war.[22]

The message is clear. The Board of Governors is deceiving itself if it believes it is dealing with one obstreperous radical whose loss no one would mourn. He is part of a brotherhood of veterans that crosses political lines, and these are fearless people. There will be hell to pay if he is fired.

But it is a cleverly political document as well as an emotional one. Innis is giving Cody the ammunition he needs to counter the arguments of the ultra-conservative elements on the board. He continues: 'We are menaced continually by the ravaging effects of ideologies which have spread across Canada since the last war. On all sides we are slightly encouraged, both through the support of the President and the general alleviation of bitterness, to believe that we can eventually secure the confidence of labour. If the proposals regarding Professor Underhill are carried out, our task is hopeless and whether we like it or not we encourage the growth of beliefs in the class struggle and foster the movements with which we believe democratic civilizations are at present threatened.'[23] He concludes by asking a rhetorical question: 'Can the University make a contribution to this war by dismissing a veteran of the last?'

While Innis defended Underhill on the principle of academic freedom, it is likely that the case was decided by much more geopolitical considerations. These centred on the position of the United States vis-à-vis the European war and the role of the American media. In this regard, Innis's role in the Underhill affair was significant not so much because of his position regarding academic freedom but because it fell to him to circulate the following letter from J.B. Brebner, who was viewing the controversy from New York City:

> I am distressed to hear that Underhill is under fire and in danger of dismissal and I am wondering whether there is much general recognition in Toronto of how serious an effect that would have down here. In influential circles American admiration for Great Britain is habitually connected with the contrast between the public expression in the British Isles and under the dictatorships. Already American periodicals have made a certain amount of capital out of Canadian restraints on public expression, and anti-British groups have pointed to Canada in support of their allegations that the British can be as dictatorial as anyone. Inevitably American journalists are already in possession of information (good, bad and indifferent) about the attempt to dislodge Underhill in September, 1940. As you perhaps know, the New York Times has already used this, and I gathered a couple of weeks ago from acquaintances on the Times that it was prepared to do so again. The Times would be judicious and restrained, but other papers would not. They would pick out what they wanted from the Times story and use it for their own ends.
>
> Is this the time, then, to allow a Toronto professor to be forced out for exercising the ordinary rights of a British citizen? Above all, can Canada risk widespread dissemination of extracts from the Canadian press charging Underhill with the sin of being pro-American? There is an immense amount at stake, and on the whole Americans are normally inclined to be more friendly towards Canada than Great Britain. Lord Halifax comes here in a few days facing an extraordinarily difficult task, and if Canada furnishes his opponents with any kind of weapon you can be sure that they will use it to the limit. All this may seem to you strangely Realpolitik, coming from me, but I believe that in this matter I know what I am talking about.[24]

In the end, the crisis seems to have been settled by the intervention of the prime minister and the Department of External Affairs citing Canada's wartime national interests, rather than by the efforts of faculty at the University of Toronto.[25]

While celebrated for his consistent defence of academic freedom, Innis was often judged to be naive in his beliefs, even by those on the left whom he defended.[26] I do not believe this is true. Innis was fighting these battles on the basis of a longer-term perspective than that of the other protagonists. As in so many other areas, he recognized that universities in Canada found themselves influenced by two great traditions of academic freedom. On the one side was the British tradition stressing the self-governance of the university and, because of this governance, the freedom of scholars 'to follow and express political and social views in opposition to government or convention.'[27] On the other side was the Germanic scholarly tradition, taken up by American universities, which stressed the freedom of the scholar to pursue his research even if it challenged conventional belief. Neither the German nor the American universities had the autonomous governance structure of their British counterparts, however, and so the freedom to pursue research was coupled with more obvious limitations on the scholar's freedom to express his views outside the university.

Innis recognized that crises concerning academic freedom broke out in Canada when scholars exercised the British ideal of freedom of expression outside their university. While this may have been possible in Britain, it raised hackles in Canada where university governance structures were dominated by non-scholars. This was the case with the vast majority of North American institutions, whether denominational or state-run. For instance, the University of Toronto's Board of Governors at the time of the Underhill crisis was largely dominated by businessmen appointed by the provincial government.

These two different positions on academic freedom were reflected in divergent philosophies regarding the main purpose of the institution.[28] For the British, it was the teaching of a moral or philosophical approach – in a nutshell, the formation of the cultured man. For the German and American institutions, it was the pursuit of independent research at the edge of knowledge by individual scholars. Innis's academic formation had exposed him to both of these traditions. At Chicago he had experienced the research-oriented, Ph.D-driven style of the Germanic tradition. This had found fertile ground in the United States and spread widely after the establishment, in 1878, of the first institution incorporating the Germanic approach – Johns Hopkins University. At McMaster, Innis had been exposed to the alternative British ideal of centring teaching on a moral and philosophical foundation for young men and women. This same ideal dominated his own department under its British-born leadership.

As with so many other areas of his thought, Innis's vision of the university was an amalgam of these two traditions. From Chicago he took a dedication to research on the edge of knowledge, actively demonstrating what this meant in his great studies on fur and cod. From the British tradition, particularly under the influence of Urwick, he borrowed the concept of the university's role in inculcating a generalist, philosophical approach – the teaching of culture, if you like.

This element of the British tradition became a lifelong theme for Innis. It was an explicit critique of the trend to specialization that seemed so readily to accompany the establishment of the research-based institutions of the Germanic tradition. He was much more aware than his colleagues that threats to academic freedom could come from inside as well as outside: from the self-referring realms of narrow compartmentalized knowledge that defined work beyond their boundaries as illegitimate. He hoped that the broad research approach that he applied to his communications studies would yield a practical example, thereby demonstrating the limitations of a more specialized research strategy.

On the issue of academic freedom he was won over to the British ideal of the defence of scholarly independence based on autonomous self-governing institutions. But he was also acutely aware that the contemporary governance structure reflected no such history of autonomy. That is why he argued that scholars should behave in ways that were deemed appropriate in the Germanic tradition – insisting on scholars' control of the course of their research while accepting limitations on their freedom of expression outside institutional boundaries.

The contradictions and naivety that he is so often accused of in his stance on scholarship and scholarly institutions arise not from Innis himself but from a lack of appreciation of the depth of his vision and its political nature. For others, his ideal seemed to be unattainably pure and, therefore, unrealistic. For Innis, the stance represented a complex and multi-level but practical project that Canadian scholars were currently engaged in developing. Take, for instance, his love-hate relationship with university presidents in Canada. Unlike his critics on the left, he recognized the presidents' key positive role in protecting scholarly independence from the interference of governance structures dominated by non-scholars. At the same time, he hoped that this was a transitional structure, and in his institution-building efforts he doggedly insisted on establishing organizations that were self-governing. On the one hand, he could write: 'A University president in Canada is required to have some of the qualities of the superintendents of lunatic asylums or of ring masters in circuses. But they should be appointed from those concerned with the protection of scholars against colonialism, imperialism, nationalism, ecclesiasticism, academic nepotism, political affiliations and the demands of special groups and classes, and with encouragement of scholars concerned with the search for truth.'[29] On the other hand, in establishing the CSSRC, he would insist on excluding university presidents from membership. In short, he had a clear idea both of where the university as an institution was and where he wanted it to be in the future. This explains the paradox that he was viewed as being close to Cody and Sidney Smith and yet would exasperate them with his threat of resignation in defence of colleagues such as Underhill with whom he had profound disagreements.

Our inability to credit the coherence and realism of Innis's vision has more to do with the bias of our era than with an inconsistency in Innis's thinking. We are currently as far away in time from his struggles of the 1940s as he was, then, from the establishment of the first North American institution founded on the Germanic research-based model (Johns Hopkins University). Innis looked back on the example of the impressive institution-building efforts that borrowed from European models, leading to the development of a vibrant higher-education sector in the United States. We look back on a period in which universities evolved in such a way as to become synonymous with ever more specialized and compartmentalized knowledge, with the result that today we judge their performance in terms of the immediate practical impact of their research; their governance in terms of their inclusion of a wide range of stakeholders rather than a scholarly elite; and their finances in terms of their private-donor recognition programs rather than their scholarly achievements. Rather than discount Innis, we would be better to mourn the loss of what might have been.

Innis's Growing Influence

Innis's decision to persevere with his university work during the war had the effect of greatly augmenting his reputation and influence. While Canadian intellectuals left en masse to do 'war work,' Innis not only remained but redoubled his efforts to build scholarly institutions. By deciding to stay at the university, to continue his scholarship, and to plan for the post-war reintegration of the veterans, Innis became a pivotal personality. He was a key contact both in terms of recommending staff to keep the universities functioning and in terms of recommending current academic personnel for war-work postings. His perspective on the world, as well as his influence on it, became immeasurably enlarged as former close colleagues took up important posts in wartime administrative agencies. Finally, the fact that he was one of the few intellectuals *not* mobilized to deal with day-to-day problems of wartime administration meant that he was in demand as a speaker with a long-term perspective. It is not an exaggeration to say that Innis during this period became the chief spokesman for the interests of scholarship in Canada. As Anne Bezanson of the Rockefeller Foundation puts it: 'With his prestige, the advice of Innis was sought without his initiative, because he was thought of as "Innis of Canada."'[30]

During and immediately after the Second World War, Innis's increased reputation and influence was accompanied by a renewed fascination with the effects of contemporary media. One of Innis's most bitter memories of the First World War period was the propaganda that had sent idealistic Christian youngsters into the horror of trench warfare. While this had largely been a wartime phenomenon, he now confronted a situation where U.S. propaganda was to be cranked up and continued as a permanent feature of peace under the banner of the Cold War. In this climate, Russia was not a subject to be diligently researched and fully understood before policies towards it were adopted. Rather, it was to be contained and isolated behind its Iron Curtain by American military power. Innis viewed the world differently and called for maximum efforts to maintain the institutions of free thought between the Iron Curtain, where Marxist propaganda dominated all, and the 'Gold Curtain' of commercialism that surrounded the United States. He believed that the world was being trapped between the two 'primitive imperialisms' of the United States and Russia. During the same period in which the Gouzenko revelations were substantiating post-war paranoia of Russia, Innis was opining, in one of his rare newspaper interviews: 'Some of the best work in English economic history is being done in Russia ... [Canadian and Russian social scientists] have much to learn from each other, and probably no countries have more to learn through an exchange of information than Canada and Russia.'[31]

It was a view that was increasingly out of sync with the times. If he had developed such a view as a young scholar, he probably would have had a very short academic career indeed. As it was, he had a well-established reputation and was beginning to function as an international scholarly statesman. His odd position on the Cold War and American imperialism seem to have been simply discounted by his colleagues while they increasingly celebrated his contribution to Canadian economic history. Yet, it did lead Innis, during the last years of his life, to focus much of his attention on scholarly individuals and institutions in Europe rather than in the United States.

It is also probable that Innis recognized the overwhelming and capricious influence of the media in the latest Underhill crisis. After all, these crises were triggered not by what Underhill had said but by what he had said earlier and was now appearing in print (1939) and by what a journalist paraphrased him as having said (1940). While the Underhill affairs must have reinforced Innis's conclusions on the influence of the media and his decision to reorient his research to deal with the area of communications, he did not choose to increase his influence through use of media techniques. As he put it, 'Generally, I avoid newspapers and the radio like the plague.'[32] He was never recorded and even seems to have declined an invitation by the Canadian Broadcasting Corporation (CBC) to participate in broadcasts on a subject of special interest to him: post-war reconstruction.[33]

Yet, while he maintained an almost paranoid fear of being used by or caught up in the media, he developed a number of practical media contacts. Moreover, he exhibited a great deal of respect for these people. George Ferguson, a veteran journalist, became his close friend during this period, despite the fact that Ferguson never understood why they were friends. 'I couldn't understand how a boob like me could be found interesting by Innis.' In fact, Ferguson came to view himself as a kind of a guinea-pig for Innis – one of the few links Innis had with the working world of the media.[34] Interestingly enough, Ferguson was not convinced that the media had anything like the influence attributed to them by Innis. 'When I would say to him that I didn't think journalists had any influence he would shake his head and say, "You're wrong. You're wrong about that."'[35]

Innis seems to have concluded that his personal influence, if it was to be consistent with his theoretical perspective, had to be primarily oral. His preferred format was lectures and speeches: the point of origin of virtually all of his communications publications. He also seems to have determined that his influence could best be exerted behind the scenes rather than through extensive media coverage. Thus, during the war years, he began to give talks to organizations such as the American Association of Advertisers and the Canadian National Association of Newspapers and Periodicals.[36] He also developed personal contacts with a great number of important non-academics who were generally sympathetic to his views on the liberal university. Innis would send out his communications essays to contacts such as R.A. McEachern, editor of the *Financial Post*, and in return would receive information that could be of use to him in academic politics. McEachern, who was responsible for printing Innis's series of articles on his Russian trip, would brief Innis on university-related matters appearing in his newspaper.[37] McEachern also once wrote to announce that the CBC president had confidentially intimated that his position was not permanent but only a stepping stone to the University of Toronto presidency. McEachern added, 'Surely, that post should go to a Canadian.'[38]

Innis seems to have relished operating quietly and behind the scenes. Typically, he denied any suggestion of his influence on academic affairs. To Anne Bezanson he replied equivocally, 'You make me smile in thanking me for making you become President of the E.H.A. [Economic History Association].'[39] It was not a matter of passively recommending people when he was asked. Innis made certain that the 'right' people applied for grants even if they themselves had not thought of doing so. Mabel Timlin wrote in surprise: 'About ten days ago I received a letter from Dr. Moe of the Guggenheim Foundation suggesting that I might want to make an application for a Fellowship ... I suspect that you are the one responsible for the suggestion and thank you very much.'[40] The war years gave full scope to Innis's penchant for manipulating the career opportunities of others – always with a view to strengthening higher education in Canada. In 1941 he wrote to his wife from summer school in Stanford: 'I ... spent most of my time reading and catching up and writing letters. It seems impossible to keep one's hand on people. Vincent [Bladen] wrote this morning of people resigning and wanting men. Alex Brady threatens to go to B.C. but I hope this has been stopped. Graham is going to Manitoba – I have tried to stop this too. My hair has gone perceptibly greyer.'[41] Innis became a virtual placement service for academics moving into or out of government service. Hubert Kemp put it concisely: 'As everybody that wants to employ an economist sooner or later applies to you, I am enclosing a [curriculum vitae] just received ... for your information and any action you may see fit to take.'[42]

Innis, by staying at the university, became a clearing house for career plans of a wide range of people in a variety of organizations. A review of his correspondence in the mid-war period indicates that he recommended people for awards from or appointments to, or communicated with former colleagues in, the following institutions: the federal government departments of External Affairs, Labour, Finance, Fisheries, and National Defence (navy and army), as well as the Dominion

Bureau of Statistics, the Wartime Prices and Trade Board, the Joint War Production Committee, the National Selective Service Commission, the Committee on Reconstruction, the Board of Grain Commissioners, the Canadian Legation (Washington), the U.S. Office of Price Administration, the Guggenheim and Rockefeller foundations, McMaster university, and the Universities of Pennsylvania, Saskatchewan, and Manitoba. This is in no sense a complete list but only a sampling of the institutions that appear in the correspondence surviving from this period which is preserved in the Innis Papers at the University of Toronto Archives.[43] In the post-war era, this level of influence continued to increase. When in the late 1940s Ferguson, his journalist friend, was given responsibility for the supervision of Nuffield Foundation grants to Canadian scholars, it was to Innis that he turned to set the guidelines for the selection committee.[44]

This summary of Innis's scholarly network and influence may give the impression that he was becoming a consummate insider who made quiet, unchallenged decisions about the future of higher education in Canada. There is much truth in this perspective, yet Innis always felt like an outsider and was often obliged to resort to his ultimate strategy of threatening resignation to get his way. Sometimes it worked. Sometimes it did not. S.D. Clark reports, for instance, that Innis wrote to the chair of a CSSRC committee who had taken a decision of which he did not approve. 'I don't like what your God damned committee is doing and I hereby resign from your God damned committee.' That was the entire content of Innis's letter. In short, these were not polite discussions but political struggles that built the framework for scholarship in Canada. In the case of the CSSRC, it is unlikely that the organization would have been constituted as a representative body of working scholars independent of government without Innis's action.[45]

Yet, despite his influence in making administrative appointments or recommendations, Innis retained his scholarly disdain for administrative matters. Connie McNeill, a secretary in the department during the 1941–6 period, recalls that the administrative paperwork was left to Professor C.A. Ashley and to the departmental secretaries. When she would approach Innis concerning a matter not involving policy or appointments, his reply would be: 'You run the office, you do it yourself.'[46] Ev Smyth, who was a junior faculty member at the time, recalls departmental meetings as once-a-year affairs convened without an agenda. Innis, with a pained expression indicating he was wasting his time, would go round the table asking each person if he had anything to say.[47] A similar low level of administrative attention was devoted by Innis to his duties as dean of the School of Graduate Studies. He was interested only in policy issues, particularly questions concerning the maintenance of high academic standards or the introduction of new programs. All the day-to-day correspondence of the School of Graduate Studies went, not to Innis, but directly to the two chairmen of the school or to the administrative secretary.[48]

On the basis of his war experience, Innis adopted an attitude that colleagues who left the university to take up wartime administrative positions were 'deserting the ship, though, as already noted, he was prepared to help them if that what we as they decided to do.'[49] Colleagues who stayed, however, did not fare much better since they were subjected to his austere, monk-like attitude towards scholarship. As far as academic salaries were concerned, Innis's parsimonious nature was legendary. Bladen, Karl Helleiner, Dallas Smyth related anecdotes concerning the setting of salary levels. Donald Innis recalls one case in which a lecturer's salary was cut in half because his level of scholarship and teaching failed to meet Innis's standards.[50] Innis's attitude seemed to be that young scholars were privileged to have the opportunity of working at the University of Toronto and therefore low salary levels should not be an issue.[51]

When the war ended, it became clear that the returning veterans would find more opportunities than Innis had in 1918. Probably the most significant factor in this was the payment of tuition and the provision of a living allowance for veterans pursuing their education. (There was no equivalent commitment in the First World War.) As a result, the student population exploded. At the University of Toronto, enrolment rose from 7,000 to 17,000 during the first post-war school year. This increase at Toronto represented one-quarter of all those Canadian veterans who went on to university work.[52] Innis's efforts to maintain functioning liberal arts programs during the war can be credited for laying part of the foundation on which this tremendous post-war expansion was built.

Although time was at a premium during this period, Innis kept his promise to the veterans. He turned down generous offers to pursue his research at Chicago and continued to teach his fourth-year course in economic history. He also led a graduate seminar. The memories of students who passed through his tutelage during these years are clear on one thing: Innis was a terrible lecturer. He would arrive at the lecture hall no matter what the state of the weather, slump down in a chair at the head of the room, and begin reading his notes in an almost unintelligibly low voice. It was the opposite of a template course in which the professor would deliver the same standard lectures year after year. Innis did not condescend to his students. It appears that he was presenting elements of his communications research as a work in progress. The links he made between discrete and seemingly unrelated facts baffled and fascinated them as much as they do us today. Many of those recall that, while they did not often understand what he was getting at, they were very much aware that they were in the presence of a great thinker. They revered him. After the last badly delivered lecture in 1947, Innis was surprised to hear the class break out into a standing ovation, a rare occurrence indeed. Given the high proportion of veterans in that class, it was probably one of the honours that Innis most appreciated.[53]

'To Have Much Insight and Power over Nothing'

The same wartime forces that heightened Innis's position in the university undercut his project in other ways. The typical experience of the First World War had been that of trench warfare. The trenches had exposed soldiers in that conflict to a commonality of experience that did not exist to the same degree for Second World War veterans. (The traumatization of individuals facing long periods of imminent death was the experience of Holocaust survivors of the Second World War, and indeed the work done on the massive psychic traumatization of this group helped us to understand better the effects on First World War veterans.) For the budding Canadian intellectual class, the dominant experience of the two wars was quite different. The widespread slaughter of promising young minds in the First World War and the sharing of the experience of trench solidarity found its counterpart in the Second in the wholesale recruitments of young academics into the wartime administration apparatus and weapons-development programs. In other words, the general experience of Innis and other veterans of the First World War was replaced by the more specialist role of young intellectuals in the Second. The First World War reinforced a generalist trend in the post-war context. The war strengthened Innis's project insofar as it weakened links with the dominant metropolitan area (Britain). The Second World War, however, weakened the project insofar as it strengthened links with the dominant metropolitan area: the United States. Therefore, while this later war highlighted

Innis's leadership in the university and society in general, it also subverted the potential for the non-specialist and nationalist aspects of his project.

Innis continued to hold to a principled position against political interference in the affairs of the university, and to associate this principle with a resistance to the establishment of specialized programs or centres within the university, as well as to university work on government and business projects. He did so at a time when universities and the social sciences in general were moving inexorably in this direction. He attempted to exert his imprint on the university and yet remained aloof from the day-to-day tasks of administration through which such an imprint had to be promulgated. In addition, he had less and less time to devote to administrative duties as the area of his scholarly investigations continued to expand. As a result, his efforts to impose his influence took on the air of interventions in an ongoing administrative process – interventions more often than not designed to veto some appointment. He extended the tactic of the personal ultimatum that he had developed in the 1930s to the appointment of others, even going so far as to intervene in the appointment of the president of University of Toronto. Innis was livid when he learned in 1945 that Sidney Smith was to be appointed president. He felt that Smith's political ties with the Progressive Conservative Party should disqualify him from the top position of an institution that should be militantly non-political. Innis went so far as to write to the leader of the federal Progressive Conservative Party, John Bracken, in an attempt to stop Smith's appointment. It was an immature and immoderate action that drew the following, somewhat piqued response:

> Mr. Bracken ... was inclined to think that there was little if anything that he should do about it. The fact is, as you probably know, that Mr. Bracken worked hard for several months to get Smith into our ranks and it was only about two months ago that Smith finally gave a 'no' answer. Mr. Bracken naturally felt that for him to interfere now might look like trying to prevent Smith from accepting the job that he had chosen rather than accept Mr. Bracken's offer. In other words, since Smith had obviously declined to come along with him, because of this offer of the University Presidency, Mr. Bracken did not wish to appear to be standing in the way of Smith's free choice.
>
> In addition to all this, Mr. Bracken definitely felt that, regardless of possible political consequences, this whole matter of Smith's appointment

> was definitely a provincial one and that he really should not interfere in any way. He is, however, quite willing to agree that this appointment may produce some political results which may not be too good. At the same time, I doubt if he thinks they will be as serious as you suggest.[54]

While Innis was unable to block Smith, the new president continually had to reckon with Innis's opinion. Innis received academic appointments only partially on the basis of his obvious scholarly reputation. At least as important as the latter was the magnitude of the crisis that Innis could cause if he was passed over in the selection process. As in the earlier phase of his career, nothing came automatically to Innis. Smith, for instance, had intended to bring E.K. Brown back from Cornell University to take up the position of dean of the School of Graduate Studies. Innis's decision to fight this plan led Smith to rescind the offer to Brown and offer it to Innis as the only means of defusing the situation.[55] Innis's tactics were, more often than not, somewhat less than subtle. The university was always on notice that their leading social scientist could, and might, leave at any time. He appended the following afterthought to a personal administrative note to the president in 1943: 'I have had an intimation from the University of Chicago that you are to be approached with regard to possible negotiations (re) my appointment as a professor at that institution. This will warn you ahead of time of their designs and my confusion.'[56]

Innis had neither the time nor the inclination to develop the administrative skills that would have allowed him to propagate his vision of the university in a more effective manner. Even if he had not launched himself into a major new field of research, he would have found the last ten years of his life exhausting. As it was, he became more frustrated than ever with holding the trappings of power within the university while the substance of what he wished to accomplish with that power eluded his grasp. His report to a colloquium on graduate studies as dean of the School of Graduate Studies is telling in this regard for it is essentially a long complaint concerning the limitations on his power.[57] After citing the article of the School's statutes pertaining to the duties of the dean, Innis said: 'If I may speak more plainly the powers of the Dean are limited. To "maintain and improve the quality of graduate scholarship in the University," he can intervene at very few points.'[58] Feeling himself caught in a figurehead position, he went on to identify the power of appointments and promotions to graduate faculty, over which the dean had no control, as the central failing of the system:

> There is nothing so frustrating as to see what should be done and having no power to do it. Anyone familiar with the problem of appointments and promotions will be aware of the considerations involved which have little to do with the graduate school: marriage, the birth of children, particularly of twins, rising prices, age levels, an offer from another institution, the hope that some other institution may appoint him, are among the factors which must be taken into account and are taken into account. All this involves unbelievable pressure for promotion or advancement and eventually leads to hearty backslapping.[59]

In such a context, Innis believed that the 'processing' of students was replacing genuine scholarship as the purpose of the university: 'The process of supplying standard erudition in uniform packages guaranteed under the pure food law fully sterilized and sealed without solder or acids to which it is only necessary to add hot air and serve is capable of indefinite extension. Education ... becomes a series of memory tests or at best mazes ... The maze is used to test the capacity of the student. Examination papers are systematically studied over a long period and the best teacher becomes the one who has the best mastery over the mazes.'[60] Innis attributed this deterioration in the quality of higher education directly to the increasing role of university administration. 'The administration becomes concerned with the need for a larger teaching staff and consequently is apt to emphasize mediocrity,' he argued. 'In a recent visit to a conference of administrators of Graduate Studies in the United States, I found myself overwhelmed by ... the devastating effect of administrative bureaucracies ... concerned (primarily) with professional education and discussions of income tax details, government contracts and the like.'[61] His main criticism of administrative personnel was that they were thoroughly incapable of assessing scholarship: 'Scholars can only be judged by their peers. Scholarship is a tough activity and involves unremitting toil over a long period with no prospect of immediate recognition – indeed with the prospect of decreasing recognition as the university is attracted to publicity stunts or becomes concerned with media such as radio and television which are impossible for scholars.'[62] Innis ended his presentation by relating his critical remarks to his overall concern that the university demonstrate an independent direction in the face of metropolitan interference. 'I have no apology, speaking unofficially as I have done throughout,' he said, 'to make for stressing the distinctive Canadian tradition of this institution. Again and again, among English speaking institutions we have shown the power of our independence in providing ultimate direction in the face of pressures from Great Britain and the United States and from the Canadian government itself.'[63]

Innis's disclaimer to the effect that he was speaking unofficially is disingenuous. He was, after all, the dean of the School of Graduate Studies making a presentation to a public forum. The paradox that this speech emphasizes is that, no matter how great Innis's reputation became, no matter how senior a position he held, his personal project never permitted him to view himself as part of the academic establishment.

Against the standard of the British ideal of scholarly autonomy, the one important judgement in Innis's eyes was the level of scholarship of which he viewed someone capable. He would manipulate colleagues according to his assessment of their scholarly abilities. Administrative detail could be delegated to those not up-to-the-mark as far as scholarship was concerned, but Innis would never countenance giving these individuals formally recognized positions of authority. Such posts went to those he judged to be solid scholars whether or not they had administrative abilities. Nor did close friendships interfere with this logic. During his later years, Innis tried to shut out his old friend, Vincent Bladen, from the departmental chairmanship in favour of S.D. Clark. Innis did so in spite of the contrary opinion of virtually all his colleagues, including Clark himself.[64]

Royal Commissions

One of the apparent contradictions in Innis's position that is often cited is his dim view of scholars being employed by the government while he himself agreed to serve on three royal commissions. From Innis's perspective, this was not a contradiction but two entirely separate matters. On the issue of government employment, Innis believed that one of the key functions of the university was the training of a talented civil service. Innis believed that the formation of a cadre imbued with a sense of culture and the limitations of public policy in the face of social complexity was the sine qua non of fostering an effective public service. The inculcation of these individuals with a longer-term perspective during their university years was one of the best ways to give them a lifelong immunity against the pretensions of social engineering.

He did not think it unnatural that students should move on to a career in the public service. Nor did he think it irregular for scholars to move on to become public servants in a one-time-only mid-career change. He did object to scholars who believed they had found policy

024074 No. 16

When Born.	Nov 5th 1894
Name.	Harold Adams Innis
Sex—Male or Female.	Male
Name and Surname of Father.	William A. Innis
Name and Maiden Surname of Mother.	Mary Adams.
Rank or Profession of Father.	Farmer
Signature, Description and Residence of Informant.	William A. Innis Hawtrey Ont
When Registered.	Nov 20th 94
Name of Accoucheur.	Dr A. J. Collver

1 Harold Adams Innis's birth registration. (Archives of Ontario)

2 The Innis family ca. 1912. Back (left to right): Lillian, Hughena, Harold; front (left to right): William, Samuel, Mary Adams. (Norwich Archives)

3 S.S.#1 South Norwich, 2003. Innis's elementary school, now a private residence. (A. John Watson)

4 Harold Innis (fifth from right, back row) at S.S.#1 South Norwich, ca. 1906. (University of Toronto Archives)

5 Network of railways around Otterville in the early 1900s. Innis commuted daily to Woodstock Collegiate Institute by train. At the time, there were twenty railway stations within twelve miles of the town. (Otterville Museum)

6 Otterville train station, 2003. Moved after the railway was closed, this station now serves as the local museum. (A. John Watson)

7a Hawtrey General Store as it appeared in an advertisement at the height of the town's prosperity. Harold worked for his Uncle Sam here ca. 1912. (*Oxford County Atlas*, 1876)

7b The same store in 2003. Hawtrey became a virtual ghost town after the local railway line closed. (A. John Watson)

8 Harold teaching in Landonville, Alberta, summer 1915. (University of Toronto Archives)

9 Professor Ten Broeke's Philosophy Club, McMaster University, Toronto, 1915–16; Harold seated, far left. (University of Toronto Archives)

10 Innis went directly from being a graduate to being a soldier. (University of Toronto Archives)

11 Canadian Officers Training Corps (COTC), 1915–16. Innis – back row, fourth from left. (University of Toronto Archives)

12 'Are we downhearted "No"': Shorncliffe Camp, England, basic training. Innis – tallest in the back, sixth from the left of standing men. (University of Toronto Archives)

13 'Harold in war with gas mask. Just came off duty. Very tired.' (Note written by Mary Quayle on the back of this soldier's postcard.) (University of Toronto Archives)

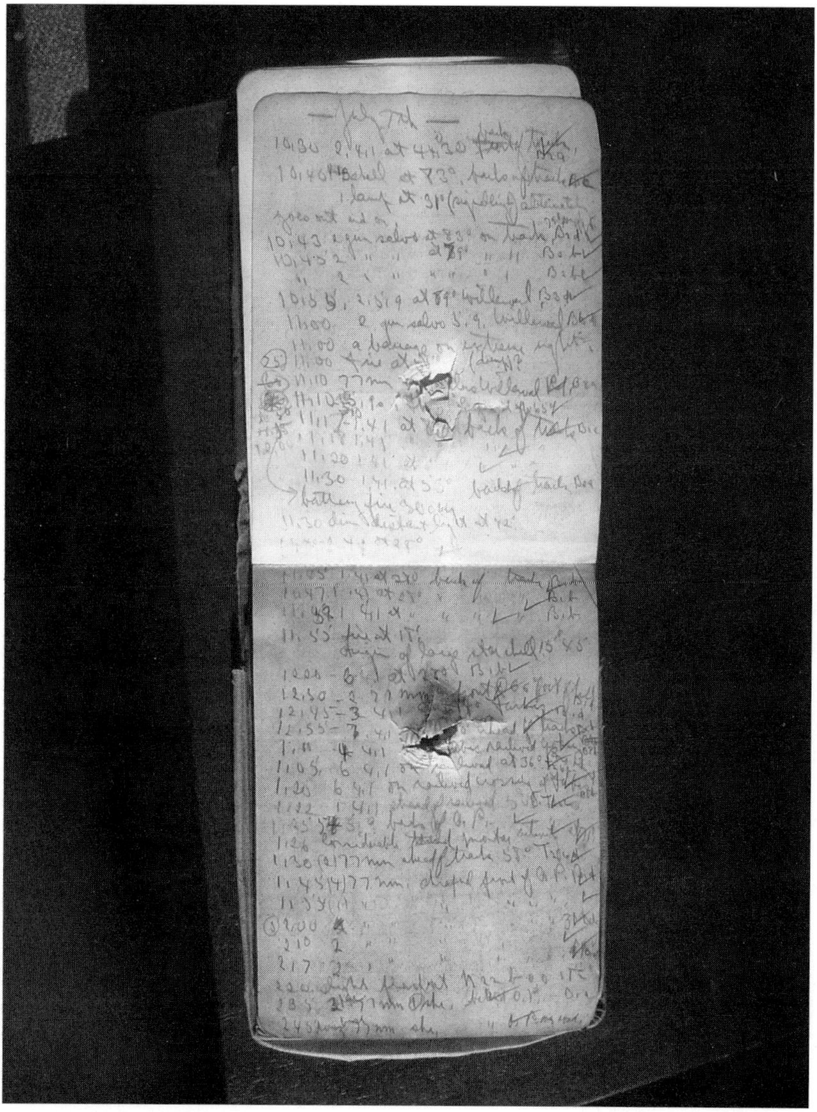

14 Innis's army field notebook with shrapnel hole. Injuries sustained in this incident ended his artillery-spotting entries on 7 July 1917 (and almost ended his life). (University of Toronto Archives)

15 The movie-theatre lantern slide used to present Innis as a '4 Minute Man' – a speaker for war bonds in Chicago. (University of Toronto Archives)

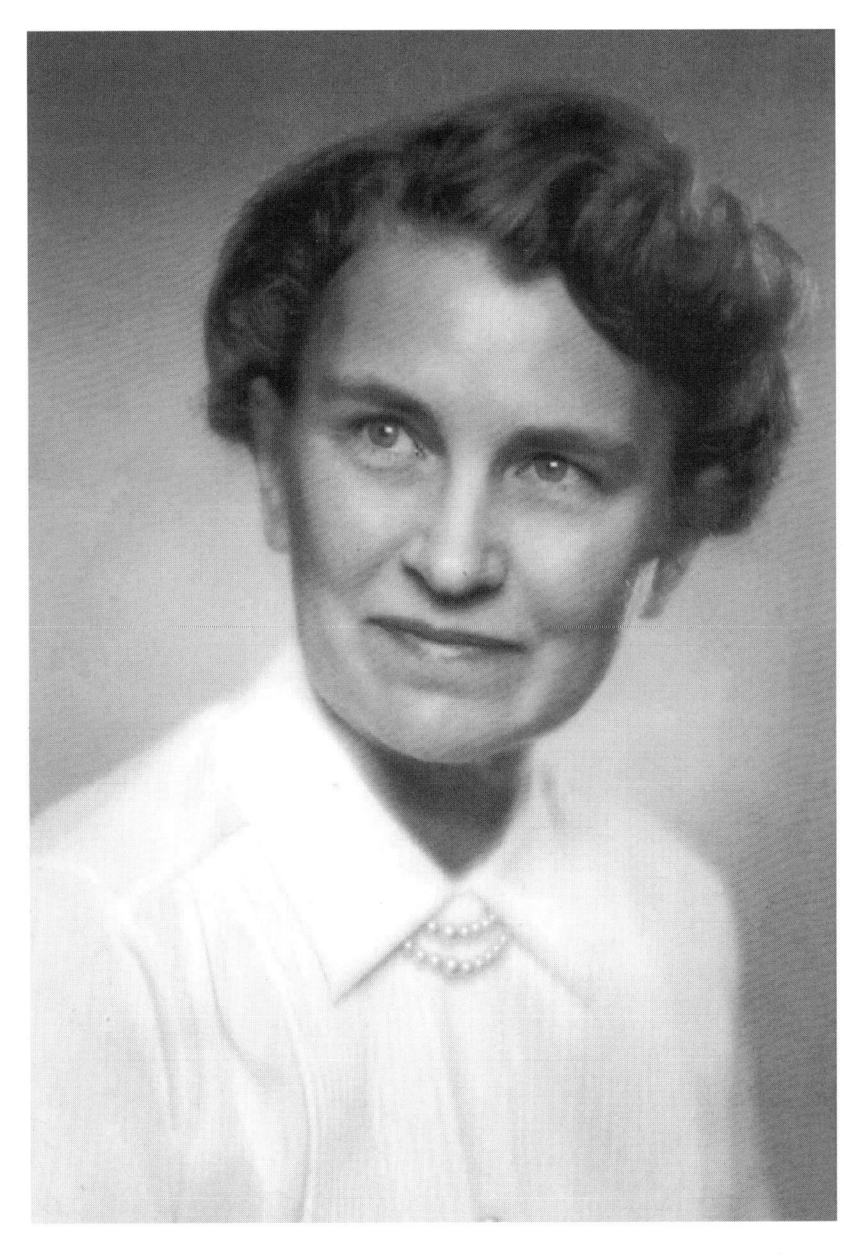

16 Mary Quayle Innis in 1942: Harold's partner, confidant, and emotional support, and an author and editor in her own right. (Anne Dagg)

17 High Park, Toronto, mid-1920s. Harold, Vincent Bladen, Mary Quayle Innis. (University of Toronto Archives)

18 By river steamer down the Mackenzie River in 1924, enjoying his 'dirt research.' Innis seated, far right. (University of Toronto Archives)

19 More 'dirt' research on the Peace River, 1924. (University of Toronto Archives)

20 A Man's World: Department of Political Economy, Baldwin House, University of Toronto, ca. 1930. Innis front row, third from left. (University of Toronto Archives)

21 Irene (Biss) Spry, August 1941. (Lib Spry)

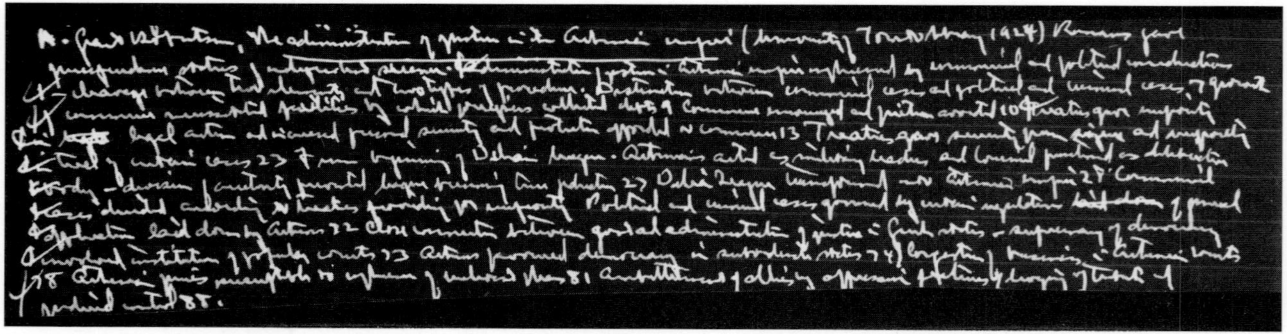

22 Photocopy of a clipped photostatted copy of Innis's reading notes on Grant Robertson's *The Administration of Justice in the Athenian Empire*. (University of Toronto Archives)

23 Photocopy of a typical full-page photostat of Innis's reading notes (Jaeger's *Paideia: The Ideals of Greek Culture*). Original photostat size: 10½ × 14¼ inches. (University of Toronto Archives)

GREEK CIVILIZATION 93

able in an inexpensive and widely read book.[1] Xenophanes used poetry and developed the *silloi* which was satirical in character. Poetry was recited and the rhapsode was held in high esteem. He attacked Homer as a source of errors and denied that gods had human forms.

> But if cattle and horses had hands and were able
> To paint with their hands and to fashion such pictures as men do,
> Then horses would pattern the forms of the gods after horses
> and cows after cattle, giving them just such a shape
> as those which they find in themselves.

In the words of Jaeger,[2] by his influence in the dissemination of philosophy he transfused philosophical ideas into the intellectual blood-stream of Greece. Xenophanes was the first to formulate religious universalism.

The Dionysian tradition had retreated in the face of restraints imposed by Delphi, legal reforms, and advance in philosophy, but it advanced from the courts of the Tyrants to the artistic outburst of the fifth century. Stone-cutting had been used in the publication of laws and in the making of records as Greek epigraphy attests. Sculpture escaped from the traditions of imperialism in the East. Polytheism and the art of statuary based on it checked the development of a divine unity as a dogma. 'The cause of myth and plastic art are really one' (Dill). After the defeat of the Persians, when the festival and the worship of Zeus became stronger bonds among the Greeks, Olympian victors became heroes of the first rank and were celebrated in statues. Sculpture ceased to be exclusively the handmaid of religion and emancipated itself from architecture. Pindar the Theban (502–452 B.C.) wrote hymns celebrating the greatest moments in the lives of athletes and pointed to the advantages of the

[1] Werner Jaeger, *The Theology of the Early Greek Philosophers* (Oxford, 1947), p. 155.
[2] Ibid., p. 42.

24 Photocopy of page 93 of Innis's Oxford Press edition of *Empire and Communications* (now lost) showing his holograph additions. (Innis College Archives)

25 Innis at the height of his influence. Academic conference ca. 1946, Innis at centre with cigarette. (University of Toronto Archives)

26 Royal Society of Canada meeting, ca. 1946, Harold fourth from right in lower group, hands clasped. (University of Toronto Archives)

27 Royal Commission on Transportation (1949), Innis holding coat, seventh from left. (University of Toronto Archives)

28 Building a campfire with the children. (Left to right: Mary, Anne, Hugh, Harold, their dog Tigger, and Donald.) Probably Foote's Bay on Lake Joseph, ca. 1940. (Anne Dagg)

29 The Man in the Background: Harold and Mary Innis with children (left to right) Donald, Anne, Mary, Hugh, ca. 1950. (Anne Dagg)

30 Otterville war memorial, 2003. (A. John Watson)

solutions in the paradigms they had been exposed to in an academic setting moving to the civil service to implement these solutions. He was also very much against individuals returning to the university after having served the government. Innis would reject the argument that their practical experience would strengthen the scholarly milieu. For him, the more likely outcome of the reintegration of these individuals was the distortion of research to deal with short-term problems, thereby undermining the university's central role of concentrating on a longer-term perspective and the development of a philosophical approach. It would also compromise the university's independence from governments and politics.

Above all, he was against talented scholars taking up government positions because scholarly work was of ultimate importance in a new country like Canada that lacked adequate scholars to work through an understanding of its problems from an indigenous perspective. This attitude was based on his belief that a much smaller proportion of the population had the capacity to become scholars than to become civil servants, businessmen, or politicians. The occurrence of these individuals was random, not class-based, in Innis's view. Their early identification through a functioning public-education system was paramount, their mature defection to a non-scholarly calling a social tragedy.

Innis did not equate service on a royal commission with employment in government positions.[65] He felt that it was right and proper for scholars to deliver their best advice to government when called upon to do so *as scholars*. The royal commission format, he felt, permitted this. The corollary that inevitably followed from this judgment was that Innis, and any scholar who shared his views, was destined to become the royal commissioner from hell.

These commissions, after all, were a peculiar institution in Canada that empowered a panel of experts from a variety of backgrounds to examine a complex area through exhaustive interviews with a wide range of stakeholders. They were expected to filter the information derived from these interviews so as to devise a set of recommendations that could serve as the basis for practical changes in public policy. Innis's view of the function of royal commissions was much less grandiose. He was fond of citing the 'Rt. Hon. Lord Kennett, P.C., G.B.E., D.S.O., DSC,' a veteran royal commissioner, who maintained:

> That the functions of the Governors of a democracy were very largely the functions of a medicine man. They had to distract the attention of democracy as a whole from really vital matters by displays calculated to keep

> them happily occupied – a function in which they received the greatest and most constant assistance from the press ... one of the favourite devices of the medicine man was the tribal dance ... to persuade his tribe that something very important was going on in the way of activity ... that was the first function of a Royal Commission – a tribal dance to persuade the general public to believe that something very active was in progress.
>
> ... another device [was] the medicine hut. This was a hut shrouded by curtains, into which he retired for a long period with the object of persuading the tribe that something very important was going on and that it was essential that they should wait for his emergence. This was the second function.
>
> ... a third function ... was the promotion of the dog-fight. When a government finds itself extremely hard put to it to distract the attention of the public from one of the fundamental human ills for which the public expects a remedy from the Government, and for which the Government is sorry it can find no remedy, it promotes a dog-fight between the people with different views, and for starting a dog-fight there is no method so valuable as that of a Royal Commission.[66]

Following Lord Kennett, Innis defined his role on these commissions as that of indigenous 'medicine man.' He brought over the historical, long-term perspective, which he thought was the only one appropriate for a scholar, into his work on the commissions. His scholarly scepticism did not fit easily with an institution designed to analyse present day problems and make practical recommendations for public-policy changes. He was more at ease undercutting the certainty of the analysis and casting doubts as to whether the recommendations being put forward would yield the desired results. While Innis' sociability led him to get on well with his fellow commissioners, the closer it came to producing the final report, the more tensions arose as a result of his scholarly scepticism.

The usual threat of resignation was resorted to on several occasions. Resolution came about invariably by Innis being allowed to write a separate section of the final report over which he could claim ownership distinct from the joint product of the commission. For the 1946 report of the Manitoba Royal Commission on Adult Education, Innis drafted a section that is a rambling defence of the role of the university in the face of the challenge posed by mechanized communications. It is distinctly sceptical of the prospects of discrete adult-education initiatives: 'Adult education, appealing to large numbers with limited training, can be disentangled with difficulty from the advertising of large organizations concerned with the development of goodwill ... To reach lower levels of intelligence and to concentrate on territory held by newspapers, radio, and films, adult education follows the pattern of advertising.'[67] Innis's prescription was to concentrate on strengthening the university to produce better graduates and to give refresher summer courses for teachers. It was not a formula that went much beyond Innis's own practical experience teaching for the Workers' Educational Association in the 1920s. The readers of the report must have been perplexed by Innis's 'contribution.' (He did, however, contribute a much broader overview to the commission's field of investigation. He also, reflecting his own background, argued passionately for a more open access to educational opportunities so that the 'best minds' of the generation could be identified early, even if they came from rural areas.) Similarly, in the case of the 1934 Nova Scotia Royal Commission on the effect of federal fiscal and trade policies on the provincial economy, Innis drafted what was termed a 'complementary' report that, while it did not contradict the conclusions of the main report, added a longer-term historical perspective on the effects of the National Policy and the tariff.

This sense of being odd man out as a scholar carried over into his work on the Royal Commission on Transportation (1949–51). At one point, he wrote to the secretary of this commission: 'I am writing to submit to you for transmission to the proper authority my resignation as a member of the ... Commission ... as an economist ... I can scarcely be accepted in a system in which I am almost an isolated exception. My status is one that can hardly be expected to command attention and my influence will inevitably be negligible.'[68] As in so many other instances, the threat seems to have accomplished its end and Innis completed his work by producing a separate section in the final report to the commission.[69]

Creighton maintains that Innis's appointment to the Transportation Commission 'may be regarded as a misfortune in that it delayed and, in the end, prevented the completion of studies that were both more congenial and more permanently valuable.'[70] He also notes that '[Innis] probably did less than his best work in these circumstances.' While there is some truth in these observations, they miss the mark on two counts. First, the scholarly mission that Innis was pursuing was a political one with many levels of activity. One of these was to respond with the best advice possible when called on to do so by the provincial or federal governments. In other words, Innis, unlike Creighton, did not view his royal commission assignments as diversions that took him away from his central calling but as an integral part of that calling. Second, the fact that his transportation report was 'less than his best work,' which would have been avoided if he had only stuck to his scholarly research and writing, is misleading. The note-taking, file-making, text-writing patterns Innis used for his work on this commission demonstrate the same questionable methodology of research and writing that he employed in generating his *History of Communications* manuscript. I would argue that it is precisely the flawed nature of this methodology of research and composition (examined later in this book) rather than the subject matter of the Transportation Commission that led to Creighton's correct judgment that this was 'less than his best work.'[71]

To a great extent, Innis was the author of his own sense of marginalization at the height of his career. He continued to harbour a deep-seated belief that scholarship *demanded* a lifestyle with as few personal duties and material desires as possible. Certainly, just as he always had, he devoted a minimum of time to his family. In Mary Quayle Innis' semi-autobiographical novel *Stand on the Rainbow*, the father figure is virtually absent. While this novel is ostensibly fiction, it is so obviously drawn from Innis family experience that it probably represents the best insight we have into this area. It was certainly viewed as such by her acquaintances. It is also consistent with Hugh Innis's description of his father as a 'one-quarter parent.'[72] In light of his personal situation, it is not surprising that Innis often tended to greet the announcement of marriage engagements of young colleagues with open hostility. S.D. Clark, poverty-stricken at the time, was flabbergasted by Innis's cynical retort to the news of Clark's engagement: 'Well, I suppose the next thing you'll be doing is to buy a house!'[73]

All efforts to pinpoint Innis's sense of isolation at a time when he was enjoying his greatest success inevitably lead back to the basic research he was pursuing during this period. As already explained, with the coming of the 1940s, Innis's personal research switched from mainly Canadian staples industries to communications studies far beyond Canada. His point of crossover was his work on the pulp and paper industry for a volume in the Canadian-American relations series. He had tried to entice Irene Biss to work on this volume, his intention being that she would write the section on the related development of hydroelectric power. Her decision to get married and therefore be forced, as the rules of the time dictated, to leave her university career in 1937–8 meant that this joint book never appeared. Instead, Innis contributed entries on both subjects to the *Encyclopedia of Canada* in 1938. Consideration of pulp and paper led naturally to an interest in the advent of the mass-circulation daily newspapers in the United States and their effect on public opinion. His first communications essay per se, 'The Newspaper in Economic Development,' appeared in the *Journal of Economic History*, in 1942. Just as his work on the CPR had led Innis back to studies of the earlier industries based on fur and cod, so his work on newspapers in the United States led him back into older forms of media and their effect on earlier empires.

He was alone in making this changeover in subject matter. His old colleagues in economic history kept to their field. He did move closer to a group of classicists whose interest more closely paralleled his new tack. However, as we will see later, they had no idea of the importance of their work to a man they so admired.

Innis had started the decade working alone on relatively modern media, probably hoping, through a superhuman effort, to produce a magnum opus on communications. He could not convince his colleagues in economic history that the next stage in Canadian intellectual development was for them to apply the methodology they had developed at the margin of Western civilization in such a way as to make a new contribution to knowledge, and hence to the revitalization of that civilization. If they could not travel with him on this intellectual odyssey, perhaps he could lead – just as he had with *The Fur Trade in Canada*.

By the late 1940s, he probably knew that his big-volume strategy would not work. The topic was too wide-ranging, and he sensed that he was running out of time. And so he adopted another strategy, which had two parts. First, as an authority on Canadian economic history, he would be asked to give a guest lecture or to give a scholarly speech when receiving an honour. Innis would respond with a sweeping and mystifying analysis of the ebb and flow of empires and the media that they employed. It was as if a theatre audience had arrived expecting 'Death of a Salesman' only to be treated to a long and obscure Greek epic. The reviews were at best condescending. They were almost uniformly negative. Yet, because of the reputation he had developed in the previous two decades, he kept getting the invitations and he kept delivering his interpretation of the impasse in Western civilization.

At the same time, to make the best use of the time he felt was left to him, he abandoned plans for one big book, deciding instead to publish smaller, less comprehensive volumes. He made a virtue of necessity by observing that modern scholarship seemed to favour shorter treatments. We can see the effect in his publications of the period. *Empire and Communications* was published in 1950 and probably represents a condensed version of three chapters of 'A History of Communications,' the magnum opus he had hoped to publish. *The Bias of Communication* came out in 1951 and *Changing Concepts of Time* in 1952. These books were collections of essays and speeches that drew heavily on the masses of material collected for 'A History of Communications.' These publications were 170, 214, and 147 pages in length respectively. Compare this with *The Cod Fisheries*, at 508 pages, and *The Fur Trade in Canada*, at 419. He had embarked on an undertaking that could not be completed in a decade, and no more time would be available to him to invest in its completion.

The Contemporary Reaction to the Communications Works

The increasingly cranky way in which Innis exerted his influence during the communications period reflected the fragmentation of the original symmetry of his personal project. Until the end of the Depression, his personal research, publications, teaching and administrative practice, growing reputation and influence, and the positions of authority that he held had all been related to his pioneer work in Canadian economic history. After 1940, a severe element of incongruity entered his career.

It would be difficult to exaggerate the degree of intellectual isolation in which the communications works were composed. In our era, where the name 'McLuhan' is internationally recognized, it requires some effort to imagine the extent to which Innis's essays of the late 1940s and early 1950s fell outside of the acceptable bounds of scholarship. Yet, without the precedent of Innis's communications essays, it would have been difficult for McLuhan to write, and for the University of Toronto Press to publish, his seminal *Gutenberg Galaxy* in 1962.[74]

Innis was working in virtual isolation. His acknowledgments in the communications works refer to source material, whereas in the staples studies they made reference to the help of colleagues. Those individuals with whom Innis did discuss communications invariably had no sense that they were influencing the direction of his research. Grant Robertson, a classics professor at one of the University of Toronto's colleges and a minor source for Innis, was frankly puzzled as to why he should be asked to vet the *Empire and Communications* manuscript. He was even more perplexed by its content.[75] Colleagues who had been close friends during the economic-history phase often reported 'losing touch' with Innis during the 1940s because of his new research interests.[76]

Innis felt that his communications essays were vulnerable to attack from specialist academics. This suspicion, coupled with the almost complete lack of positive response to the communications works, contributed to his sense of overwork and depression. Innis repeatedly would send out copies of his essays to colleagues, but, unlike his experience with the economic-history essays, he received no replies other than polite acknowledgments of receipt.[77] His son, Donald, reports that he was particularly depressed by the lack of reaction from colleagues at Chicago – John Nef, Chester Wright, and Frank Knight – whose interests Innis felt most closely paralleled his own. Indeed, the only significant feedback Innis received on his work came from his own son.[78]

Nor can it have been very satisfying to realize that the invitations Innis took advantage of to present his communications works were extended not for that purpose but for lectures on economics in general, and on colonial economic history in particular. Francess Halpenny, an editorial assistant at the University of Toronto Press in the post-war years, recalls that the conventional wisdom of the time, not surprisingly, was that Innis 'was out of his depth' in dealing with communications.[79] General puzzlement from the audience greeted his essay 'Minerva's Owl' at the Royal Society of Canada. The Beit Lecture Series – the high point of his career and the basis for *Empire and Communications* – was received in much the same way. It appears that the audience was expecting a detailed examination of some aspect of British imperial history. According to Graham Spry, who attended one of the initial lectures, the lecture was not received favourably. After Innis had set out his rendition of ancient and classical history, the general reaction of the audience was that such topics 'should best be left to Europeans.'[80]

Innis's degree of intellectual isolation can perhaps be gauged by his choosing of S.D. Clark, who was not all involved in communications studies, as a colleague to talk to about his misgivings vis-à-vis his research. Innis expressed to Clark his 'discouragement,' his feeling that 'he was not getting anywhere,' that he 'wasn't pulling things off.' It was the first time he had ever talked to Clark about his work, and Clark believed that he had never talked to colleagues about his staples research in such a negative manner. Innis was particularly hurt by the tremendous fall-off in audiences during the week of the Beit Lectures at Oxford.[81]

It is doubtful that Innis ever felt fully prepared or at ease with a communications presentation. At times, this sense of disquietude brought on impossible demands and paranoid reactions. One week before he was to give the Cust Lecture at Nottingham, Innis asked that it be postponed so that he could be more adequately prepared.[82] When he was informed that this was impossible since the lecture had already been widely publicized, Innis asked that he be allowed to present two lectures on the same day. He added: 'As you suggest I shall be compelled to use notes and this gives me some worry as to the Press. Perhaps I am unduly fearful because of my experience with the Canadian Press but speaking from notes one is apt to be a little less cautious than from manuscripts and I should like to guard against difficulties which may arise.'[83] Innis feared that press paraphrasing of his presentation might generate an Underhill-type crisis.

As it was with his lectureship invitations, so it was with his published materials on communications. If they had not been preceded by a reputation founded on Canadian economic history, they would have been much less likely to make it into print.[84] The editor's role in these later Innis publications, as we shall see, was restricted by the strength of the author's reputation to the checking of his footnotes and references.[85] In terms of print runs and sales, Innis's communications works were not even modest successes. *Changing Concepts of Time* was restricted to a run of 600 copies because it was used to incorporate the remaining 600 copies of 'The Strategy of Culture' that had already been printed in pamphlet format.[86] By March 1955, the University of Toronto Press still had not sold out even this modest number. Only one thousand copies of *The Bias of Communication* were printed, of which 180 still remained in stock in September 1959. Clarendon Press declined to reissue *Empire and Communications*, as its original sales had been so meagre. *Political Economy in the Modern State* had not sold its first edition run by the mid-1950s.[87] These statistics are all the more depressing when we recall the significant number of author's copies Innis sent out and the number of sales that were made in the early 1950s to undergraduates in commerce, economics, and political science who used these books as texts in Innis's compulsory fourth-year economics-history course.

The poor sales reflected the generally savage treatment that the communications works received at the hands of book reviewers. *The Economist's* review of *Political Economy in the Modern State* complained:

> This is no consecutive treatise but a collection of essays and addresses, published and unpublished, written over the last dozen years for a vari-

> ety of purposes and occasions and covering (various) subjects ... Professor Innis is a connoisseur in borderlands ... The title essay, for instance, is less a study of political economy, in the modern state or elsewhere, than a stimulating but inclusive study of the mental climates prevailing in western Europe and America since the seventeenth century, in which no coherent thesis is discernible – at least after the second careful reading.
>
> Incoherence, indeed, is Professor Innis' besetting sin ... He suffers from an extraordinary lack of organization. Facts and quotations are less marshalled than hurled pell-mell at the reader, who is never certain whether the juxtaposition of two statements is meant to imply a causal connection, an illustrative parallel, a whimsical contrast or the chance neighbourhood of two items in Professor [Innis's] card index. Coupled with a style even more atrocious than is normally to be expected of North American academics, this characteristic goes far to make his essays totally unreadable.[88]

The Bias of Communication received this treatment in the *Canadian Historical Review*:

> It is doubtful if any reader will rise from their perusal with a feeling of much satisfaction, for these papers have serious defects. They suffer from endless repetitions of the same evidence, often in almost the same words; this is especially true in regard to the material in the first two papers. They are utterly lacking in historical proportion; for example, the paper of forty pages entitled 'The Problem of Space' devotes twenty-one pages to the pre-Christian era, twelve pages to the Christian era before 800 A.D., four pages to the later Middle Ages, and four to the period since 1450. This leads to an even more serious difficulty: Mr. Innis is not an expert in ancient or medieval history; it is true that he has consulted numerous authorities, as his many references show; but with only very incomplete evidence at their disposal, ancient historians can rarely come to certain and unqualified conclusions. Too often, however, Mr. Innis takes such conclusions, strips them of their qualifications, boils them down to single statements, and so presents to the reader what is at best a mere series of half-truths. To add to the reader's bewilderment, there is also Mr. Innis' rather unfortunate literary style.[89]

Innis was much affected by negative reviews, particularly when they came from those specialist scholars for whom he had particular respect. One such individual was V. Gordon Childe, who reviewed *Empire and Communications* in an extremely condescending manner.[90] Childe wrote: 'The argument is distinctly complicated and seems to be interrupted by lively characterizations of various aspects of Egyptian, Asiatic, classical, and medieval European civilizations, the relevance of which ... is not always obvious in this printed version.'[91] Childe then proceeded to demolish Innis's argument by showing, detail by detail, how it did not correspond to existing archaeological evidence. Even when Childe agreed with Innis, he gave him no recognition for anything resembling an original contribution. Instead, he concluded by giving Innis credit basically for having a great deal of scholarly nerve: 'In brief, though handicapped heavily by remoteness from first-hand sources and expert advice, Professor Innis has courageously pursued with great erudition a promising line of research, but unluckily it has not led to such reliable results as might have been anticipated.'[92]

Innis was so disturbed by this particular review that he took the unusual measure of writing a rebuttal. However, the ambiguous and provisional nature of Innis's reply[93] coupled with its overly polite tone serve to convey his feeling of insecurity in the face of a critique by specialized scholarship rather than his self confidence in the logic of his argument.

The scholarly assessment of the communications essays was quickly reflected in the attitude of his academic colleagues. Rudolf Coper, who shared a work area in the library with Innis during the 1940s, has described how Innis was viewed by his colleagues:

> I held all of those gentlemen in high esteem. However, if I had relied solely on them, I might have formed for myself a wrong opinion of Innis, the person. They felt towards him, and talked about him, rather condescendingly, but without animosity, for his withdrawal, at least outwardly, from current events (the outbreak of war), as evinced by his immersing himself in topics ... that could not have been further from the concerns of the day. He refused to be drawn, at least in my presence, into political discussions. It is impossible that I was stand-offish in that respect myself, as I was painfully aware that I knew little about Canadian affairs.
>
> In those days I had ... a dozen conversations with him ... I remember him from those conversations as a man of exceeding gentleness, although once in a while he would throw in a scathing remark about someone or other we both knew, and then make a laudatory remark about the same person, as if to apologize to him and to tell me I should forget the first dictum.
>
> Always he struck me as a man suffering from anxieties ... He was not a happy man. He obviously enjoyed talking to me, an outsider, in a manner in which he perhaps would not talk to his colleagues without, however, taking me into his confidence in any way.

> In my lifetime I have met few persons as paradoxical as he, withdrawn yet outgoing.[94]

This thumbnail sketch of the 'communications Innis' corresponds with one provided by his wife: 'I don't think he ever thought of himself as having achieved academic success – he just felt more and heavier academic burdens on his shoulders. That and his anxiety about all the work he wanted to do. All he had done was a mere prelude to what he planned to do and when he became discouraged it was sometimes because he thought of all the work he was so eager to do, all the academic routine that stood in the way and his own fatigue as he grew older and illnesses were more frequent.'[95]

Innis's sense of anxiety and depression are only partly explained by his impossible workload. During the communications period, he developed a new methodology of research and exposition that allowed him to increase his scholarly output and avoid attacks from specialist scholars while working on eras and subjects in which he had little background. He was undoubtedly aware that this methodology exposed him to the possibility of being charged at any moment with the ultimate academic sin: plagiarism.

CHAPTER SEVEN

A Telegram to Australia: Innis's Working Methods

The principle of oral tradition ... is, that what was once well said shall not be changed, but repeated in the very words of the original author.

Solomon Gandz[1]

What was once well said was not to be changed but repeated in the very words of its original author.

H.A. Innis[2]

In the portraits of the graduating class of 1916 in the *McMaster Monthly*, Innis's entry describes his 'ambition' as 'to invent a new loose-leaf system.' This early concern with how the content of research was processed through index and file systems remained a lifelong obsession. From the early 1920s, we find Mary Quayle Innis being drafted to work on an index-card system to give order to the mass of material being generated by his research. In the communications period, Innis's methodology for processing the content of his research changed significantly – so much so that it is difficult to grapple with the meaning of the communications works without understanding the methodology by which they were produced.

Before examining the details of Innis's methodology during the communications period, I will summarize key elements of the context and content that led to these changes in the 1940s. I will also examine how the conclusions he was reaching, which his research led him to, fed back into the manner in which he pursued that research and its presentation.

Innis entered the 1940s at the height of academic success. He had become the first Canadian to head the department of political economy at the University of Toronto, on the basis of his pioneer work in Canadian economic history. His leadership on publications and in developing academic institutions was recognized both in Canada and internationally. He had become a scholar statesman.

And yet his research path was taking him on a more solitary route than he had ever experienced. In the 1920s and 1930s he had been engaged in a grand collective endeavour to rewrite Canadian history. None of his colleagues viewed these efforts as a prelude to reinterpreting world history using the insights gained during the staples phase. In the 1940s, he set out, through his own exhaustive research efforts, to prove this could be so. Similarly, he had failed to convince his more progressive colleagues that the best way to have a positive impact on Canadian public policy was to stick to the work of the apolitical, monk-like scholar.

It was an unprecedented period of 'isms' offering contradictory political world-views to society's problems. The struggle of these world-views had led to a Second World War which, like the First, Innis feared, would destroy great numbers of promising young minds, deform common sense through organized propaganda, and undermine the universities by stripping staff and cutting budgets. Innis recognized early on that this war would be even worse: that the pernicious effects of wartime propaganda would not end with the peace but would become a permanent feature of the post-war world, which indeed happened under the banner of the Cold War.

In short, it was a time during which all the elements of Innis's project – his individual research, his efforts to maintain and build academic institutions, and his contributions to national public policy – were demanding increasing amounts of his time and energy. In such circumstances a reasonable person might be expected to focus on one or at most two of the elements of his overall project. Not Innis. Through working harder and increasing his productivity, he still hoped to make a contribution to all three levels of his project. In this regard, changes in his methodology of processing his 'dirt' research were central to his efforts to increase his productivity.

The changes in his methodology were not just an issue of productivity, however. The acrimonious debates of the 1930s had left their scars. Innis had seen first hand the viciousness of attacks on such diverse individuals as Underhill and Urwick. His success and seniority offered him no comfort. He felt more exposed to attack because his works would be looked at more carefully by vested interests outside the university as a result of his attaining a position of authority in the scholarly world. This lack of security was made more intense by his recognition that many of the promotions and appointments he had achieved in the course of his career had been given grudgingly and under his threat of resignation. His fear of being attacked for public pronouncements was a carry-over from the rancorous debates of the Depression. Added to it now was his fear of attack by specialist scholars in the fields of research that he was only now entering. Innis quite consciously adjusted his methodology of research and presentation as a result of his perception of these potential attacks. As he put it at the time:

> I am largely compelled to avoid making speeches in public and to resort to the careful preparation of material to be made available in such guarded fashion that no one can understand what is written or using quotations from the writings of authors who stand in great repute. I have often envied the freedom of my colleagues in other subjects. On the rare occasions on which I read the reports of their speeches I am always impressed by the ease with which they make statements largely because no one will pay much attention to what they have to say, or because they speak about subjects which do not affect people's direct interests. I am unhappily too aware of the fact that I am the first Canadian to be appointed to the position which I have the honour to hold [head of the department of political economy] and that such an appointment coming at so late a date reflects the very great fear of pronouncements made by the holder of my chair.[3]

It was not only the context but also the content of his research that had changed. His work in the 1920s and 1930s had been founded on a thorough examination of virginal archival material available largely in languages he understood. This was coupled with extensive field travel and wide-ranging conversations with individuals currently active at all levels of the staples industries. In the 1940s, his research led him far back in time to periods in which the primary source material had survived in a selective fashion and could be deciphered only by those with skills in ancient languages. Innis sought to overcome this impasse through exhaustive reading of authoritative secondary sources: historical surveys, translations, biographies, commentaries, and commentaries on commentaries. Given the period he was researching, the work of the classicists became of central importance, as we will see in the next chapter. But it was a one-way influence. His classicist colleagues had no idea of the importance of their work to his research. Nor did Innis's research have any perceptible influence on their own.[4]

There was no easy way that the conversations and field travel of the staples period could be duplicated in his communications phase. As we will see again in more detail later, Innis attempted to compensate by reading large numbers of trashy biographies of people connected with early media work. At best, this provided him with elements of conversation with non-intellectuals who had lived through the periods he was researching. All in all, this new research material provided much thinner ore than what he had mined during the staples phase. A new methodology of work was necessary to process the vast amount of material he was reading and to focus on the occasional insight imbedded in the vast stream of dross.

Finally, there was the issue of how the results of his research fed back into his methodology and its presentation during the communications period. I say 'presentation' rather than 'publication' because virtually all of his works in the 1940s had their debut as speeches, were later published as monographs, and then still later appeared in books that were, in fact, collections of these speeches. As a scholar statesman, he was receiving many more invitations to address institutional gatherings of select audiences. He invariably accepted these invitations without, as we have seen, being thrown off his lonely path of research by basing the subjects of his talks on his research in progress. His hosts might have been expecting some elaboration of his work on Canadian or imperial economic history. What he invariably delivered was an often puzzling overview of media and their relationship to the rise and fall of empires. Alternatively, he would present an eccentric interpretation of current affairs based on the perspective he was developing through his communications work.

His changeover to releasing his work through speeches went far beyond the convenience of getting multiple uses out of the research at hand. Innis was having serious second thoughts about the tradition of presenting scholarly thought through the large-book format. In this regard, we should take at face value what he was saying at the time. He pointed out in his preface to *Empire and Communications*: '... All written works, including this one, have dangerous implications to the vitality of the oral tradition and to the health of a civilization, particularly if they thwart the interest of a people in culture and, following Aristotle, the cathartic effects of culture. "It is written but I say unto you" is a powerful directive to Western civilization.'[5] In short, Innis delivered his communications works in speeches rather than in one big book (as he had with *The Fur Trade of Canada* and *The Cod Fisheries*) not just for practical reasons but as a defence of the oral tradition.

There is what appears to be a 'big book' manuscript from this period in Innis's papers: 'A History of Communications.' This manuscript has resisted four separate attempts over the years to wrestle it into publishable form. I would argue that these efforts were fundamentally misguided in that they were predicated on the belief that Innis intended eventually to publish a big book on communications similar to his tomes on fur and cod. He may well have started his communications work with this in mind, but I believe that he had jettisoned the idea by the time of the Beit Lectures in 1947. The manuscript from that time forward served as a mine for extracting material to be released as oral presentations and scholarly articles. These in turn would eventually be collected and published as small books.

This strategy relieved Innis of one task that he viewed as a nightmare: the editorial work involved in preparing a large volume for publication. His experience with *The History of the Canadian Pacific Railway* and *The Fur Trade in Canada* had been trying. The editorial challenge of *The Cod Fisheries* was even worse. Irene Biss had gone over the final manuscript twice without much success. Finally, Shotwell of the Carnegie Foundation had sent up A.E. McFarlane, an experienced editor, who stayed in Toronto working with Innis until the manuscript was reduced to publishable form.

In any case, Innis must have been well aware that his big books, although celebrated for their academic excellence, were not selling well. A corollary of the overall reduction in the length of his works is a severe compression in writing style. Innis may have reacted to criticism of his manuscripts' excessive length by becoming fanatical about their not containing any padding. As Mary Quayle put it: 'Somewhere he read that writing should be as stripped and spare as a telegram to Australia. His writing was as stripped and compressed as that. He often quoted the saying about the telegram but he would probably have written with the same constriction if he had never come across it.'[6] The result is that the tortured prose of his 1938 presidential address to the Canadian Political Science Association, 'The Penetrative Powers of the Price System,' became the rule rather than the exception during the communications phase. It is as if Innis were attempting to reduce significantly the length of his works without in any way compromising on the amount of ground covered. What is more, although Innis initially prepared the majority of the communications works for public presentation, their length, moderate in terms of academic essays, was still quite excessive for a platform address. He appears to have read them from beginning to end and to have lost his audience as a result. It is unlikely that Innis, at any time, felt relieved of this pressure to compress his ideas.

I have argued that the context of the period, the content of what he was researching, and the feedback from the research itself led to significant changes in Innis's process of intellectual production. We must return to the archival documentation to find further clues to what this entailed.

Once we do so, it immediately becomes apparent why so little work has been done on the methodology as opposed to the content of Innis's work. Relatively little survives from the initial drafts of either the staples or the communications phase of his research. His handwriting was so crabbed and crowded that his secretaries tended to scrap the holograph copies of anything that they had transformed into typescript.[7] This general practice has led to the preservation of different types of holograph material from each phase. During his staples years, Innis followed a standard pattern of research and composition made exceptional by painstaking attention to detail (including site visits) and by a faculty for pattern recognition. He travelled not only to obtain a first-hand view of the milieu and technique of staple production but also to plumb the oral tradition of those involved in the trade. He used this knowledge to offset the bias of the written documentation. Since virtually all the material Innis was examining during these years was virgin primary-source material, it was incorporated in his steady stream of published works. What was worth saving from each research project was published and the original holograph material was scrapped as Innis moved on to a fresh start on a new project. The result is that, while many of the travel notes, correspondence, and typescripts have been preserved, this is so for relatively few of the reading notes, holograph drafts, and corrected manuscripts.

Innis's 'Banks'

All this changed in the 1940s with the advent of work in the communications field. His claim to scholarship during this phase was entirely different. Publication and commentary on primary sources was replaced with wide-ranging eclecticism applied to secondary sources. Beginning with his work on the press, Innis established what might be called 'banks' of documentation from which he would draw again and again for material that would form the basis of his written work. There were three distinct banks. One contained, in roughly alphabetic order by author, the quotations that Innis had found interesting in the course of his reading. The second contained ideas, aphorisms, anecdotes, and insights that occurred to him or that he picked up, usually outside his reading. The third and most important bank was composed of the reading notes of the large numbers of books he consumed during his later years.

The existence of these banks of quotations, ideas, and reading notes represents the main difference in the type of documentation surviving from the two phases. As in the staples period, the original holograph drafts of many of the communications works seem to have been destroyed as the typescript version became available. The principal exception to this rule is the massive, 2,400-page manuscript of the unpublished 'A History of Communications,' mentioned above. However, since it was never completed by Innis, it is unclear where this handwritten document lies on a continuum between the initial compilation of reading notes and the final draft of a text before its publication.

The significance of the appearance in Innis's archival papers of these banks of documentation from the communications phase has been overlooked. After all, there is nothing exceptional about a quotations bank per se. Nor does the roughly alphabetic structure of the quotations file give one a clear notion of its accretion over the years.[8] It remains much as it was, apparently produced directly in typescript from the original books according to Innis's pencil markings.

The ideas bank, in contrast, is so clearly remarkable that recompiling and microfilming it became the first project concerning the Innis Papers after his death. To return to its original format and use, we must sort out the various scrambled versions of the file produced to date. The tortured history of this one bank indicates just how difficult it is to reconstruct Innis's methodology.

Originally, the format of the ideas bank was a set of file cards accompanied by a cross-referenced index on the same type of file cards.[9] These cards were in holograph form and have since disappeared. As with other Innis holograph material, they may have been jettisoned after they were transferred to the typescript version on standard-format bond paper. Innis seems to have had this done during the last years of his life. The result was a 339-page manuscript. The notes seem to have been made in blocks corresponding to different periods and research projects from approximately 1944 to 1952. These blocks were typed in roughly chronological order beginning with those most recently composed. This ordering left no place to fit in new or overlooked blocks, and it appears that two sections of this kind have been tacked on to the end of the typescript. They, therefore, appear to be out of order.

As William Christian has pointed out, it is quite possible that the card file still existed at the time of Innis's death for there is a second, partially completed typescript in existence.[10] This typescript may have been started and then aborted once it was realized that the cards had already been transferred to typescript, a discovery that in turn may well have led to the scrapping of the cards. A fourth version of the file was produced by the committee set up to review his unpublished work just after his death. The material was reorganized alphabetically under headings never used by Innis. Sections of notes were broken up. Repetitions, indiscretions, and all sense of chronology were avoided. Microfilm and limited bound versions of this effort were circulated and it came to be viewed as 'The Idea File' per se. The circulation of this fourth version was rationalized as being more accessible in format, and indeed it was. However, it masked the original purpose of the material as a slowly accumulating working-research file.

A fifth version, *The Idea File*, has been produced and published by the University of Toronto Press (edited by William Christian). Its presentation follows the reasonable course of reproducing Innis's original typescript version but in as nearly chronological an order as possible. Since the original index cards and cross-reference file have been lost, this approach to editing Innis's idea bank gets across, as much as possible, its purpose as a research tool. Innis added to it in sections corresponding to his progress in moving from one research project to the next. While he made new notes with each new project, the accumulated file was always on hand to serve as a bank of material of a certain type (ideas, aphorisms, anecdotes, definitions, and so on) to which he could return in the course of composition. Given the tortured history of the documentation and the loss of the original version, it is not surprising that its full significance has seldom been grasped.

As mentioned, the third bank of material to be found among the archival documentation for the communications phase is Innis's reading notes. This is the only bank available in his own handwriting. Even so, the original material has disappeared and we find primitive white on black photostats in its place. Innis's crabbed handwriting is difficult to decipher at its best. Add to this the problems posed by stains, sheets sticking together, and out-of-focus copy, and it is understandable why the exceptional nature of the reading notes has not been pointed out.

These three banks of material – quotations, ideas, and reading notes – are unique to the documentation of the communications phase.[11] It appears that Innis began all three around the same time, towards the middle or end of the Second World War. Accordingly, they represent a change in his method of work as he entered the communications phase. While Innis never directly wrote about his methodology, we are able to reconstruct it in some detail based on these files and the recollections of his colleagues, family, and secretaries.

Innis's shift from primary to secondary-source material was dictated both by the pressure of time and by the availability of material. He did not have the time required either to master ancient scripts and languages or to develop a philosophical background. Following the practice of his staples work, he could and did gain some first-hand experience in the functioning of modern media through such contacts as his friend George Ferguson of the *Winnipeg Free Press* (and later of the *Montreal Star*). However, Innis faced fundamental limitations that were inherent in this type of research. For instance, observing the drying of cod or the grading of beaver pelts today will tell you much about these trades in the past, whereas visiting a modern newsroom tells you little of an English broadside of the eighteenth century.

Innis attempted to overcome these difficulties through a particular strategy of reading. He would listen with great interest to the mealtime conversation of 'specialist' scholars concerning the current debates in their fields. While they would see his earnest manner as a demonstration of politeness, Innis took the conversations far more seriously. He would follow up with a visit to the library to pick up the source material that had been mentioned.[12] He also attempted to make up for the lack of primary-source material by the sheer weight of his reading. One of the tragedies of the Innis documentation is that the original library checkout listing for Innis, which would have allowed us to sketch the chronology of his reading, has been destroyed.[13] Moreover, as with work on his ideas bank, the documentation of what Innis was reading has been terribly scrambled by researchers working on the papers after his death. There are at least four different versions of his reading lists in existence. Although it is virtually impossible to determine how each of these lists came into being, they do leave us with a general impression of the scope of what he read.[14] At the very least we are confronted with a ninety-page bibliography. The 'Classical' section of one of these bibliographies (and this is a relatively small section) lists 179 volumes, and reading notes exist for each of these volumes.[15]

His reading strategy was evident in the quality as well as the quantity of the material. As mentioned above, it has often been observed that Innis read a lot of 'trash' books during his communications period.[16] It would be difficult to understate the extent to which he did this. Two of the largest book lists are found under the section headings 'England 19th Century' and 'America 19th Century.' The approximate proportion of mundane biographies and autobiographies found under these headings is 80 per cent and 65 per cent respectively.[17] The influence of such works in the published text of Innis's communications works is almost invisible, in the same way that direct reference to his travel experiences is absent from the staple works. Nevertheless, both these banal biographies, like his trip notes and conversations of earlier times, were essential ingredients of the material he was producing in the communications and staples periods respectively. The biographies were the closest Innis could come to direct conversations with ordinary people who had been engaged in media work but were long since dead. He felt obliged to read them to avoid being blinded by the great names of the specialist scholars working in these new fields. As long as a reasonable supply of these unscholarly popular biographies was available for a period, the quality and precision of Innis's published work is high. When he was forced to rely on more general secondary source material, the scholarly quality of his work deteriorates. In short, this biographical material was a quasi-primary source.[18]

Innis had apparently mastered what would now be called speed-reading techniques. The amount of time he devoted to a book depended on his identification of themes in it that were relevant to the communications work. He would read several books at a time. He attacked, line by line, those of great interest to him. He concentrated on promising sections in others after having looked over the books' indexes and/or tables of contents.[19] In the course of reading, he pencil-marked the actual book and inserted small flags of paper to indicate these marked pages.[20] Quotations demarcated with square brackets seem to have been retyped directly onto plain bond sheets that were then filed in the quotations bank. Items appropriate to the idea bank were either underlined, noted in margins, or written onto the small flags of paper inserted in the source book. It is unclear whether these were then transferred onto the (now lost) handwritten card files or typed directly onto the 339-

page typescript version of the idea file. The material in Innis's 'Idea File' was the least dependent on his reading. Often he gathered ideas from conversations and travel experiences, or just as they occurred to him.

The notes in Innis's readings bank do not represent his commentary but rather his telegraphic reduction of the author's key sentences. Some words are changed or reordered, but mostly words and phrases are simply dropped from the original version. For instance, the original passage

> *The administration of Justice in the Athenian Empire* was *influenced by* various factors, but above all by *commercial and political considerations*; and it is the existence of these two particular influences which explains a remarkable *cleavage between two* distinct *elements* in the judicial system *and two* corresponding *types of procedure*. In general the *distinction* was *between commercial cases* on the one hand *and political and criminal cases* on the other[21]

is reduced in Innis's reading notes to:

> Administration of justice in Athenian Empire influenced by commercial and political considerations – cleavage between two elements and two types of procedure. Distinction between commercial cases and political and criminal cases.[22] [See figure 23.]

This note was most likely developed as follows: Innis would underline key words in the original text as he read the book. He would then return to the books he had read and transfer all the bracketed passages, underlining, and marginal glosses to his three banks. The pencil marks would be obliterated with an art gum eraser before the book was returned to the library. The underlined passages were then transferred as reading notes to large sheets of foolscap.[23] Each sheet was packed with notes. There were no margins. Two lines of text were fitted between each ruled line and the back of each page was also used. Innis's extremely frugal use of space seems simply to have been a matter of habit, one that went back to his days as a poor undergraduate at McMaster and as a private in the First World War when he kept a tiny (clandestine) daily diary.[24] These notes were not retyped but a special grant was obtained to photostat them.[25] They emerged as enlarged white on black copies. Many of these copies have been destroyed through discolouration or pages sticking together, or through disintegration as a result of their high acid content. It is ironic that Innis, the media expert, should pick this manner to pass on his reading notes, thinking that the copies produced a more legible script than the originals. (For an example of a typical reading notes page, see Figure 24.)

Virtually none of the original (that is, pencil on foolscap) reading notes have survived. They were either jettisoned after having been copied or were clipped up and used as the raw material for Innis's publications. Using a pair of twelve-inch scissors, Innis was able to separate tiny sections of the reading notes by cutting across them at one stroke. (In various interviews, three of the Innis children appear to have been impressed by these oversized scissors.) Whether he literally cut up and juggled sections of the notes, or copied them onto a new sheet, or made a rough paste-up and then recopied them in a more coherent style is not clear, for the original manuscripts no longer exist. However, reminiscences of the family indicate that pasting up the black photostat strips was one stage of the work. The 'History of Communications' manuscript offers little help in sorting out this question since the original handwritten manuscript is not available. The archives version – a typescript containing Innis's marginal notations – has less coherence than the other published works. It is usually thought of as a draft based on a compilation of reading notes that would have been later reworked. The traditional view is that Innis would have added his all important 'bridges' during the reworking phase.

Direct evidence of all the different stages of the work process has not survived. It may also have changed somewhat from subject to subject, and as the last years went by. Nevertheless, we can be certain of the broad outlines of Innis's method of working. First, he did establish and work from the three banks of sources to which he was constantly adding and on which he constantly drew to compose his articles. Second, he did use verbatim a massive quantity of telegraphically copied source material. Third, the further he strayed from Canadian economic history and related subjects into ancient history and philosophy, the more pronounced these methods became.

The key to this process was the effective organization of the banks and the work area itself. Whatever happened to the original copies of reading notes, it is clear that Innis kept one copy only of each item in the quotations, ideas, or reading notes files. (Some repetition did occur in the ideas file but this was not of crucial importance since this material was generated by Innis himself.) Where Innis used a section from any of these files, he noted this through the expedient of placing a mark – ✗ – beside the appropriate line or lines. It was a simple credit/

debit accounting system applied to text.

The work area became important, therefore, because Innis would be carrying on these transactions from dozens of text 'accounts' in the course of writing an article. He was obliged to juxtapose note to note, book to book, to obtain the result he wanted. His ability of dealing with this extreme externalization of material used in composition accounts for the phenomenon that everywhere he worked – whether his table in the periodicals room of the library, his office in the department of political economy, or his study at home – was piled high with papers and books. There are many anecdotes recounting the amazement of individuals at how Innis could locate documents in this seeming chaos. He seems to have done this by treating the piles of paper themselves as filing cabinets. Each referred to a particular subject and was kept in order vertically. The top paper was an index of what was in each pile. Joyce Wry, former secretary in the political economy department, recalls, 'One time he was home and he telephoned me and said, "Will you go to my office and count over nine columns [piles] and go down to the thirteenth letter and bring me the letter from Professor So and So," and I did and, Lord, it was the letter!'[26]

His study at home was his most important work space, and virtually every evening he spent four or five hours reading, making notes, and writing there.[27] It was organized around two points: an armchair and a work desk. Books were spread out in on the floor in concentric circles from the armchair where Innis read, four or five deep closest to him and one deep at the margins. It appears that he would read and make marginal notes or insert notes in the text while at this chair. He would move to the desk to transform the scattered notes into foolscap reading notes, to clip, file, or paste up these reading notes or their photostatted copies; or to write. As well as books, piles of notes littered the room. Yet, despite appearances, the chaos made sense to Innis; a filing cabinet, which his wife had given him in a vain attempt to get him to adopt a more orthodox approach to classification, sat unused in a corner.[28]

His methodology in the use of secondary-source material accounts for the inordinately large number of 'passim,' 'ff,' and 'see ...' notations that occur in the footnotes of the communications works. It also accounts for Innis's habit of following a short quotation in the text with the author's name in brackets. Almost invariably these quotations do not come from the original source material but from excerpts included in secondary sources. If Innis had decided to tighten up his references while still using this eclectic methodology, he would have spent more time verifying footnotes than producing new text, since the majority of its lines would have had to be referenced.

The very complexity of the methodology and its concurrent light referencing makes it difficult to retrace Innis's process of composition. We do not have paste-up first drafts or rough copy to help us trace the sources of non-footnoted copy in the final published material. Furthermore, Innis, in his text, often mixes several secondary sources in any one passage, juxtaposes completely different parts of each source in adjacent sentences, and now and again throws in bridge material to connect the passage to his overall theme. He often reorders or changes words in his sources. Finally, no original reading notes but only scarcely legible copies exist for much of this secondary-source material. Despite these difficulties, however, it is possible to probe certain promising passages and, by so doing, to obtain a good notion of his methodology.

Grant Robertson was the only classicist to comment on Innis's *Empire and Communications* manuscript in its final draft form. Innis cites Robertson's work in this book, and legible reading notes exist for it. The unique citation of Robertson, in addition to the brevity of the corresponding reading notes, makes it a promising point at which to probe Innis's incorporation of secondary-source material in his final copy. Given the general nature of the citation, one would expect to find a précis of Robertson's argument at this point in Innis's text. Yet both Robertson and George Grube (whose work was referred to in a similar manner) were puzzled by Innis's references to their books. Since there appeared to be no clear connection between Innis's line of argument and what the two men had written, they interpreted the references as a form of compliment paid by Innis to their scholarly efforts.

Robertson's puzzlement is understandable for Innis has used his work in a way that is exactly opposite to the standard scholarly citation. Instead of reproducing the gist of the original argument in his own words, Innis has used Robertson's very words to create text. We can trace back this selection and compression of source material by comparing Innis's final text to his reading notes and to the original-source material. The passage we are concerned with is the second and third sentence following Innis's citation of Robertson in *Empire and Communications*. It reads in the published text as follows: 'Political and criminal cases were decided by regulations of general application laid down at Athens with the result that the courts suffered from congestion and juries were suspected of susceptibility to irrelevant pleas. The allies protested against oppressive features in judicial control and the levy of tribute.'[29] This is a reduction of the following reading notes (the numbers indicate the pages in Robertson's book from which the notes were extracted):

> Commercial cases decided according to treaties providing for reciprocity – *Political and criminal cases* governed *by* certain *regulations of general application laid down by Athens.* 72 close connection between government and administration of justice in Greek states – supremacy of democracy involved institution of popular courts. 73 Athens favoured democracy in subordinate states. 74 *Congestion* of business in Athenian *courts.* 78 Athenian *juries susceptible to* influence of *irrelevant pleas.* 81 embitterment of *allies* by *oppressive features, judicial control* and *levying of tribute.* 88.[30]

These reading notes were, in turn, Innis's compression of the following sentences taken out of context from over sixteen pages of Robertson's book:

> *Commercial cases* with most, if not all, of the allies were *decided according to treaties providing for reciprocity* ... More important as regards their influence upon imperial relations were the *political and criminal cases* which Athens transferred to her courts without any system of reciprocity. It has been shown that these were *governed by certain regulations of general application laid down by Athens* ... 72
>
> There was a *close connection between government and the administration of justice in Greek states* and hence the *supremacy of democracy involved in institution of popular courts* ... 73
>
> Systematic persecution of the oligarchs ... was quite in accord with *Athens'* general policy of *favouring democracy in the subordinate states.* 74
>
> In the first place, the *congestion of business in the courts* was so great that litigants who had made a long voyage to Athens might be compelled to wait for months before their cases were decided ... 78
>
> *Athenian juries* were peculiarly *susceptible to* the *influence of irrelevant pleas,* and nothing was more likely to affect their verdict than the claim that the speaker was a supporter and benefactor of the democracy ... 81
>
> The mistake of the Athenians, which was certain to lead to the ultimate failure of the system, lay in the *embitterment of the allies by* the *oppressive features* of the *levying of tribute and* the *judicial control.* 88.[31]

This short example makes obvious the extent to which Innis's methodology brutalized the source material. Whole sections of original material are reduced to the telegraphic notation of one of their sentences. This is further reduced in constructing text to be published. Finally, disparate elements of the already highly compressed material are welded together into the specious unity of a single sentence. The overall impression conveyed by text produced in this way is that the author knows what he is talking about. One feels automatically on the defensive, unable to follow the line of argument. In fact, however, this putting of the reader, rather than the author, in question is indicative not of the strength of Innis's argument but of the vulnerability he felt at this stage in his scholarly career. As he had earlier put it, 'terminology becomes a defence against the inquisitive.'[32]

The task of retracing Innis's methodology through lines of reading notes back to the original source is difficult enough in the case of a minor reference for which limited reading notes exist. The more important the source becomes the more difficult it is to trace its use. It may be cited in a general or a specific way at many different points in Innis's text and any one author may provide numerous books to serve as grist to Innis's mill. The jumbling of the material becomes more thorough the more Innis finds it of interest.

The classicist Werner Jaeger could serve as an example of an important source for Innis. (I am speaking here only in respect to Innis's methodology of work. C.N. Cochrane, E.T. Havelock, and E.A. Owen had a far more profound effect on the structure of thought in the communications period, as we will see in the next chapter.) We can see how Jaeger gives content to the later Innis because many pages of reading notes still exist, with checked-off lines indicating the incorporation of his material at some point in Innis's writings. For example, Innis wrote eight oversized pages crammed full of notes from a reading of Jaeger's *Paideia: The Ideals of Greek Culture*.[33] (See Figure 24.) These pages alone contain approximately 350 lines of text used by Innis either verbatim or in a somewhat reworded fashion. Looking at the process from the other end, we find that Innis, in *Empire and Communications* and *The Bias of Communication*, makes at least fifteen references to Jaeger drawn from five of his books. More significantly, seven of these references are of a general nature (for example, 'passim,' 'ff,' 'see ...').[34]

The following passage from *Empire and Communications* provides a typical example of Innis's compression and incorporation of Jaeger's material (with my numeration added):

> 1. Philosophy had its impact on large numbers of the population 2. The work of Anaxagoras was in prose and made available in an inexpensive

and widely read book. 3. Xenophanes used poetry and developed the *silloi* which was satirical in character. 4. Poetry was recited and the rhapsode was held in high esteem. 5. He attacked Homer as a source of errors and denied that gods had human forms.

> 6. *But if cattle and horses had hands and were able to paint with their hands and to fashion such pictures as men do, then horses would pattern the forms of the gods after horses and cows after cattle, giving them just such a shape as those which they find in themselves.*

7. 'Men imagine not only the forms of gods but their ways of life to be like their own' (Aristotle). 8. In the words of Jaeger, by his influence in the dissemination of philosophy he transfused philosophical ideas into the intellectual blood-stream of Greece. 9. Xenophanes was the first to formulate religious universalism.[35]

The section above has been put together as follows (emphasized words and phrases indicate borrowed text):

1. Innis begins with a bridge sentence stating the theme: the communicating of a new philosophical perspective to the populace.
2. 'It is not easy to form an idea of *Anaxagoras'* views on nature in general; for the remaining fragments of his *prose work,* which at the time of his Athenian sojourn *was an inexpensive and widely read book* comprehensible even to the layman, must obviously have been culled from the whole with the intention of demonstrating the theoretical basis of his doctrine of elements and explaining it in the philosopher's own words.' Jaeger, *The Theology of the Early Greek Philosophers,* 155.
3. 'But the personal character of *Xenophanes'* work is most clearly revealed in his invention of a new type of *poetry – the silloi.* These poems were *satirical in character.'* Ibid., 39.
4. 'But [Xenophanes] also *recited* his own *poems* like a *rhapsode.* At that time the rhapsode's profession was *well esteemed.'* Ibid., 40.
5. '... *Xenophanes* felt compelled to *attack Homer as the mainstay of the prevailing errors* ... He ... makes way for a philosophic conception by *denying that God's form is human.'* Ibid., 42–3.
6. This quotation is found verbatim in ibid., 47.
7. This sentence does not appear in the original text but was borrowed from another source, and it was added in Innis's handwritten gloss in his original Oxford Press copy of *Empire and Communications.* Mary Quayle Innis incorporated this gloss as supernumerary footnotes in the revised (University of Toronto Press) edition. This one sentence was incorporated

directly into the body of the revised text because it appeared in this position in Innis's Oxford Press copy. (See figure 25.) It is likely that Innis would have incorporated the better part of his new marginal notations into the body of a revised text. There was simply no room to do so in the Oxford Press copy. In the case of sentence 7, he was able to do so only since it was preceded by a direct quotation that provided the necessary blank space in the body of the published text.

8. 'And so it was with Xenophanes that the work of deliberately *transfusing* the new *philosophical ideas into the intellectual blood-stream of Greece* began.' Ibid., 42. This source is noted in a footnote in the original text.

9. '*Xenophanes was the first to formulate* that *religious universalism* which, both in later antiquity and more especially in the Christian era, was deemed to be an essential feature in the idea of God, indispensable to any true religion.' Ibid., 48.

The overall effect of this use of source material is similar to that in Robertson. Each sentence seems to make sense. It sounds authoritative but where is it all leading? Innis tries to impose an order on the material with the introductory sentence. Popularization of an anti-Homeric cosmology through the two media, prose and poetry, is the subject. But this very imposition of order leads to a jumbling of the source material. There is no good reason for putting Anaxagoras together with Xenophanes except that both were anti-Homeric and one seems to have used prose while the other used poetry.

Innis's compression of Jaeger's commentary on Anaxagoras produces an awkward sentence. This is because the verb forms do not agree and a book cannot be 'widely read' before it is 'made available.' More important are the shifts in meaning caused by the methodology. It is true that Anaxagoras has come down to us by way of prose fragments. However, these are couched in dialogues so that the impression that Innis gives us of dissemination through books is misleading. The books would have been merely a supplement to a primarily oral dissemination of his ideas. Furthermore, that prose fragments have survived does not mean, as Innis would have it, that Anaxagoras used only prose. In fact, the source points out that his surviving commentary on *Nous* or 'Mind,' the notion that he substitutes for the Olympian pantheon as supreme in the world, is written in the style of a hymn. His use of the poetic form here is extremely important for it would indicate an association that is not explicit in any of the documentation that has come down to us. Using a form reserved for praise of the old gods would indicate that Anaxagoras did indeed view the 'Mind' as the 'Divine' in a far more traditional sense than is generally supposed. Innis ignores this.

Following Anaxagoras' prose work (circulating as a book), Innis introduces Xenophanes' poetry, which one assumes is also in book form as opposed to typical orally transmitted poetry (sentence 4). At best, the connection between Xenophanes, books, and the dissemination of poetry is not clear. What is clear is that Homer as a source of errors is Xenophanes' key point of attack. Again we find that Innis's methodology has twisted the sense of the source material. Sentence 4 in the original is not a general statement about poetry but a specific one about the status of Xenophanes. He was 'like a rhapsode,' a public reciter of Homer who added his own satirical poems to traditional recitations of the epics. As such, he did not view Homer as the 'source of errors' but as the man 'from whom all men have learned since the beginning.' However, just as he respected Homer as the 'overwhelming authority throughout the realm of Greek culture,' he was obliged to criticize the Homeric tradition as the main source of prevailing errors. These changes of meaning brought to the material by Innis are significant. His reading of the sources misses the whole point of why it was Xenophanes that so successfully transfused the new ideas 'into the intellectual blood-stream of Greece.' This was so because these new ideas came to the Greeks through the guise of the traditional educator, the Homeric reciter. In overlooking this point, Innis further distorts the sense of Jaeger's passage.

The deformation of source material through this methodology is the prime factor producing the obscurity of Innis's style during the communications phase. A jumble of secondary sources replaced the mountain of detail on geography, cultural modes, and the pricing and characteristics of staples that filled his earlier works. But the two data sources were fundamentally different and so was Innis's grasp of them. In the staples works, he could produce an insightful conclusion that succeeded in clarifying the plethora of facts by recognizing the pattern running throughout. In the communications work, in contrast, one often has the feeling that he is forcing unifying final paragraphs onto material where they do not readily fit.

This methodology was used to construct approximately the same proportion of the communications essays as was taken up by primary-source material in the staples works. Furthermore, as Innis's banks of quotations, ideas, and readings grew and he became more adept at using the technique, his ability to produce publishable material increased. This (along with exhaustingly long hours) accounts for the phenomenon that puzzled his admiring colleagues: his ability to produce an explosion of scholarly essays at a time when he was saddled with an unprecedented amount of work with the department, the School of Graduate Studies foundations, royal commissions, professional associations, and so on. In this regard, too, one of the key factors that came into play was his ability to mobilize far more secretarial support at this stage in his career than had been possible when his reputation had not been so thoroughly established.

The technique also accounts for a related phenomenon found in the communications phase. This is the impression we have of Innis repeating himself without actually repeating himself. In several essays he will deal with the same era and similar themes while using different historical examples to support his argument. He is, in fact, returning to his consolidated banks of secondary sources as he makes a new attempt to articulate the underlying theme of all the communications works: the long-term unfolding of the classical dialectic of intelligence and power throughout human history.

The question is, given its importance and ingenuity, why was this methodology not recognized? In fact, it was. The attributes that it exaggerated in Innis's writing – abstruseness, discontinuity, a rambling style – were noted and criticized from the beginning. But his critics focused on the syndrome rather than its source.

The only significant work done on the methodology itself was carried out, well after Innis's death, by the Canadian Radio-television and Telecommunications Commission (CRTC) editorial group working on the 'A History of Communications' manuscript. This group produced virtually a line-by-line source analysis of some parts of the text.[36] Their findings, however, seem to have been minimized by the manner in which they viewed their task. The material on which they were working was assumed to be different from Innis's published work. It was thought to be a rough first-draft manuscript heavily dependent on reading notes that would later have been transformed into a final product by Innis through the addition of significant amounts of commentary introductions, bridges, conclusions, and other material. This view of the document had been formed largely as the result of the comments of readers to which the manuscript had been sent. In contrast, I am suggesting that the most complete chapters of 'A History of Communications' were far closer to being Innis's final version than is generally supposed. What started out as editing notes for 'A History of Communications' have, therefore, a more general relevance for understanding Innis's methodology of work throughout the entire communications period.

In a similar manner, the significance of Innis's marginal notations in his original edition of *Empire and Communications* has been minimized. In the revised edition, these notations have been incorporated as additional footnote material (and, indeed, many of the notations would have been turned into footnotes by Innis). Yet seven of the notations quite clearly represent a continuation of Innis's habit of incorporating new secondary-source material directly into the text.

It is understandable, that both with the 'A History of Communications' manuscript and with the revisions to *Empire and Communications*, an opportunity to deal, in depth, with Innis's method of work was missed. A metaphor for the old and new styles of archaeological excavation is appropriate here. The old style concentrated on the finding of the most precious artefacts for display well out of their context. The new concentrated on the logic of the site itself and the social history indicated by placement of the artefacts as found. Once a site has been dug up using the old approach, much of the valuable insights that might have been gained by the more careful attention to cataloguing the entire geography of the site are forever lost. In similar fashion, the effort to look for publishable material by massaging the material Innis left behind when his life was cut short has resulted in the loss of important evidence regarding how he worked during the communications phase. We would know more about Innis if his files had been left exactly as they were at the time of his death.

But, given that the method led to some questionable results from the scholarly point of view, why is it that during Innis's lifetime no one seems to have recognized how he was pursuing his work? If something was basically wrong, why was it not pointed out to him? There are several answers. First, compared with the process the staples books went through on the way to publication, a minimum amount of editorial criticism was given to Innis on the communications works. By the later 1940s, he had become such an oracular presence that his books virtually went straight to press. This was partly because they were reproductions of speeches, many of which had already been published in pamphlet format. Editorial vetting amounted to a junior editorial assistant checking the accuracy of each of the references contained in the manuscript. Francess Halpenny held such a position. She recounts discovering a discrepancy in the sense of a certain reference and having to build up her courage for three days before phoning the 'Great Man' to suggest he might have made a mistake. (Halpenny also recalls finding Innis's art-gum eraser tracings in many of the references she followed up in library books. Not having anything but the final typed draft to go by, however, this told her little of his methodology of work.)[37] Certainly, to have sent Innis's work out to readers at this stage in his career would have been considered an insult. If this had indeed been done, there might have been considerable difficulties in getting approval for publication. Nor did Innis receive significant criticism from his peers on manuscripts before they went to press. I remain astounded that only his son Donald, two junior colleagues who knew little about ancient history (W.T. Easterbrook and R.H. Fleming), and Grant Robertson read *Empire and Communications* 'in whole or in part' while it was in preparation.[38] Robertson, the only one whose training suited the task of a critical commentator, read only the section on Greece. His comments are of a minor kind (spelling mistakes and matters of detail). It is quite clear that he is overwhelmed by Innis's reputation and, consequently, discounts from the start any criticism that he might make: 'What you probably want to know is whether there are any places where I do not entirely agree with or do not quite follow you. One of the latter (probably due to *my abstruseness)* is the section on the early philosophers.'[39] The impression one gets is that Robertson cannot follow the overall drift of the argument yet is awed by the tremendous scope of the ostensibly correct comments on various aspects of ancient Greece.

The most perceptive of his colleagues guessed that something was wrong but also sensed that the field Innis was working in was of tremendous importance. They attempted to articulate this but without great effect. For instance, Brebner writes to Innis, on reading *Empire and Communications*:

> The lectures are remarkable in their great breadth of scholarship and pungent originality. I don't know enough to comment helpfully in the former and, as I see the latter, you are still at your old habit of 'factoring' in ways not thought of before. I think I see one change in that, where formerly you opened yourself too often to the charge of exclusive attributions by rather casual use of the definite article, you now use the indefinite more often or specifically number a factor 'among others.'
>
> I feel that you loaded yourself with some unbearable problems of exposition and perhaps made them worse by not being able to resist incorpo-

> rating facts and ideas that had more importance to you in other connections than in this. In fact, I'd be unhappy if you left your communications studies at this point. I think your terminus is probably 'Power and Communications,' including transportation, and I believe that you could write a book of about 100,000 words which would be a distillate, encumbered as little as possible with quotations and scholarly apparatus, and be addressed with great art and simplicity of logic to a developing theme, i.e., a kind of criterion for a great deal of history, economics, and political science. The developing theme would solve the problems of general and particular relevance, indeed of the relation of relevance to space and emphasis, that cannot be avoided when you are conscious of saying new things and having to prove them.[40]

The tragedy was that Innis was not in a position to take a new tack. Time was running out, and the methodology allowed him to produce, in a short time, a great deal of material using reorganized reading notes. Furthermore, he had made such an investment in the 'overhead' (that is, the consolidated banks) of this methodology of composition that the adoption of a different approach was impossible. And then there was the inescapable fact that he could not make a new start. Furthermore, he was a generalist operating in an increasingly hostile specialist environment. The abstruseness and the proximity to the work of specialist 'authorities' (secondary sources), which his methodology entailed, offered a modicum of intellectual protection. Finally, his methodology was in tune with his belief that a genuinely profound and new intellectual vision was likely to occur on the margins of Western civilization. Instead of working within a paradigm elaborated on the foundation of a Great Name, whether Keynes or Marx or another, Innis used his methodology as a way of incorporating all schools, paradigms, and greater or lesser known intellectuals into his work. The textual scrambling of secondary sources was the technique through which he sought to avoid the bias of his times, which was, as always, the bias of the imperial centre.

'On the Neglect and Non-Publication of the "Incomplete and Unrevised" Manuscript of Harold Adams Innis'[41]

In recent years, William Buxton has argued that 'A History of Communications' represents an overlooked source that could offer new insights into the meaning of Innis's communications phase. As he puts it: 'My intention is to shed light on how it is that the vast bulk of the material placed in archives remains undisturbed and unexamined ... The lack of attention given to the document is a scandalous state of affairs and a sad commentary of the state of Canadian scholarship.'[42]

There are several elements to Buxton's position on 'A History of Communications.' The first is that the document has not been subjected adequately to scholarly scrutiny. To explain why this document remains unpublished despite several attempts, Buxton posits a bias among the executors of Innis's scholarly estate towards the economic-history/staples period. In considering the later attempt by the CRTC to publish the manuscript, Buxton identifies George Ferguson as the individual who worked behind the scenes to sabotage the publication project because he had his own personal agenda for the material. Buxton implies that the decision not to publish was based on a fundamental misunderstanding of the importance of the manuscript.

None of these arguments holds up in the face of the primary-source material. First of all, far from lying 'undisturbed and unexamined,' A History of Communications' has been reviewed and analysed by many scholars over a number of decades. As Buxton himself points out, it was considered for publication on at least four occasions. It is true that the executors of Innis's estate[43] did have a greater familiarity with his economic-history work, but their main bias was towards preserving and making accessible Innis's heritage *in toto*. For this reason, although neither the idea file nor 'A History of Communications' was published immediately after Innis's death, both documents were recompiled, reviewed, revised, microfilmed, and distributed to Canadian university libraries by the committee responsible for his estate.

Buxton further asserts that 'within the voluminous material pertaining to the publication project, there is virtually no detailed discussion of the volumes' content.'[44] This is simply not the case, as even a cursory glance at the Elspeth Chisholm Papers at the National Archives will show.

Chisholm was hired by the CRTC to wrestle the material into publishable form in the late 1960s. After months spent chasing down Innis's references in 'A History of Communications' and reviewing the primary material with Mary Quayle Innis, Chisholm reported in a memorandum to her colleagues: 'The pattern emerges: it is only necessary to find the appropriate source before finding the words, in précis, in the manuscript which might be subject to correction or query. Since the ms. is a shorthand for these sources, it is time-wasting to go to general reference books ...'[45]

After many more months of research into 'A History of Communications,' Chisholm drafted 'Innis' Method of Working,' which reached conclusions consistent with my analysis presented earlier in this chapter. Chisholm wrote: 'As I discovered, however, that the citations referred to only a small part of the actual summaries, like the proverbial tip of the iceberg, I began to realize that these were *entirely* notes, not *bridges plus notes* as might have been assumed.'[46] she continued:

> His method was his own, resulting in the style of his published works. It has been said that his citations were 'quixotic.' I would substitute 'arbitrary.' Sometimes he cited a quotation, sometimes he did not. Only he knew how to recall the author. Sometimes he cited a summary, or indirect quotation, using very nearly the original author's words, usually reordered, using passive verbs instead of active, cutting out inessentials, anecdotes, illustrations and paring it down to the meat at the centre that he was after. Sometimes he would use the author's own words in this order or another, and NOT cite him.[47]

In Chisholm's papers we also find notes defining the verbs 'quote,' 'cite,' 'paraphrase,' and 'plagiarize.' Comparing the last sentence in the above quote to the following leads me to believe that no other reason is necessary to understand why the decision was taken *not* to publish 'A History of Communications' after so much time and effort had been invested by the CRTC: 'To *plagiarize* is to quote without credit, appropriating another's words or thoughts as one's own.'[48]

As for George Ferguson, far from being a behind-the-scenes saboteur of the publication project, he was a fellow veteran and close friend of Innis who, as a newspaper editor, offered him first-hand insight into the workings of the press. Ferguson understood perhaps better than anyone the motivation of the CRTC officials, who, in their plans to publish 'A History of Communications,' thought they were revealing the source of much of McLuhan's thought. At the end of 1969, he wrote to Elspeth Chisholm:

> I laughed like hell at your letter ...
>
> But my laughter was based on your mentioning the Ark-of-the-Covenant look that passes over the faces of Martin and Chiasson [of the CRTC] when the name Innis is mentioned. Imagine the look I got when I happened to say to them that the Innis mss. was 'secondary research.' They flared up and said it was original. I said I couldn't quite see it. The chapter

> for instance on China: I had known Innis well, I said, and doubted whether his grasp of Chinese was profound. He had, I said, been educated in the schools of Woodstock, thence to either McMaster or Toronto, thence to Chicago for his doctorate. Woodstock, I was sure, I said, had no classes in Chinese.
>
> Indeed, I continued, I was pretty sure Innis was a unilingualist, and his lingo was English.[49]

I would maintain that Ferguson, perhaps more than anyone, understood the limits of the methodology Innis employed in crafting his communications work and that this explains why he was against publishing 'A History of Communications.'

A key aspect of 'A History of Communications' not stressed by Buxton is that the first three chapters are missing. This is because, as Chisholm and Mary Quayle Innis knew, they had been used as the basis for *Empire and Communications*. In other words, Innis was mining the manuscript as an intermediate source between the 'banks' of quotations, ideas, and reading notes, on the one hand, and final text of the published communications work, on the other. There is no doubt in my mind that its publication, with a description of the methodology that lay behind it, would have damaged Innis's reputation and the esteem in which *Empire and Communications* was held by later commentators.

Buxton believes that 'A History of Communications' is important because it represented an interpretation of 'the onset of modernity.' In this he is correct but in an unexpected way. What the manuscript anticipates are the pressures and opportunities that modern scholarship would be subject to as a result of the introduction of new communications technologies such as photocopying and the Internet.

Innis and the Computer Age

Innis expected that new contributions to Western civilization would take place on its margins at times of imperial reorganization, where new media technologies being introduced, and indeed this belief sustained his work for the strengthening of higher education in Canada. It would be surprising, therefore, if he did not appreciate the extent to which his methodology of work was made more efficient (although it did not depend on) the application of new photostatting techniques. When he obtained the grant that allowed for the complete copying of his reading notes in the late 1940s, he was able to scramble the source material more efficiently by cutting it up rather than recopying it a portion at a time. Innis must have had an inkling of how the new technology was affecting the work he was producing. Its labour-saving aspects were insignificant compared to the obscurity of style that it fostered.

Yet, even if we consider Innis's methodology of composition to be a failed aspect of his work, we should also recognize the audacity of his attempt. At the end of his life, Innis was often said to be 'like an oracle' or 'like a computer.' In a sense, both of these observations are true.

Innis, in the methodology of the communications period, was anticipating the impact that technological developments in the field of communications would have in changing the way we view the world. Working before the invention of the transistor, the microchip, the personal computer, and the Internet, Innis was approximating their effect on thought and research in his use of primitive photocopying, a paste-pot, and a long pair of scissors. His research led him to recognize that changes in technology ushered in periods of cultural florescence during which everyday human consciousness shifted irrevocably and fundamentally and a new common-sense was formed.

In mining the work of hundreds of authors, paraphrasing and juxtaposing their most profound thoughts, and creating new works at the end of the process, was Innis not anticipating the most profound cultural change that would occur in the generations following his death? I am referring to the switchover from the biological storage of information in the brain and in non-interactive physical records banks such as libraries and archives to the storage of far greater amounts of information in digital form and its interactive retrieval and manipulation through software and search engines.

If this seems far-fetched, try a simple experiment – a kind of Rorschach test, probing the common-sense of our time. Ask anyone under twenty-five years of age what the expression 'cut and paste' means to them. Most will describe it in terms of computer commands. Few will describe a process involving scissors and paste-pots. Some will even be unaware of the primitive technologies that gave rise to the language of these everyday computer commands.

This also explains the difficulties that have arisen in trying to wrestle the massive 'A History of Communications' document into book form. I contend that, in the end, Innis accidentally constructed his own personal incipient Internet through the development of the research methodology of the communications phase. The search engine 'software' for this Internet was in Innis's brain itself. With Innis's death, the only copy of that software was lost, and 'A History of Communications' is destined to remain the ruin of a historically interesting but dead database. Innis, in his methodological work during the communications phase, was rediscovering the technological underpinning of the oral tradition – the ancient art of memory – and attempting to revive it. It was an extraordinary effort. Again it is Brebner who recognizes the centrality of Innis's ability to use the art of memory and his banks of notes to mobilize secondary source material. But he does not fully realize the problems the method entailed. Brebner asks:

> Who can tell us how he kept available his enormous learning? Innis' explanation to me was that he took 'very full notes' of his reading, but my ultimate impression was that, although it seems incredible, most of those notes were available in his head. At any rate his master calculating machine was inside his own brain-box, where a powerful mind digested knowledge into meaning and yet remembered the ways that would, when necessary, lead back to that knowledge in note-books and on scraps of paper, in printed sources, and in monographs of the utmost variety. His bulging little office, with its cascades of seldom-disturbed books and papers, were almost hilarious commentaries on the card-indexes, carbon copies, manila folders, filing cabinets, and bookshelves, whose endless management keeps the rest of us safe from thinking and writing as stamp-collectors at their albums.[50]

More than any other aspect of his work, Innis's application of a new technique in the pursuit of his intellectual endeavours marks him as a transitional figure with the ambivalent reputation that that implies. While the overall insights of the communications work are brilliant, the bulk of the text is a meandering and abstruse scramble of secondary sources. As I have indicated, I believe that 'A History of Communications' was not a pre-publication draft in the usual sense but a semi-processed body of ore from which Innis intended to mine more books like *Empire and Communications* and *The Bias of Communication*. Nevertheless, even the final published products suffer from the oddness of the research methodology on which they are based. Not surprisingly, then, many scholars have shied away from serious consideration of the communications works. During Innis's life, they were little read. They did come into vogue again as a corollary of the interest in the work of McLuhan. At this point the methodology of their composition was not recognized, but the abstruseness it engendered was viewed as a hallmark of the profound. A kind of cult has arisen with its own vocabulary for the initiated. It is all tremendously interesting for those on the inside and terribly boring for those on the outside.[51]

A favourite quotation of Innis from Socrates in *Phaedrus* provides an appropriate commentary on his method of composition and the apotheosis of the obscure to which it led in the McLuhan era: 'This discovery of yours will create forgetfulness in the learners' souls, because they will not use their memories; they will trust to the external written characters and not remember of themselves. The specific you have discovered is an aid not to memory, but to reminiscence, and you give your disciples not truth but only the semblance of truth; they will be bearers of many things and will have learned nothing; they will appear to be omniscient and will generally know nothing; they will be tiresome company, having the show of wisdom without the reality.'[52]

CHAPTER EIGHT

Innis and the Classicists: Imperial Balance and Social-Science Objectivity

The classical tradition precludes the possibility of selling the classics, and is overshadowed by subjects with little other concern than that of selling themselves and even threatening to improve the classics by selling them. We must be aware of those who have found the truth.

H.A. Innis, 1946[1]

Innis's shift to communications studies is usually described in two ways: in terms of the internal logic of the research path of the individual intellectual (that is, Innis's pursuit of the staples approach until he came to consideration of the implications of pulp and paper), and in terms of the influence of the Depression and war on the Canadian academic world.

Although there is a great deal of truth in these interpretations, I believe that they are only partial explanations, for they have a metropolitan bias that tends to overlook the importance of collective efforts originating in the periphery. We have seen that Innis, as part of his overarching project, was greatly preoccupied with strengthening the Canadian scholarly community. Nowhere has his transition to the communications studies been examined thoroughly from the perspective of how he benefited from the collective scholarly support available to him during the transition. The work of certain classics scholars at the University of Toronto seems to have been a major source of this sort of influence. Innis found in the work of the classicists on ancient empires a possible way of addressing the issue of social science 'objectivity' which had so troubled him during the Depression years.[2]

Similarly, the example of Adam Smith, the eighteenth-century economist who developed classical political economy on the periphery of

Britain (Scotland), inspired Innis's obsession with developing a new approach on the margins of empire. Innis believed that this peripheral perspective could have such critical power that it would lead to 'political' effects without requiring involvement in party politics. Innis was basically a liberal who saw in classical political economy an exemplary intellectual project.[3] He considered the development of political economy as the ideological underpinning to the extension of the price system. There is no doubt that he believed the extension of this system to be one of the key advances in human social organization.

> The relatively impersonal character of the price system has enormously enhanced the efficiency of industrialism.
>
> It appears to be the most effective system for introducing freedom and efficiency into hierarchical systems. It has largely avoided the hierarchical limitations of Chinese civilization with its dependence on written examinations, the costs of the ecclesiastical system shown, for example, in celibacy in the Roman Catholic Church, and the dangers of autocracy based on the divine right of kings.[4]

Innis saw the price system as an essential mechanism for permitting the entry of intelligence and knowledge into the hierarchy of power. And he viewed the extension of the price system and the advent of liberal-democratic society as a single coherent social project.

Yet Innis also believed that the very success of the price system was leading to its dissolution. In like manner, classical political economy's broad concerns had dissolved into a fetishism of administrative technique. Moreover, the university, just when it most needed to serve as a crucible for a new world-view, was itself thoroughly caught up in this degradation of thought. Innis wrote that the spread of the price system and free-trade ideology

> provided the prosperity which enabled large-scale organizations to extend their activities and compelled the state to restrain them.
>
> [The result has been] overpowering demands of administration [which] have been reflected in the decline in emphasis on philosophy in the study of political economy. The curricula of universities are concerned to an increasing extent with the routine and details of administration, and students are taught more and more about less and less.[5]

Innis's studies of the price system and his conviction that the thought system associated with it had succumbed to day-to-day detail led to his belief that a radically new paradigm must be developed. It was at this point that Innis left his colleagues in economic history behind, setting out on an individual quest to develop just such a paradigm.

The years of the Second World War mark the gestation of Innis's new approach. During this period he was particularly concerned with three areas that would be essential prerequisites for the new perspective. The first was the university tradition: his spirited defence of this tradition emphasized his belief that the institutional foundation for a new system of thought was still most likely to be located in the university. The second area was the history of communications: Innis's next most important activity during these years was a wide-ranging, basic 'fact gathering' project on communications topics, similar to the 'dirt' research that had supported his staples works. Finally, there was a new set of intellectual architectonics: he began a serious examination of the humanities, and in particular classical studies, in an attempt to find the philosophical approach that would provide a framework for interpreting the multitude of raw materials. In all three areas he was to a great degree working alone and against the grain of the times. As a result, he was developing what would become an intellectual synthesis that was far more singular than that coming out of the staples studies.

Innis's defence of the university tradition goes far beyond a facile plea for the principle of academic freedom. He adopts an *intensely political* stance.[6] Remember that Innis was involved in a continuing practical struggle to control appointments and the disbursement of research funds. He was an administrator, as well as an intellectual, and he wasted no words in directly denouncing the individual cases of what he considered to be the interference of 'vested interests' in the functioning of the university. Nor was this stand merely a case of faculty politics. Innis's position was supported by a thorough understanding of the origins and role of the university as a key institution in Western history. As a result of his research into the history of the university, Innis maintained that

> the university must play its major role in the rehabilitation of civilization which we have witnessed in this century by recognizing that western civilization has collapsed. This University, which represents the great tradition of freedom from state control, offers a platform on which we may be able to discuss the problems of civilization. We stand on a small and dwindling island surrounded by the flood of totalitarianism.

> For all universities ... [the interference of vested interests] is a crime against the traditions of western civilization for which men have been asked to lay down, and have laid down their lives.[7]

Considering the formative influence on Innis of the Great War, we know that these are the words of a highly engaged scholar.

Innis was absorbing tremendous quantities of basic data during these years. He seems to have made a concerted effort to read the tomes of encyclopedist historians such as Oswald Spengler, Arnold Toynbee, and Alfred Kroeber, as sources of information rather than for their methodologies. He was also reading survey studies of many fields: law, language, music, art, drama, religion, and poetry. Finally, he read massive amounts of literature related directly to communications and communications technology. All of these efforts were directed towards accumulating a general 'sentiment of experience' that later could be organized for presentation. The intellectual approaches of these sources tell us nothing about Innis's methodology, and, indeed, he seems to have looked on them, especially the encyclopedist historians, as uncritically imbued with the thinking of their times.[8] As we saw in the previous chapter, the manner in which Innis handled such sources tells us a great deal about *his* methodology.

The fruit of this extensive research activity is the unpublished, massive manuscript 'A History of Communications.' This seminal manuscript, as already explained, is actually a detailed fusion of reading notes that formed the basis of Innis's later published works on communications, and, logically, it would be an obvious place to look for clues to the framework that would later appear in his published communications writings. One might even say that 'A History of Communications' and *Empire and Communications* are to Innis's intellectual development what *Grundrisse* and *Das Kapital* were to Marx. But, whereas in *Grundrisse* there are important, albeit short, passages on the method of research and exposition that Marx used to develop *Das Kapital*,[9] a search for similar passages in *A History of Communications* proves futile. In fact, as far as I am aware, Innis never wrote a detailed account of the methodology he used in his research work and expositions. He repeatedly insisted that what was needed to cure the bias of present-day economics as a discipline was a philosophy of history. Both his methodology of research and his efforts at developing a related philosophy of history are found in implicit rather than explicit form in his work.

What one does find in 'A History of Communications' is an incredibly dense collection of detailed facts on communications technology that winds back in time and spans continents. Unlike Innis's publications in his earlier staples studies, this manuscript contains no key passages knotting the multitude of facts into some kind of structured whole. But this in itself tells us a great deal about Innis's method. Its primary prerequisite was an incredible amount of painstakingly catalogued research. Having collected an enormous amount of facts, Innis seems to have 'leapt' to a conclusion based on the sudden realization of a pattern in these mounds of facts. We can now see how this method would be a difficult one for individuals of a 'school' to pick up and make use of, even in a deformed manner. Neither the exhaustingly intense level of research nor the epiphany that sealed the process would lend itself to duplication.

While the previous chapter dealt with how Innis marshalled these armies of facts, here we are concerned with understanding the gestalts that lay at the basis of the integrating framework of the communications studies. To gain an inkling of the patterns that Innis used to order his data, we must focus on the third area in which he began working during the war years: the field of classical scholarship. It seems that Innis, from the earliest phases of his transition to communications studies, tended to see in the current problems of the university the recurrence of a dialectic originally set in motion in classical times.

Back to Classics

This habit of seeing relevant parallels between contemporary and classical situations is by no means a habit only of Innis. In Canada, classical studies, in much the same way as political economy, had been affected by an approach transplanted from Oxford and Cambridge. The mainstream of scholarly activity in the field of classics was dedicated to the textual interpretation of esoteric pieces of Greek and Latin literature. However, in the 1930s the University of Toronto was a major centre of classical studies in North America. There, a new generation of classicists was working against the bias of the time in much the same way as Innis.[10] They stressed a more professional, research-oriented approach, and their efforts were aimed at casting a new perspective on modern problems by elucidating classical themes.[11] Three classical scholars who fall into this category and who are specifically mentioned by Innis as having had an influence on his work are C.N. Cochrane, E.T. Owen, and E.A. Havelock.

Cochrane seems to have been a transitional influence on Innis's work. Both men had had similar backgrounds: both came from rural Protestant Ontario, attended university in Toronto in the pre-war era, and served with the Canadian army during the Great War. After the war, both had participated in the collective effort of Canadian intellectuals to develop a popular historiography that would serve as an authentic basis for a growing sense of national identity.[12] The acquaintance of Cochrane and Innis was founded at an early date on the basis of common interests in scholarship and a similarity of beliefs concerning the role of the university.[13] In the early 1930s, when Innis was engaged in departmental in-fighting with Gilbert Jackson over the commerce course, Cochrane weighed in on Innis's side with a densely argued essay, 'The Question of Commerce.'[14] Cochrane put the case against increased specialization in the arts faculty, against an independent commerce school, and against dropping classical arts subjects as prerequisites to the commerce degree. Moreover, with his classics background, Cochrane was more capable then Innis of situating the debate in a philosophical framework. In 1932, he summarized the situation as follows:

> The spirit and methods of the [commerce] schools are not merely 'objective' and 'scientific,' but pragmatic to the last degree; and the end of the study is exclusively the cultivation of power.
>
> Not that I wish to put myself in the position of repudiating entirely this ideal of education. In fact, I feel very strongly that in most discussions, the antithesis has been too sharply developed between the cultivation of power and the power of culture. These two conceptions are not so much opposed as complementary and, like the positive and negative poles of the magnet, are always present together in any true philosophy of education. But to aim at either exclusively is certainly to mutilate the ideal and perhaps to miss both.[15]

Innis and Cochrane decried the overwhelming tendency of their times to stress specialization and the 'cultivation of power.' They responded by accentuating a more general or philosophical approach in scholarly work. In this respect, both men shared a 'political' attitude towards their intellectual activities.

The two men complemented each other. While Cochrane supplied the philosophical perspective that Innis lacked, Innis supplied encouragement to Cochrane, helping him overcome his block on putting his work into publishable form. Both scholars were animated by a non-parochial, anti-colonial spirit. When Cochrane died in 1945, Innis stressed in an obituary the commitment of his colleague to fostering a high level of scholarship at a hinterland university, pointing out the difficulties Cochrane faced in producing his celebrated study *Christianity and Classical Culture* in such a colonial context.[16] Cochrane had been forced to fight to retain his affiliation with University College, Toronto, on the title page of the book, against the wishes of the publishers who protested that such a 'colonial' identification would have a depressing effect on sales; they wanted to use his Oxford affiliation. Cochrane's scholarly success substantiated Innis's belief in the role of the peripheral intellectual who was aware of metropolitan paradigms without being enslaved by them. In the obituary Innis summarizes Cochrane's significance as follows: 'In a sense his contribution represents the flowering of a long tradition of classical scholarship in University College and in Oxford. His roots were in Ontario soil. His robust independence of word and phrase reflected his background. His concern not only with the role of thought in Graeco-Roman civilization, but with its reflection in the work of the great historians of that civilization, enabled him to make the *first major Canadian contribution to the intellectual history of the West*.'[17] Upon Cochrane's death, it was left to Innis himself to live up to the standards of the profound and wide-ranging scholarship that had been realized in *Christianity and Classical Culture*.

Cochrane is a transitional influence in the work of Innis in two ways. Most obviously, he helped Innis extend the spectrum of his research, which now ran from the history of the fur trade to classical studies. More significantly, the classics scholar, in all his work, attempted to write history by placing the historian of the times (and his biases) at the centre of his analysis. Innis was primed to adopt this approach. During the transitional period to the communications studies, he was becoming increasingly alarmed at the effect of social conditions on social-science thought. Cochrane's study 'The Mind of Edward Gibbon,' produced during this period, offered a seminal example of how thought and history might be related.[18] In Cochrane's presentation of Gibbon, we can read direct echoes of Innis's strategy in the communications studies. Like Gibbon, Innis 'asserted that the historian should be, in the best sense of the word, a philosopher, because the first qualification for his task is the power of perceiving the relative importance of facts ... The need for philosophy is the need for a principle of discrimination.'[19] And like Cochrane, and as opposed to Gibbon, Innis was developing the position that (as Cochrane put it):

> it is quite possible to reject the findings of experimental science and to make a fresh beginning from a less inadequate starting point. From this standpoint historical investigation will (on the factual level) take the fullest possible advantage of available techniques for discovering and assembling its data. On the level of presentation, it will seek with Gibbon to measure up to the most exacting standards of logic and artistry. But finally, on the ultimate level of interpretation, it will abandon conventional illusions of scientific objectivity and will seek with the aid of sympathetic imagination, disciplined and controlled by the comparative study of peoples and cultures, to enter into and recover what it can of past experience, so far as this is possible within the narrow limits of human understanding; and this experience it will seek to 'represent' in such a way as to convey something, at least, of its meaning to contemporaries. In this formidable undertaking the historian can ill afford to neglect any possible assistance; he will ignore at his peril the rich resources of language and literature.[20]

This is a concise summary of the approach taken by Cochrane in writing *Christianity and Classical Culture*. Commenting on this work, Innis, in one of his rare statements about methodology, singles out precisely Cochrane's perspective as being important in the current crisis of the social sciences:

> The significance of the volume for social scientists is in its philosophical approach. In classical civilization reason asserted its supremacy and in doing so betrayed its insecure position with disastrous results. 'Such perversions of intellectual activity,' Augustine called 'fantastica fornicatio, the prostitution of the mind to its own fancies.' Classicism was indicted 'in the fact that it acknowledged the claim of science to be architectonic and therefore, entitled to legislate with sovereign authority for the guidance of human life' ... History written from the philosophical background of classicism differs sharply from history written from the Augustinian point of view with its emphasis on will, personality, and unpredictability. Paradoxically classicism assumed the unpredictable in the incalculable, in fortune or in chance, whereas Augustine admitted the possibility of understanding the unpredictable by emphasizing personality or individuality. A society dominated by Augustine will produce a fundamentally different type of historian, who approaches his problem from the standpoint of change and progress, from classicism with its emphasis on cyclical change and the tendency to equilibrium. The doctrine of original sin

> became the basis of a philosophy of progress in contrast with the philosophy of order of classicism.
>
> The sweep of the Platonic state in the nineteenth and twentieth centuries and the spread of science have been followed by the horrors of the Platonic state. The social scientist is asked to check his course and to indicate his role in western civilization. His answer must stand the test of the philosophical approach of Cochrane.[21]

The important factor to note in the work of Cochrane, and especially in Innis's interpretation of Cochrane, is the close relationship between the writing of history and the making of history. This is fundamentally a political vision of the role of intellectual activity.[22]

Thus, Augustus, classical understanding, and the *Romanitas* as a coherent human social project are juxtaposed with Augustine, Christian belief, and the Holy Roman Empire. In terms of writing history, these two polarities are presented by related dichotomies in the areas of *logos*, unpredictability, and 'ends.' For the classical historian, the *logos* is found in the concept of order and its corollaries of a tendency towards equilibrium and an identification of change with recurrence or cycles. For the post-Augustine writer, the *logos* is presented as the concept of progress and its corollaries, a millennial theory of change, and an apocalyptic vision of the day of judgment. Ironically, in the 'rational' approach of the classicist, the element of unpredictability is located outside human ken in fortune, whereas, in the post-Augustine vision, human personality itself becomes what is the radically indeterminate in history. Meanwhile, the 'end' of the classical ideologue's work is the foundation of a balanced policy that realizes the common good. The Christian vision, in contrast, is one of fulfilment of the potential for good of individual personalities. The state is significant in this vision only in the negative sense of 'an instrument' for regulating the relations of what Augustine calls the 'exterior man.'

The emphasis of the classical pole had much to do with Innis's approach and purpose. But, as a scholar true to the credo of classicism, his aim was not to launch a frontal attack on the notion of progress and the role of personality in human history. Rather, he sought to incorporate these elements in a balanced approach that stressed the origins of the dominance of the Augustinian vision in Western civilization. The end result of this colossal attempt at synthesizing the two streams of thought was a strange amalgam of classical and Augustinian elements. For Innis, the essence of order, or a concern with balance, is found in that individual par excellence, the intellectual. At the same time, the notion of progress is coupled, not with the action of personality through history, but with the limitations imposed through the impersonal succession of different material media of communication.

Although there is a clear connection in Innis's communications work between the re-emphasis of the importance of classical culture and his treatment of societies, it is implicit rather than explicit. We can trace this link in greater depth by making use of certain clues offered by Innis in the above-mentioned obituary. There, Innis makes reference to what he considers to be important passages in Cochrane's work, and it is useful to look in more detail at the original passages in Cochrane to which Innis refers. Commenting on the synthesis carried out by Augustine, Cochrane says:

> The order of human life is not the order of 'matter,' blindly and aimlessly working out the 'logic' of its own process nor yet is it any mere reproduction of a pattern or idea which may be apprehended *a priori* by the human mind. To think of it as either is to commit the scientific sin of fornicating with one's own fancies; in other words, of disembodying the *logos* in such a way as to rob the *saeculum* of all possible significance. For the Christian, time, space, matter, and form are all alike in the words of St. Ambrose, 'not gods but gifts.' They thus present themselves, not as causes but as opportunity. As such they may be said to 'unite' as well as 'divide.' This they do by giving us our status as individuals in the '*saeculum*.' But this status involves its specific limitations, not the least of which is the difficulty of communicating with our fellows. This difficulty is intensified by the confusion of tongues ... from it not even the saint can claim to be exempt: 'Moses said this and passed on,' remarks Augustine, 'what did he really mean?' The difficulty in question is that of creatures whose limitations of *mind* and *sense* compel them to adopt such expedients as that of 'making noises in the air.' Yet, he accepts these limitations as inescapable ...
>
> From this standpoint human history presents itself as 'a tissue of births and deaths,' in which the generations succeed one another in regular order. In this context of generations, each and every individual has his own times and spaces, so that the notion of a man 'out of his age' is a vicious and irrelevant abstraction.[23]

This passage could very well have been written by Innis. Here we find Innis's concern with charting an analytical course between crude materialism and simple idealism. We also find the description of the solution Innis will attempt to work out in the communications studies. Essentially, he was concerned with the limitations on individual action imposed by the matter and form of systems of communications. He traced the biases that these systems imposed on how societies conceived of time and space and the way in which they tackled their problems on the basis of these conceptions. Finally, Innis was concerned with a self-critical approach that entailed nothing less than seeing men, including the social scientist, as inevitably tinged by the mentality of their times. For Innis, as for Cochrane and Augustine, the freedom of the individual does not spring from a recognition of a human, transhistorical essence but from grappling with the limits to action of a being 'in his own times and spaces.' In general, Innis stressed a classical bias in his work that presented this dialectic of limits and action in terms of human communities (empires) rather than individuals. However, behind these imperial projects in Innis's work stand those monopolies of knowledge constructed by individual intellectuals. It is here that the dialectic is played out on a personal level. The same is true of Cochrane's treatment. It now becomes clear why Innis notes in the preface to *Empire and Communications* that 'an interest in the general problem was stimulated by the late C.N. Cochrane.'[24]

It is not so evident why Innis credits E.T. Owen with having played a similar role in his transition to the communications studies. If one looks for the evidence in Owen's scholarly work, one finds only articles on textual criticism of the Greek tragedians and Homer. There seems to be little relationship between the substance of these articles and Innis's work. As far as I can determine, Innis cites Owen directly only once, and this is in an insignificant footnote.[25] But Innis seems to have fastened on one published article by this classics professor that falls outside the category of textual criticism. This is a complex philosophical argument found in his papers at the time of his death and published posthumously. Entitled 'The Illusion of Thought,' it contains strong echoes of themes developed by Innis in the communications studies.[26] The article is a long exposition on the saying 'the wisdom of a learned man cometh by opportunity of leisure.' Innis often stressed this necessity of security and free time for intense scholarship; in fact, it was the basis of his defence of the university as the essential tool of the social scientist. But Owen's article also reveals more subtle aspects of this theme in Innis's work that are often overlooked.

The above saying is usually taken as being the basic credo of an elitist vision of knowledge and education. But in Owen, as in Innis, the interpretation is fundamentally democratic. The saying 'is of serious import, not in so far as it touches the learned few, but in that it condemns us all to share, in however small a degree, the burden of philosophical speculation.'[27] This is because man, whatever his basic ability and in whatever social context, is unable to saturate his mental capacity with perceptual and motor demands. As Owen puts it, 'the eye is satisfied with seeing, and the ear with hearing, but the mind, being not fully occupied by its practical functions, frets over the incomprehensible.' Owen is referring to a dialectic of thought similar to that developed by Innis. But, whereas Owen is dealing with man in general, Innis is concerned with how the historically specific techniques of communication set channels in which perception runs and how, therefore, the thought of the times is related to the structures of the channels of perception.

Owen goes on to say: 'In a sense man has through his power of thought got out of touch with the world he lives in. The delicate machinery by which life becomes conscious is running somewhat loosely in him, and out of the waste energy thus generated he creates a world of thought which he cannot quite distinguish from nor quite identify with the sensible world. He sees through a fog of reflective thought, which subdues everything to its own colour, and where it reveals nothing seems to be concealing.'[28] This is a statement similar in character to the philosophical basis of Innis's communications works. Innis's concept of the biases inherent in media is a way of describing the external or historical determinants that fix the composition of the 'fog of reflective thought' for a particular time and place.

Owen continues: 'For what actually is the Unknown that thus keeps ever beyond his reach? It is surely but the shadow of his own thought, the obverse side of knowledge. It is the Known unknown he is puzzling over, i.e., the Known imagined as apart from its being known. The limitation man frets at and calls the limitation of his knowledge is knowing. Knowing is the impenetrable curtain that hides the Unknown from his sight.'[29] In effect, Owen offers here a philosophical basis for Innis's intellectual strategy. For Innis, the search for knowledge is not the search for final truths, for the Known, but for an understanding of the limitations of knowledge. This is the only real knowledge.

At this point in the argument, the positions of Innis and Owen diverge. Still, even here, we may look at Innis's efforts in the communications studies as an attempt to overcome the central problem of thought as posited by Owen: 'You cannot by thought get beyond thought, for beyond thought is part of thought ... If there is a reality independent of all the forms in which we can apprehend it, thought's discovery that it is undiscoverable is no more than an arbitrary assertion of the superior "reality" of the appearance presented by thought. In acknowledging its limitations it is asserting the validity of its authority beyond those limitations.'[30] Owen rejects, then, the objectivity of an approach that takes as its basis an examination of the limitations of thought. Not surprisingly, he concludes in a way similar to Urwick that 'life in its deeper aspect, life significant, is nothing but life thought about, life as it is rendered by thought. Its problems and mysteries are the fruit – not the failures – of the musings of men.'[31]

A social *science* is impossible. However, a social *art* is important if not objective. Unlike Urwick or Owen, Innis never gave up the struggle for a genuine social *science*. Yet neither did he satisfactorily conclude that it was possible. His efforts may be seen as a long endeavour to develop a position between the artful humanist critique on the one hand and the false science of the price theorists and statisticians on the other. In the end he was to develop a position closer to the former than to the latter. The results of Innis's work and Owen's conclusion are hauntingly similar. Owen writes: 'We have all the facts before us. They do not change or increase. We merely express them differently as we elaborate our mental processes.'[32]

An anecdote from Ernest Sirluck may help to place Owen's influence on Innis in context.[33] Innis used to dine regularly with faculty colleagues at Hart House. (It was an ongoing complaint of Irene Biss that she was unable to enjoy the discussions at these social gathering during the Depression because of Hart House's men-only policy.) At this period, there was much discussion among classicists concerning the use of epic poetry as a technique for the inter-generational communication of the 'cultural baggage' of a non-literate people. Sirluck recalls a stimulating conversation with Owen on this subject, with Innis as a quiet, source-noting witness. Since Innis had contributed little to the conversation, Sirluck was taken aback to see him that same afternoon borrowing from the library all the authorities Owen had cited. When Sirluck expressed his surprise that Innis should be interested in this area, Innis replied emphatically that he thought the subject was of fundamental importance.

On the basis of this anecdote, it seems clear that we would be looking in the wrong place for the influence of Owen on Innis were we to focus on the former's published works. The real influence may have come via oral, not written, channels through lunchtime conversations that can never be recaptured. At the very least, we know that Owen (and other classicists), without being aware of it, provided Innis with a guide to current scholarship in the field. These sources were then tracked down and used as raw material for the elaboration of Innis' 'banks' of quotations, reading notes, and ideas.

A.E. Havelock was another classicist who was interested in, among other things, the techniques of the Homeric epics and their transformation into the basis of Western civilization through the invention of the alphabet. I will argue that Havelock had a profound influence on Innis's communications period that went far beyond that of Owen. Yet in his 1984 address, 'Harold A. Innis – A Memoir,' Havelock denies that this was so:

> To this attempt of mine to reconstruct Innis' intellectual journey I add a postscript on the relationship between his investigations and my own. This is more slight than some may have supposed.[34]
>
> Studies of mine bearing on conditions of orality and literacy in antiquity had not appeared, and when they did, many years later, they were written, to tell the truth, without reference to Innis' work.[35]
>
> The contiguity between Innis and myself seems therefore to have been as much as anything else, a matter of happy coincidence.[36]

Havelock reaches this conclusion by focusing on the timing of his scholarly publications that deal with themes of interest to Innis. Only one of these appeared before Innis's death (*Prometheus Bound* [1950]). As for the possibility that he may have influenced Innis through public lectures, he dismisses the idea as unlikely: 'I recall having given one or two public lectures in Toronto University during the war years, on the oral character of the Homeric epics and the society which produced them, and it is barely possible that Innis attended them. As I recall some other senior colleagues did.'[37] But Havelock overlooks another possibility, namely, that the influence took place through such less formal 'oral' channels as that of overheard conversations of the classicists reported by Sirluck in the case of Owen. Certainly, this type of opportunity was readily available. Havelock taught at University of Toronto for seventeen years (1930–47). In the same memoir in which he dismisses the idea that he had an influence on Innis's work, he outlines a number of occasions on which he had dealings with Innis on issues of some importance: the establishment of the Hart House group of progressive intellectuals; the defence of Underhill; and the recruitment of young economists for the British government at the start of the Second World War.[38] There were also strictly social occasions. 'I recall one pleasurable session with Innis on a summer afternoon in Muskoka – more precisely on the shores of Fairy Lake north of Huntsville; he had brought members of his family with him.'[39] So it appears that Havelock would have had ample opportunity to talk to Innis about his research.

Moreover, it is unlikely that Havelock would have been aware of the extent of his influence on Innis. During this period of social interaction with Innis, Havelock would have viewed him as a rather conservative economic historian while Havelock himself was decidedly on the left and in a different discipline. Given their political differences, Havelock remained clearly puzzled that 'there was a certain liking and respect between us.'[40] If they did talk about ongoing research, it is likely that Havelock would have read this at the time as the polite conversation of a friendly colleague from an unrelated discipline.

In the end, it is Havelock's own understanding of the oral tradition that blinds him to the profound effect his work had on Innis. He says: 'I suppose, if I were asked to add any corrections of my own to Innis' work, it would be to suggest that his moral preference for the oral word is colored by a certain romanticism which history fails to justify. It is all very well to stress the oral component in Greek culture, but after all, it was mainly the alphabet which released the energies of this culture into history. Without this technology, how much would the Romans, not to mention ourselves, have known of the Greek mind?'[41] Havelock's point makes sense regarding history. But, in the present the two of them shared, Innis could have known a great deal of Havelock's mind without this knowledge being passed on by print.

Because he has no inkling of influencing Innis orally, Havelock is surprised that, five years after he left University of Toronto for Harvard, he was contacted by Innis. He reports: 'One morning in the fall of 1951, in Cambridge, Massachusetts, I received, somewhat to my astonishment, a personal phone call from [Innis] to come up to Toronto as soon as I could, to speak to his seminar in the old McMaster building. He did not even make it clear to me precisely what I was to speak about.'[42] Havelock comments no further on the seminar other than to report his shock at Innis's wasted appearance.

In 1952 Innis's collection of essays *Changing Concepts of Time* was published with a preface citing the argument of Havelock's book *Prometheus Bound*. Havelock, in his 1984 address, simply repeats Innis's words without further speculation as to why his material would crop up in the preface to Innis's last essays. Not having been influenced by Innis, Havelock finds it difficult to conceive of influence travelling in the opposite direction to the older, more established scholar.[43]

And yet, in his attitude towards specialization, towards the need for going outside the confining boundaries of his discipline, and towards the relation of ancient to modern, Havelock seems to have mirrored, in classical studies, the same sort of renegade position adopted by Innis in economic history. The similarity in tone and intellectual strategy of the two scholars is immediately apparent in this passage from the foreword of Havelock's translation of *Prometheus Bound*:

> It is particularly unlucky for the classics that administrative convenience, or else snobbish gentility, has tended to accentuate the artificial division between the humanities and the social sciences. This has killed classics as a vehicle for the history of ideas, by encouraging classicists to strike attitudes in defence of cultural traditions, rather than attend to certain continuities between let us say, Aristotle's *Ethics* and modern behaviorism or instrumentalism. If the essays of this present volume seem slight in substance or too absorbed in the modern at the expense of the ancient scene, let it be pleaded in their defense that, so far as they compete with specialists, they aim at only amateur status.[44]

However, Havelock's similarities with Innis went far beyond their attitudes to their respective disciplines. The dominant objective of Havelock's research was to trace the rise of Greek literacy following the introduction of the alphabet and to situate personalities, institutions, and literary works according to the logic of social change that is derived from this understanding of communications innovation. It is not an exaggeration to say that Havelock's work was the equivalent of a detailed extension of Innis's treatment of Greek civilization. At the risk of simplification, it seems that Havelock picked up Innis's concept of media supersession as a focus for looking at Greek literature and society. Innis, on the other hand, appears to have been influenced by Havelock's presentation of the classical dialectic of power and intelligence. This becomes transferred to the communications studies in the form of monopolies of force and knowledge.

The importance of recognizing the genealogy of concepts is brought out particularly well in comparing the two men's work. If we were to summarize the theme that Innis was dealing with in the communication studies, it would bear strong resemblance to a subject that the intellectual left has been concerned with under the name of reification: how the products of man come back to him in an increasingly horrifying and alien form the more his level of knowledge and material productivity is extended. However, this theme seems to have developed not from a dialogue between Innis and the scholarly left but from classical studies, where it is present in mythic form. The unlikely source of this theme is found in Havelock's translation of *Prometheus Bound,* which he subtitled *The Crucifixion of Intellectual Man*.

During his last years, Innis included this work in the reading lists for his economic-history course. In the preface to his book of essays *Changing Concepts of Time,* he summarizes their theme by saying that 'the problems of understanding others have become exceedingly complex partly as a result of improved communications.' He goes on to point out that:

> this general argument has been powerfully developed in the *Prometheus Bound* of Aeschylus as outlined by EA. Havelock in *The Crucifixion of Intellectual Man* (Boston, 1951). Intellectual man of the nineteenth century was the first to estimate absolute nullity in time. The present – real, insistent, complex, and treated as an independent system, the foreshortening of practical prevision in the field of human action, has penetrated the most vulnerable areas of public policy. War has become the result, and a cause, of the limitations placed on the forethinker. Power and its assistant, force, the natural enemies of intelligence, have become more serious as 'the mental processes activated in the pursuit of consolidating of power are essentially short range.'[45]

It is important to recall these elements, isolated by Innis as significant ones, in Havelock's argument as we turn to an examination of that argument in more detail. Havelock begins by examining the paradox that man's mastery of the scientific method has exploded the limits of time and space in which he lives to the extent of bringing on a horror at the insignificance of all human values. According to Havelock, the relevance of examining the ancient myths is that they present, in an encapsulated and powerful form, the essence of the circumstance of modern man. 'It is as though the collective consciousness of the human species were a continuous thing, living outside the confines of time, able to guess the dark meaning of its history before that history has been realized.'[46] Whether or not Innis would have agreed with this general statement, it is clear that he viewed the purpose of the study of the classical past in the same way in which Havelock did: as an attempt to develop an overall perspective that might counter somewhat the myopia of the modern mentality. Havelock's introduction to *Prometheus Bound* is a scholarly attempt to do just that through an interesting counterpoint of modern circumstance and ancient myth.

As Innis studies innovation in communications technology, Havelock studies innovation in myth. He interprets the great myths that arose in the classical period and links them with the advent of science. The first is the Eden myth, in which the bitter truth of the tree of knowledge is seen as the scientific perspective that first allows man the breadth of vision necessary to realize his own insignificance (nakedness). The second, and more important for our purposes in that it directly relates to Innis, concerns the legend of Prometheus. Prometheus, the immortal who stole the fire (technology) of the gods and gave it to mankind, is interpreted as the fountainhead of Western civilization. In Aeschylus' treatment, fire becomes an instrument of the applied sciences and the 'Fire-Giver is made over into the figure of the great inventor and teacher and thinker.'[47] The play is seen as a tragedy that is in essence the tragedy of our civilization, for the very source of that civilization – technological advance – ends up undermining its basis. What could have led (and perhaps eventually will lead) to self-conscious freedom ends in the powerless enchainment of the intellect.

At this point, Havelock returns to an examination of the non-mythic dynamics that have culminated in the ennui of modern man. In a phrase echoing Innis and foreshadowing McLuhan, Havelock points out that it was less the advent of entirely new techniques than the application of power on a vast scale that marked the beginning of the modern age. 'The resources [of power] literally extended the power of man's little finger, so that what technology had already devised and used for thousands of years could be multiplied by rapid and continuous operations.'[48] Not the creation of civilization, but its radical and enforced democratization, characterized the age. 'Civilized' ceased to apply to the cultured elite and began to refer to all within the realm of European hegemony.[49]

In Havelock's opinion, the really important development was the advent of the scientific world-view. He traces the advent of this mentality to the development of precision instruments. In striking similarity to Innis, he suggests that the transformation of this present-minded perspective into the common-sense of the times is related to the advent of literacy on a mass scale and that this, in turn, is related to the perfecting of print technology. Again in a fashion that recalls Innis, Havelock suggests that it was the very extension of Western civilization through print technology that led to the unprecedented primitivism of total war. Moreover, he points out that, for the first time, Western culture, having reached an impasse, was not about to undergo a process of regeneration following the penetration to the heart of the empire of marginal peoples untainted by the contradictions of the modern world-view. 'The Goths when they invaded Rome committed no crime against themselves. If they impaired an ancient culture it was in preparation for absorbing it. By way of contrast, between 1920 and 1940, in the cultural complexes first of Italy, then of Germany, culture attempted a form of suicide ... The fact is that these wars of the west [the First and Second World Wars] have been civil wars ... The retrospective eye of history will one day see them for what they were – a tearing apart of a single social and spiritual fabric in the heart of Europe.'[50] The current crisis is viewed as nothing less than the collapse of Western civilization.

Havelock next turns once more to the Greeks because he finds in their culture a proto-vision of the change in mentality that, much later, would occur on a mass scale, ushering in the disasters of modern times. The Greek myth deals on a far more personal level with this new mentality. It links that mentality to the discovery of science and terms it the advent of *intellectual man*. For Havelock, its presence is one of the 'fundamental reasons for dating our culture from the Greeks.' According to Havelock, the Greeks themselves were cognizant of their innovative vision of man: 'They were well aware that intellectual man had not been there in the beginning. They were therefore prepared to isolate his character and influence from the rest of human nature and human history, and to define it. Their vision of him was precise, in the sense that his vocabulary and logic were defined, and also triumphant, in that all culture and leisure and life above the savage level was seen as the creation of intellect, and also concrete ... cultural history was interpreted in terms of tools and technology.'[51] This integral vision of man depended on an understanding of the limits of intellect defined by the state of nature and history. It also posited the critical intervention of man in the world through reason at the level of culture and technology. It was a balanced understanding that rested on a subtle vision of the nature of history:

> Ancient thinkers all assumed that the universe in some guise or other was eternal. But ... they never pressed this notion as far as the vast unmanage-

> able vistas of space and time ... [Thus] they found it easy to think of the existence of the human species from two opposite points of view. On the one hand its history was finite; it had evolved from the animal level within a conceivable period of time ... A portion was spent in a savage, the rest in a civilized condition. [From this point of view, man] was surely ephemeral in terms of eternity.
>
> [However], whenever they sought to formulate the universe, it had man in it. Therefore ... this historical phenomenon of man was itself eternal. They often resolved the paradox by imagining that his history was cyclical. This mythology of history had a double advantage. It broke up eternal and unthinkable duration into manageable periods of time within which man's record could be viewed as a whole. Secondly, a proportion was established between this life of man and the ultimate reality which surrounded him. He was given the advantage on the one hand of a limited, understandable and perhaps manageable destiny, and on the other, of an eternal and significant one.[52]

Havelock suggests that the seeds of scientific mentality that extended the concepts of time and space to infinity in two directions (past and future, small and large) have thoroughly dissolved the cyclical point of view, and that this has led in turn to the collapse of our own sense of eternity. He examines this process at the point at which it became first apparent, in the myth of Prometheus as treated by Aeschylus. His version of the myth will not be a recompilation of the adventures of the gods. 'You know all [that] story. Let me rather relate to you the tragedy of man: How from the silly creature that he was I made him conscious and intelligent.'[53] Aeschylus, in the persona of Prometheus, goes on to enumerate the various gifts involved in the symbolic gift of fire. These include arithmetic and the written language. He sums up the essence both of his contribution to mankind and of its tragedy: 'One sentence short proclaims the truth unique Prometheus gave, what man received, technique.'[54] The gift of fire is the beginning of applied science and is seen as such. In this version of the myth we find: 'In the powers of the protagonist, as they are announced by himself and by others, there are united the invention of the technologist and the fore-sight of the historian'[55] – or, put another way, the modern vision at the level of the applied and social sciences.

Having established that the subject matter of the myth is indeed modern man, Havelock looks for clues to the disasters of the present in the tragedy of Prometheus. Who is responsible for the crucifixion of intellectual man? The answer is unequivocal: power and force, the minions of the supreme political ruler. (This is precisely the point selected by Innis as significant.) 'What the masque tries to say,' argues Havelock, 'is that the sheer will to control and order and compel is itself a historical force in human society and in the human soul, working in constant antagonism to intelligence which it distrusts and despises and crucifies.'[56] In the myth, Prometheus foresees the downfall of Zeus, yet he must suffer despite that knowledge. In the present he has no power to effect immediate change. In the long run, however, Zeus does not have the foresight to avoid his destiny. Prometheus is punished for disobedience but is unrepentant. 'The claim of command and obedience, once it becomes an end in itself, conflicts with all forms of science and all processes of the imaginative intellect.'[57] Zeus is finally destroyed because 'Prometheus closes up the channels of information. This becomes an allegory of the self-defeating character of power systems, which are based on power alone ... power itself cannot endure in the long run without [the] aid [of intellect].'[58] Havelock summarizes the myth and the dilemma of modern man as 'the conflict of intelligence and power.'

> If power is to be treated as an absolute, any diminution, even for temporary purposes, is to be rejected, and hence the elasticity of long range calculation is lost ... Therefore the will to power unless it mitigates itself by other considerations altogether ... cannot calculate at long range ... It cannot postpone an issue and is therefore prevented from pushing through from means to ends, and so to further ends. It therefore cannot take in that increasing area of interest, which converts itself into the area of philanthropy, where the forethinking intelligence is in charge.
>
> Therefore power always corrupts, and absolute power corrupts absolutely.[59]

One can see that there is a striking similarity in the theme and worldview that both Havelock and Innis are dealing with. Perhaps Havelock is right – he did exert little *direct* influence on Innis (although I doubt it) and vice versa. I would still argue, however, that it is important to examine Innis in light of Havelock because they are both coming out of the same milieu and thinking similar thoughts. Clearly, there was a singular intellectual fervour going on among a remarkable group of classics scholars at the University of Toronto during this period. In a smaller, less compartmentalized university, where the opportunities for talented individuals to socialize together were much greater than is the case today, Innis's exposure to this intellectual fervour was a critical factor in allowing him to make the leap to the very different subject matter he was dealing with in his communications studies. He could not have done so if he had been restricted to an intellectual dialogue with his traditional colleagues in economic history.

Sometimes it is the 'small things' that underline the provenance of new sets of ideas. For example, the one and only time that Innis used the same technique as Havelock – drawing on a Greek myth to underline the conundrums faced by modern society – was coincidental to the same period in which Havelock was using a similar expository technique in *Prometheus Bound*. Let us turn to an examination of this exercise in Innis's work, with a view to seeing how such an examination might help us to better understand what he was getting at in the last phase of his career, and how the classicists influenced him in that regard.

Minerva's Owl

Innis chose the myth of Minerva's owl to produce an essay that in many respects is the most comprehensive précis available for the communications works. Innis presented this essay in his capacity as the newly elected president of the Royal Society of Canada, and he used the occasion to expose fully for the first time the precocious sweep of his mature communications theory. The essay contains in a condensed form much of the material from the later chapters of his unpublished magnum opus, 'A History of Communications.' Innis himself underlined the significance of this piece by placing it as the lead essay in *The Bias of Communication*. 'Minerva's Owl' presents us with an ideal subject for tracing some of the linkages between biographical context and theory content mentioned above. For instance, despite the fact that Innis viewed the paper as one of his best communications essays, in its contemporary context it was an abysmal failure. A.R.M. Lower, one of his colleagues in economic history in attendance at the meeting, remembers: '"Minerva's Owl" took flight in the gathering darkness and flew off into the woods, apparently, and disappeared. Well so did his audience! He killed every audience that way, except, perhaps a few acolytes in the front row who were taking down the words of the priest.'[60]

From the start, then, there was a division between, on the one side, the economic historian (Lower) and other established colleagues who did not follow Innis into the new field, and, on the other, younger scholars who learned the new vocabulary without understanding its significance. In other words, Innis was celebrated and intellectually isolated at the same time. However, in spite of its original reception, a certain consensus has developed concerning this paper. Even those who view Innis's later work as excessively ambiguous admit that this presentation was one of his most successful, albeit still abstruse, attempts to plot the relationship between the supersession of media technologies and the fortunes of empires. The essay is seen as a summary of this quintessential 'later' Innis, the media determinist.

It seems odd, therefore, that the commentators have overlooked the obvious fact that the theme of the essay is provided not by superseding media technologies but by use of a minor classical myth, the owl of Minerva. Since Innis rarely used literary devices in his work, we might expect to find something of significance imbedded in this image. A brief reference to its use in classical times and to its resurrection in the philosophy of Hegel will serve as an introduction to Innis's treatment.

To the Romans, the image was a relatively minor one representing the 'familiar' of the Goddess Minerva and in so doing portraying certain of her attributes in a symbolic manner.[61] Minerva was known to the Greeks as Pallas Athena. Given Innis's obsession with the oral tradition and its role as a fount for Western civilization in ancient Athens, it is interesting to note that Minerva as Pallas Athena is indeed the patron goddess of that city. The synthesis of qualities found in Athens that made it a great nodal point for cultural development is reflected in the make-up of the personality of the goddess. Her two main roles corresponded precisely to those two social functions viewed by Innis as essential and complementary prerequisites to empires characterized by a high level of culture. In one role, Pallas Athena appears as a goddess of wisdom: the patron of those who search for knowledge. In her other main incarnation, she takes on the mantle of a warrior goddess capable of defending her domain by the effective use of force. At the symbolic level, she represents a society founded on a coalition of priests and soldiers, of scholars and politicians.

The owl as the familiar of Pallas Athena represents in a pure form one of her component attributes: the concern for knowledge. Innis would have especially appreciated the symbol because it was conventionally used in isolation from Minerva to represent those institutions most devoted to the search for knowledge: the universities. The owl exemplified the importance of academic freedom (the unfettered pursuit of truth in isolation from outside force) in the search for knowledge.

Hegel called on the symbol in a particularly suggestive manner in *The Philosophy of Right*. This is the direct source for Innis's usage and as such bears some investigation. I am interested here primarily in Hegel's emphasis of certain aspects of the symbol rather than the subtleties of the philosophical argument that he wished it to carry.[62] Hegel writes: 'To say one more word about preaching what the world ought to be like, philosophy arrives always too late for that ... When philosophy paints its grey in grey, a form of life has become old, and this grey in grey cannot rejuvenate it, only understand it. The owl of Minerva begins its flight when dusk is falling.'[63]

It is immediately apparent why Innis was attracted to Hegel's treatment. Long before he had run across this passage, he had championed the philosophical approach as the antidote to 'preaching' in the social sciences. Yet the Hegelian symbol is an extraordinarily rich one which goes far beyond this basic meaning. Hegel speaks of a nocturnal bird of prey. The vision of this creature is acute but it is a black-and-white vision of twilight devoid of the full range of daylight's living colours. It is not a timid migratory bird but one that belongs to the night. It gets its sustenance by feeding on carrion or by ripping the heart out of living things.

Such a symbol is admirably suited to Hegel's understanding of philosophy: philosophy flourishes when a society reaches its twilight stage, decadence. The philosopher is not caught up in the daylight colours of a living world. He hovers over a world that is half asleep and feeds on what little life remains. His art is not a creative plastic one but one of an analytic bent that requires for its raw material the cultural accomplishments of a people. His function is to understand a civilization, not to prescribe or create. Accordingly, the philosopher cannot be of any use in reviving a society that is past its prime.

Let us now turn to Innis's treatment of the image in 'Minerva's Owl.' The image recurs three times in the course of the essay. Each time it is used to mark a particular place in the imperial history of Western civilization: first, the original infusion of cultural essence from decadent Greece; second, the long decline of the West, punctuated by periods of cultural florescence achieved under certain empires; and, finally, the imminent death of this cultural heritage in the present. For Innis, the owl represents the central living traditions of Western civilization that can be traced to the oral tradition of the ancient Greeks. Its flight represents the movement of the centre of Western civilization from one people and place to another. Although personnel (refugee intellectuals) and contents (books, works of art) are involved in this flight, it is more properly interpreted as a mindset or mentality that can be transferred only through the person-to-person dialogue of living individuals.

In carrying this meaning, the symbol as used by Innis differs somewhat from that of Hegel. Innis transforms a bird of prey into a bird of passage. This transformation emphasizes, in her absence, the role of the master to whom the bird is a familiar: Minerva, the Roman version of the Greek goddess Pallas Athena. Historically, the symbol is appropriate, therefore, since it leads us back to Innis's starting point, the oral tradition of ancient Athens. But as we have seen above, on the plane of theory, it also suits his purposes since Athena as goddess embodies both wisdom and force in an essential unity. This Innis regards as a necessary combination at the heart of Athens and all its successor civilizations of the West. Without the protection of intelligence (or scholarship) by force (or the political structures), the realms in which the former is operative (the oral tradition and science) are not available for application in their counterpart domains of public opinion and technology where force holds sway.

The flight of the owl, representing one side of the unity embodied in Minerva, symbolizes the collapse of an empire and this correlative project to link intelligence and power:

> With a weakening of protection of organized force, scholars put forth greater efforts and in a sense the flowering of the culture comes before its collapse. Minerva's owl begins its flight in the gathering dusk not only from classical Greece but in turn from Alexandria, from Rome, from Constantinople, from the republican cities of Italy, from France, from Holland, and from Germany ... In the regions to which Minerva's owl takes flight the success of organized force may permit a new enthusiasm and an intense flowering of culture incidental to the migration of scholars engaged in Herculean efforts in a declining civilization to a new area with possibilities of protection. The success of organized force is dependent on an effective combination of the oral tradition and the vernacular in public opinion with technology and science. An organized public opinion following the success of force becomes receptive to cultural importation.[64]

Innis's treatment of the symbolic does not involve a pacifistic utopia that views the element of force as the central antithesis of culture. The owl cannot have an independent existence for long. It may go on scavenging flights, but in the end it remains the pet of Minerva. It must return to the protection offered by the war goddess to rest between flights. In like manner, those who are most cultured, who are most caught up in the search for knowledge, must continually seek out a new protective alliance with a sympathetic but barbarian force. Otherwise, their existence and that of the cultural heritage they carry is in danger. In other words, the conversion, through non-coercive methods, of peoples formerly peripheral to Western civilization becomes the key to the survival of that civilization. By 'non-coercive' I mean that Innis believed that a cultured mind-frame could not be imposed by force, nor could it be mimetic. It had to be absorbed slowly and differentially over a period of time into the common-sense of a marginalized people. They had to work it through for themselves. Innis clarifies this by stressing that the 'cultural importation' symbolized by the flight of Minerva's owl is not the transfer holus-bolus of the cultural heritage of one people to another. On the contrary, the strength of Western civilization depends on the loss of parts of its heritage in the transfer of its centre through time from one society to another. 'Perhaps it may not be for the advantage of any nation to have the arts imported from their neighbours in too great perfection. This extinguishes emulation and skinks the ardour of the generous youth.'[65] To put the matter another way, the oral tradition, as the living tradition of human societies, must be given room to flourish once again in a new context. There is no such thing as an authentic renaissance. 'When the arts and sciences come to perfection in any state, from that moment, they naturally, or rather necessarily, decline, and seldom or never revive in that nation where they formerly flourished.'[66]

And so Innis traces the necessary transfer of the cultural centre of the West from one empire to another, ending with the contemporary United States. Here the symbol appears for the last time but in a far more desperate form: 'Since its flight from Constantinople Minerva's owl has found a resting place only at brief intervals in the West ... [Its] hurried and uncertain flights have left it little energy and have left it open to attack from numerous enemies.'[67] At this point, Innis takes a curious jump and the fate of Minerva's owl is portrayed in an uncharacteristically passionate manner as that of the individual scholar: 'There mark what ills the scholar's life assail, – Toil, envy, want, the patron and the jail.'[68] Immediately he goes on to associate this individual repression of scholars with the most general political and economic factors. 'The Industrial Revolution and the mechanized knowledge,' he says, 'have all but destroyed the scholar's influence. Force is no longer concerned with his protection and is actively engaged in schemes for his destruction. Enormous improvements in communication have made understanding more difficult. Even science, mathematics, and music as the last refuge of the Western mind have come under the spell of the mechanized vernacular.'[69] Innis ends this gloomy scenario for the present day with an appeal to the university as the last refuge of the living or oral tradition, the final perch of Minerva's owl.

In his treatment of this theme, particularly in his collapsing of general theory into the case of the individual scholar, Innis is going beyond the many themes he had in common with Havelock. In essence, he is justifying the final stage of his life's work. Central to his project and to the symbolism of the essay is the idea that Western civilization can be renewed only by intellectual developments on a periphery that in turn becomes a new centre for cultural florescence. Innis is saying that, although these revitalized efforts in the field of knowledge may originate in the decadence of the imperial centre, they can be pursued effectively only if they become allied with a new and independent politico-economic structure or force at the margin of contemporary civilization. In the past, this force may have been styled as a tribe, court, or city, but in our day it is the nation-state. Finally, the general movement of this centre of Western civilization has been traced from eastern to western Europe and now lies in the New World. We can see that Innis's theory moved marginal man, particularly colonial man, to the centre stage of Western culture.

Innis presents a picture in which significant work in either sector – intellectual or political – can be effective in the long run only if it is coupled with work in the other. Yet, paradoxically, he sees both as distinctly different types of endeavours that cannot be pursued in unison. Hence, specific institutions are needed to provide the structural basis for this division of labour. The institutionalization of force is manifest in the army, police force, courts, the bureaucracy, the mechanized workplace, and the political structures of the society. These have as their *raison d'être* the transmission of orders, obedience, and an absence of potentially contestatory thinking. This is why power is always anathema to intelligence. The institutionalization of intellectual work, therefore, is of a totally different order and seeks to guarantee the uninhibited dialogue of the 'best minds' of a society with their peers. Other institutions are concerned with the seeking out, initial training, and guiding of these scholars towards this sheltered environment.

However, even a well-developed pre-university and university system does not guarantee the maintenance of this type of dialogue. The cultural borrowing of the new milieu from the old must be of a limited or selective nature. Without a sense of incompleteness in the cultural milieu, rote learning replaces the oral tradition. Thus, the potential for development of a new and distinctive perspective is aborted. The contemporary crisis is traced to the fact that this sense of incompleteness is less and less possible given the increasingly perfect technologies for the transmission of cultural 'baggage': the various modern media of communications.

Valid or not, this is how Innis saw his world. Moreover, he could not have fully developed this perspective without borrowing elements from the work of his colleagues in classics. The vision is an extraordinary one for, by placing intellectual endeavour linked to protection by force as the cornerstone of Western civilization, he was justifying a personal project with ramifications ranging from his most profound thoughts to the national politics of the day. The consequent sense of responsibility was enormous.

The link between what the classicists were working on and Innis's communication studies is more than just an interesting historical footnote. I contend that, unless this link is recognized, it will be impossible to comprehend the world-view that Innis was developing. To realize the extent to which this classical perspective is important to Innis, let us consider the usual interpretation of the communications studies' presentation of imperialism and then juxtapose this to a perspective that takes the influence of classical themes seriously in a re-examination of Innis's work.

Knowledge and Power

The tendency of scholars examining the later Innis has been to trivialize the communications studies by presenting them as an extension of the staples approach, in much the same way that commentators on Innis's staples period have tended to present a superficial view of his analysis of that period by overstressing the determinacy that can be attributed directly to the staples' characteristics. But the profundity of Innis' later works cannot be grasped by positing a mechanical link between a dominant medium of communication and a certain type of empire which is structurally determined by the characteristics of that medium.

While it would be ridiculous to deny any carryover from his staples works into the communications studies, it is impossible to understand his later work without being cognizant of the extent to which it was influenced by the problems and treatments present in the field of classical studies. Innis's work was not concerned just with the succession of media of communications in human society but with the broader question of the interplay of power and intelligence in human affairs. He wrote history as the formation, interaction, and dissolution of monopolies of force and knowledge. The technology of communication, to be sure, mediated the process. But, from this point of view, McLuhan's suggestive terms 'time-binding' and 'space-binding' are seen to relate not so much to media characteristics as to a concern with knowledge, on the one hand, and with immediate power, on the other. This classical dialectic of knowledge and power, rather than the simple causality of media, forms the most secure basis for understanding Innis's treatment of the human condition.

Why, then, has this dialectic not been particularly evident in previous interpretations of Innis? We can account for this partly through the observation that rearticulation of an article or book means presenting the subject matter again, in the form in which it first appeared in print, as an exposition. Doing so often has the effect of mistakenly presenting conclusions as premises and vice versa. This is to some extent what has happened to treatments of the work of Innis. If we concentrate on presuppositions or the philosophical approach that underlies his investigations, we may be able to reverse some of these assertions. For example, the concern for stability found in the communications studies may be related less to the concept of 'equilibrium' in economics and the staples works then to a new element, the concept of 'balance' as brought to the attention of Innis through the work of the classicists.[70]

A comparison of the two phases of Innis's research produces an interesting parallel pattern of regressive and progressive analysis of subject matter, clarifying how a confusion of conclusions and presuppositions could come about. The symmetry of this research pattern helps to explain why, both in terms of tendencies in Innis's exposition and in terms of the résumés of his commentators, the communications studies are presented as the staples approach writ large, with metropolitan media replacing hinterland exports as the subject of analysis. From this perspective, the principal communication medium (clay, parchment, paper) and its production tools (stylus, brush, pen) become the point of departure for the analysis. The relative durability of the medium under analysis virtually determines the make-up of the corresponding empire. The relative weight of the medium is also men-

Table 5: Characteristics of empires[71]

With durable media (stone, clay, parchment)	With non-durable media (papyrus, paper, electronic)
Heavy media	Light media
Biased towards time	Biased towards space
Duration	Expansion
Oral tradition	Written tradition
Continuity	Discontinuity
Poetry	News
Religious government	Secular/military government
Language	Armed force
Spiritualism	Materialism
Communal man	Materialism
The ear	The eye
The land	The sea
Decentralization	Centralization
Hierarchy	Democracy

tioned, but, since this is usually related to its durability, the analysis is virtually one-dimensional. The result is a deterministic presentation in which heavy but durable media such as stone, clay, and parchment are seen as the basis of empires that do not control a great deal of territory but tend to last through time. On the other hand, light media that are subject to decay produce empires that control great hinterlands but are short-lived.

The drift of this type of analysis is extremely mechanistic since the structure of empire at many different levels is related directly to the relative durability of the dominant media. As set out in table 5, we find that there are two constellations of imperial characteristics grouped around durable and non-durable media respectively.

Although virtually all commentators look at Innis's communications studies in terms of this one-dimensional but multifaceted dichotomy, some emphasize an aspect other than durability as the dominant one from which all others follow. For example, McLuhan stresses the relative importance of perception by the eye or the ear as being the determining factor. But, whatever aspect is considered dominant by different commentators, the distinction between the two constellations of imperial characteristics is seen as the essence of Innis's thought.

Indeed, this dichotomy often does appear in Innis and in precisely this form when he is making *general statements* about empires. However, the dichotomy must be looked on as a compound of the philosophical approach that underlies his analysis rather than a rigid set of opposites that he applies to history deterministically. This general dichotomy relates to Innis's analysis of historical cases in much the same way that Marx's schema of modes of production relates to his analysis of concrete social formations. As in Marx, Innis 'breaks the rule' of his own schematic dialectic wherever historical evidence dictates.

To understand this tendency to conceive of Innis's communications studies as a rather crude and deterministic approach, we must keep in mind the relationship of the communications studies to the earlier staples studies. In the case of hinterland analysis, it is quite possible for a conception of the characteristics of the succession of different resources demanded by the metropolis to provide the focus for analysis. By definition, an understanding of the hinterland context revolves around a conception of imbalance, or disequilibrium, or dependency. However, when we turn to an analysis of the centre of the empire or the empire as a whole, then *the balance of staple products* required for autonomous growth becomes the focal point. Without this balance of various staples, the empire fails in the face of less vulnerable competition. Innis himself was always acutely aware of the necessity of a more complex, multidimensional approach in focusing his attention on the fortunes of empires as a whole. His treatment of the supersession of the Spanish and French empires by the British in the North Atlantic and in North America is testimony to this fact.

In summary, then, at the time that Innis turns, in the communications studies, to consider the history of empires, two tendencies are at work in his thought. On the one hand, carrying over from the hinterland analysis of the staples works, he tends to stress at the level of *theory* the *predisposition* of the *individual* medium towards formation of a certain type of social structure. On the other hand, since he is now dealing with the history of an entire empire, he emphasizes, in dealing with *concrete historical cases*, the necessity of a balance of *various media* whose predispositions complement each other to make for a successful imperial project. While both concerns are present to a limited extent in both periods of his academic endeavours, it is legitimate to make the general assertion that in the staples period Innis is primarily dealing with 'cyclonics' or radical instability while in the communications studies period he is concerned with imperial stability or balance. The addition of the concept of 'balance' is absolutely essential in his switchover to metropolitan analysis.[72] This is why an appreciation of classical studies was so important as a prerequisite to the new stage of Innis's research. The concept of balance he uses is not primarily one of economic equilibrium. It relates more closely to the balance of power and knowledge of the classicists. Consider, for example, the following statement: 'The sword and pen worked together. Power was increased by concentration in a few hands, specialization of function was enforced, and scribes with leisure to keep and study records contributed to the advancement of knowledge and thought. The written record signed, sealed, and swiftly transmitted was essential to military power and the extension of government. Small communities were written into large states and states were consolidated into empire.'[73] Each empire is assessed in terms of how the existence of various media complemented one another to produce a solid imperial project in terms both of political administration (power) and of cultural florescence (intelligence).

This interpretation has definite implications for how we view Innis's political stance. Innis was *not* an anti-imperialist in the sense of having a prejudice against large-scale empires. On the contrary, he felt that the balanced empire represented what was best in human achievement. However, he *was* an anti-imperialist in the modern sense of being committed to opposing the imbalance (in the form of military expansionism) of contemporary empires. In many regards, the presuppositions of Innis's arguments are exactly the opposite of those of the modern analysts of empire. Both liberals and Marxists usually see the essence of imperialism in some form of imbalance (cultural, in the case of the liberals; structural, in the case of the Marxists) at the centre of the empire. Innis, in contrast, assumes that the great empires catalogued in history were possible precisely because of some sort of balance. Innis's view of empire tends to be made over to conform with the modern orthodoxy by stressing the dominant medium and its characteristics and hence the imbalance of various empires. For Innis, this process of linking a medium with certain characteristics is only the first stage, or the theoretical step, taken in an analysis that seeks to explain the complementary relationship of various media. Innis's basic *assumption* is that mutually complementing media are part of the essential nature of all successful empires.

In brief, Innis starts with the fact of empire, assumes balance, and works back to an examination of the complementary nature of media in that empire. In contrast, we tend to think that Innis's method in the communications studies is to start with the medium and its characteristics, assume imperial imbalance, and work back to the structure of the empire. The distinction is important, not only in terms of the consequent flexibility with which Innis treated the various aspects clustered around the basic dialectic of empire, but also in terms of his efforts to establish a degree of objectivity in the social sciences. The following passage summarizes Innis's view of the fundamental problem of objectivity:

> The significance of a basic medium to its civilization is difficult to appraise since the means of appraisal are influenced by the media, and indeed the fact of appraisal appears to be peculiar to certain types of media. A change in the type of medium implies a change in the type of appraisal and hence makes it difficult for one civilization to understand another. The difficulty is enhanced by the character of the material, particularly its relative permanence. Pirenne has commented on the irony of history in which as a result of the character of the material much is preserved when little is written and little is preserved when much is written.[74]

Innis goes on to say: 'I have attempted to meet these problems by using the concept of empire as an indication of the efficiency of communication. It will reflect to an important extent the efficiency of particular media of communication and its possibilities in creating conditions favourable to creative thought.'[75] The concept of empire he is referring to as the key to the problem of objectivity is precisely the idea of balance. As he puts it, 'concentration on a medium of communication implies a bias in the cultural development of the civilization concerned either towards an emphasis on space and political organization or towards an emphasis on time and religious organization. *Introduction of a second medium tends to check the bias of the first and to create conditions suited to the growth of empire.*'[76]

If this balancing of media is assumed to form an essential precondition of empire, then we can escape the (written) bias of our own time in pursuing our social-science investigations. We can do so by assuming that, for each successful empire, there exists a shadow system to the communications medium that at first sight appears most important to us. The 'dominant' medium may appear so to us because many examples of it have survived to our epoch or because it is more in tune with our present-day mentality or 'bias.' However, since we know (or assume) that by itself it does not explain the dynamics of empire, we are able to look for its complement, that is, the medium or tradition that was counterposed to it to bring about the imperial balance. Thus, while we are perhaps unable to appreciate the importance of the oral tradition in history by direct means, we can realize its importance by looking at the extent to which it counterpointed the written tradition in successful imperial projects. This was the essence of Innis's argument concerning objectivity in social-science work. The presumption of balance in the historical examples of empire was the crucial starting point of his work.

However, we have so far dealt merely with the infrastructural elaborations of the solutions Innis was proposing to the problem of objectivity. This balancing of complementary media is only one-half of the Innisian dialectic of empire. Besides 'efficiency of particular media of communication,' Innis also mentions 'creating conditions favourable to creative thought' as a prerequisite to empire. This also becomes an 'objective' probe used to investigate the superstructural elements of imperialism. In other words, Innis assumes that the fact of empire indicated at an infrastructural level the balance of media with complementary characteristics, and at a superstructural level the continued application of 'creative thought' by the power elite of the empire. The paradox of empire as seen by Innis is that power and intelligence are antithetical, and yet, at the same time, both have to be associated in the articulation of a successful imperial project.

Innis's presentation of imperialism is characterized by a double dialectic involving an historically cyclical movement relating force and intelligence to developments in communications technology:

> The relation of monopolies of knowledge to organized force is evident in the political and military histories of civilization. An interest in learning assumes a stable society in which organized force is sufficiently powerful to provide sustained protection. Concentration on learning implies a written tradition and introduces monopolistic elements in culture which are followed by rigidities and involve lack of contact with the oral tradition and the vernacular ... This change is accompanied by a weakening of the relations between organized force and the vernacular and collapse in the face of technological change which has taken place in marginal regions which have escaped the influence of a monopoly of knowledge.[77]

Instead of presenting Innis's approach as a fairly mechanistic one that links the characteristics of a medium to a type of empire, we have developed a far more complex picture of the method of the communications studies. Basically, Innis was presenting a double dialectic in which media complemented one another and were linked to the political alliance of power and intelligence dominating the empire. It was the influence of the classicists on Innis's research in the 1940s that provided the archtectonics that allowed him to develop this level of complexity and subtlety in his analysis of imperial dynamics.

CHAPTER NINE

Time, Space, and the Oral Tradition: Towards a Theory of Consciousness

A human being is part of a whole, called by us 'Universe,' a part limited in time and space. He experiences himself, his thoughts and feelings, as something separated from the rest – a kind of optical delusion of his consciousness. This delusion is a kind of prison for us, restricting us to our personal desires and to affection for a few persons nearest us. Our task must be to free ourselves from this prison by widening our circles of compassion to embrace all living creatures and the whole of nature in its beauty.

Albert Einstein[1]

The traditional view of Innis as a media determinist is deficient because it badly misrepresents his political position. A belief that the characteristics of a medium directly determine the fate of the empire with which the medium is associated seems tailor-made for a totalitarian spirit. This kind of rigid determinism calls up the image of a pessimistic anarchist railing against technological inevitability. Innis has been cast as both totalitarian and pessimistic, but these descriptions are fundamentally flawed.

In the communications works, Innis was engaged in a precocious attempt to rejuvenate the badly battered ideological foundations of liberal, pluralist society. In rejecting totalitarian world-views, he sought to develop a perspective that would reconcile the continued relevance of the role of the thinking individual with contemporary imperial structures of unprecedented range and centralization.

Much of the misunderstanding of Innis stems from the chronic ambiguity of his work, which in turn can be traced to his lifelong aversion to dealing with theory in a direct manner. In chapter 7 we saw an

Table 6: Innis's view of empire

State (directing alliance)	Monopoly of knowledge	Monopoly of force
Culture (the 'commonsense' or everyday way of thinking of society)	Spiritualism	Materialism
Infrastructure (communications technology)	Relatively 'permanent' medium	Relatively 'perishable' Medium

ingenious consistency in the communications works in that the exotic methodology employed in their composition illustrates one of the main points they transmit, a point popularized in McLuhanesque shorthand as 'the medium is the message.' To put this another way: the substance of the communications works is to be found not so much in the reams of facts they contain as in the manner in which these facts are presented. This suggests that Innis's 'theory' of communications, being consistent with itself, is found imbedded in the structure of his later works. Let us examine how this could be so.

We have seen how the influence of the classicists allowed Innis to avoid facile causality and to develop in its place a complex dialectic of empire. The components and levels of this dialectic may be presented in schematic form, as in Table 6. Innis believed that any empire indicated, by its very existence, the presence of a relatively balanced mix of these six elements. By 'relatively' I mean in comparison with contemporary competing conceptions of empire put forward by internal groups (subversion) or by external sources (the aggression of neighbouring states).

Innis's presentation of imperial history often appears confused because he hops with apparent abandon from one level to another of this dialectic. He does so to emphasize the interrelated nature of the various components and to avoid being accused of believing in simplistic causality. The weakness of the approach is that it leaves us with no clear idea of how the various levels *are* related. He was attempting an ambitious synthesis of a theory of politics or imperialism (described in chapter 8), a theory of consciousness (outlined in this chapter), and a theory of technological change (explained in chapter 10), but he did so without addressing theory directly at all. Instead, he transmitted the new synthesis by the manipulated presentation of masses of historical facts.

Innis's Synthesis

Commentators readily identify a theory of imperialism and a theory of technological change in Innis's work. In fact, they usually subordinate the former to the latter. Innis's concern with working out a theory of consciousness, however, is not so readily apparent. To understand why this is so, we must reconstruct Innis's logic of empire. In his version of political theory, the healthy empire was by definition pluralistic. Empires needed to resolve two general types of problems: duration or continued existence through time and integration or continued existence through space. To address these problems effectively, the empire had to incorporate a directing alliance consisting of two types of institutions (religious and political), each particularly adept at meeting the challenge of either duration or integration (especially during periods of expansion). The very existence of empire indicated a relative degree of pluralism.

The theory of technology can easily come to dominate this theory of imperial politics as soon as we turn to consider the nature of the executive activity undertaken by the directing groups of the empire. The communication of their decisions must be carried out within the limits of the characteristics of existing technology. Beliefs can survive over the long term only if a relatively permanent technology can be used to extend them beyond the range of personal memory. Similarly, extensive expansion through military action and the subsequent integration of new regions within the empire can be accomplished only if the means exist for the relatively rapid communication of orders. The current state of communications technology within the empire determines the potential of that empire.

Innis attempts to break out of this technological determinist bind by introducing a theory of consciousness that places an emphasis on marginal individuals within the imperial system. To put it another way, he asks the question: How do the individuals who make up the directing alliance of intelligence and force come to think in the manner in which they think?

Innis never develops a definitive answer to this question, but it is clear that merely posing it was crucial in allowing him to avoid the pitfalls of a more technologically determinist approach. The question reinforces his conviction that innovation in communication technology should be a focal point in imperial analysis, for it underlined the fact that media not only determine the limits of executive activity but also are a major factor in determining the limits of contemplative activity (or the decision making that precedes action). He believes that the media employed by an empire shape not only how efficiently the directing groups can command but also what sort of orders they would choose to issue given their mental outlook.

At first glance, this introduction of the question of consciousness, or the formation of ways of thinking, into Innis's synthesis of political theory and technological change seems to result in an even more extreme position of media determinism. Indeed, it does, if we limit ourselves to an examination of existing imperial projects. Every imperial project tends more and more to rely on its existing dominant technologies. As a result, the members of the directing alliance become more self-referring in mentality (or more 'monopolistic' or unconscious in their decision making). In time this leads to their choosing their own 'method of suicide' – that is, they make the types of decisions that accelerate the collapse of their imperial system.

Innis's approach is much less deterministic and more sophisticated. He realizes that the monopolization of certain types of technology and certain manners of thinking by the directing alliance of an empire has the corollary of leaving new innovations in military, production, and communications technology open to 'marginal' individuals or groups – those who are inside the imperial society but excluded from its directing alliance or those who lie just beyond the borders of empire (the barbarians). The marginalized individual has the ability to develop the type of critical consciousness increasingly absent in the directing alliance for two reasons: first, he is less caught up in the orthodox world-view (and its myopias) diffused by the directing alliance; and second, he can manipulate alternative communication technologies to help him to reinforce his independent perspective on events.

Again we find ourselves caught up in the terrible symmetry of Innis's position. He is writing himself, as a peripheral intellectual, into the key critical position within this theoretical structure. He is doing so not egotistically, however, but to restore value once again to the role of the thinking individual in an age of massive imperialisms. His communications studies are imbued with a liberal and pluralist stance in that they are based on two key assumptions. Innis assumes that healthy empires must contain a relative degree of *pluralism* within their directing groups. He also believes that this degree of pluralism could be retained only if the empire is periodically revitalized by the addition of new technical innovations and perspectives developed by *individuals* whose ability to think things through stems from their marginal position within imperial society. The only alternative to this renewal process is the total collapse of the empire in the face of a new imperial alliance coupling external 'barbarian' force with marginalized intelligence from the old imperial centre.

The extent to which Innis's analysis combines a theory of consciousness with a theory of communications technology and a theory of imperialism has been overlooked partly as a result of the categorization of his work as either 'early' or 'late.' The aficionados of the early Innis are for the most part economic historians who discount the later works by interpreting them as an unjustified extension of a staples determinist approach. Within this group, there is no room for consideration of the 'soft' element of consciousness. In fact, this tendency has been exaggerated as a result of the appropriation of the heritage of the communications works by Marshall McLuhan. McLuhan's contribution is precisely to emphasize the effect of communication technology on sensory organization: that is, to link technology and consciousness. In so doing he excludes the more political perspective that is made possible when a theory of imperialism is retained as a third component in the intellectual synthesis. To counter the tendency of McLuhanism to project a facile determinism of the media over sensory organization, devotees of the early Innis present his communications work as an equally facile determinism of media over imperial fortunes. To do this they portray Innis as not being interested in the question of individual consciousness.[2]

This orthodox version of Innis's communications theory could not have developed, however, if his original works had not lent themselves to an under-emphasis of the presence of a theory of consciousness in his synthesis. Clearly, Innis felt more at ease dealing with the institutional and technological levels in his examination of change, which is not surprising given his background in economic history rather than psychology, to say nothing of his extremely shy personality. What is also overlooked, however, is his remarkably persistent return to this level of sensory analysis in the communications works.[3] In fact, the direction of Innis's work at the time of his death was more and more focused on questions of changes in consciousness and their relationship to media changes.[4]

The main reason that the question of consciousness is not raised more explicitly in the communications works, however, has to do with the extent to which it was implicitly present. For instance, Innis's concern with the scholar is primarily a plea for the maintenance of a personality so far outside the power structures of a society that his manner of thinking can be significantly different from that type of consciousness directing and pervading society. The myopias of the everyday consciousness of a society can be assessed and criticized only if the culture of that society retains an institutional space for those with a different manner of thinking. Again, this should not be thought of abstractly. Innis lived out his theory of consciousness both in his insistence that the university not become caught up in immediate solutions to social problems and in his personal aversion to the popular media.[5]

Furthermore, Innis was not greatly troubled by the connection between personal and collective consciousness. A part of the Baptist heritage that he retained after his rejection of the faith was the tendency to equate institutional dynamics with the sum total of activities of the individuals associated with the institution – the Baptist congregationalist approach. Having sketched the manner of thinking present in an imperial culture, Innis would have viewed it as redundant to launch into a study of individual consciousness. In this regard, the communications works are the practical extension of his observation in 1935 in the debate with Urwick concerning objectivity that 'the habits or biases of individuals which permit prediction are reinforced in the cumulative bias of institutions and constitute the chief interest of the social scientist.'[6]

'Things to Which We Attend'

The cultural aridity of Innis's rural, 'colonial' childhood made his career take on the appearance of an extended intellectual detour through which he sought to make up for his insufficient early scholarly formation before returning to the most profound areas of knowledge to which he had been exposed as an undergraduate. The communications works, then, should be seen not so much as a radical departure by Innis as a return by him to the 'philosophical approach' of his early mentors, in particular James Ten Broeke. Innis's production of essays on law in the last few years of his life certainly represents the flowering of a personal interest in the subject following thirty confidence-building years in scholarship. This process of the turning of initial weakness into strength was one of the key reasons that Innis believed the role of the peripheral intellectual to be crucial. To take the case of law, if Innis had been more intellectually prepared in 1918, he would have taken a law degree instead of staying at Chicago. His career then would have been that of a more or less successful practising lawyer. As such, he would never have produced the wide-ranging analysis of law found in the communications works.

A similar observation holds true for the topic of consciousness, although this topic is scattered throughout the communications works without being concentrated in particular essays. The question of consciousness inspired Innis during his early studies but, as in the case of law, it was a subject he could confidently revisit only after thirty years of intensive scholarly work. Without a recognition of the early presence of this topic in Innis's background of colonial education, it is difficult to account for its sudden (re)appearance in the communications works.

Innis himself clearly notes the connection to his early schooling of his concern with consciousness in his communications essays. In the preface to *The Bias of Communication*, he writes: 'In a sense [these papers] are an attempt to answer an essay question in psychology which the late James Ten Broeke, Professor of Philosophy in McMaster University, was accustomed to set, "Why do we attend to the things to which we attend?" They do not answer the question but are reflections stimulated by a consideration of it. They emphasize the importance of communication in determining "things to which we attend" and suggest also that changes in communication will follow changes in "the things to which we attend."'[7] It may be a measure of Ten Broeke's importance to Innis that he saved the rough lecture notes from this course without preserving his own finished essays on psychology topics.[8] In any case, it is fortunate that he did, for the notes help us gain an idea of the long gestation period of a wide range of topics having to do with consciousness and sensory perception prior to their resurfacing in the communications writings.

Ten Broeke was a Christian scholar who plunged into the apparently atheistic speculation of modern science in the firm belief that, in the end, science would be reconciled with Christian faith. During this period, Ten Broeke was himself looking for answers, not furnishing his students with facile rationalizations. Hence, Innis's lecture notes contain as many questions as they do assertions. The concept of bias which Innis would later develop is merely a reformulation of the mechanism by which 'attention,' in Ten Broeke's terminology, allows us to select particular items out of the general state of 'consciousness.' The notes indicate that Ten Broeke repeatedly addressed these themes in his lectures. He also discussed the characteristics of attention, including 'duration' and 'scope': 'Why attention shifts – other things are demanding of the organism the same energies of adjustment.'[9]

Innis's later substitution of 'media' for 'other things' in this sentence produces the critical insight of the communications works. His student notes suggest that there are various levels of activity that are more or less conscious, ranging from species (or inherited) to social (or learned) to individual (or willed) activities in order of increasing self-control. This approach links individual psychology to collective and historical phenomena by putting forward the idea of neural patterning through which initially free-willed activities are reduced to the status of routine habits.[10] To put this in the terms of the later communications works, a 'conscious' innovation in media technology can have the unwitting long-term effect of changing the general structure of consciousness.

Innis's undergraduate notes examine many themes that are important in his later works: memory versus imagination, concept versus percept, the role of language in conception, the characteristics of the various senses and their evolutionary development (touch, taste, smell, hearing, sight), and the localization of sense perception in brain structure. In Ten Broeke's lectures, just as in Innis's later essays, we find 'time' and 'space' as sensually constructed, changeable concepts replacing the traditional and rigid Newtonian categories. Following upon these themes is a description of the growth of acuteness in space perception through the various senses, ending with 'Function of Space Perception – "thingness" "thinghood,"'[11] and then a discussion of the human perception of time. In general, after a careful review of Innis's undergraduate notes, one finds that Ten Broeke's concern with space/time perception is as an important factor in Innis's thinking on imperialism and changes in communications technology.

The Problem of Space and the Plea for Time

Innis's farming family might have agreed with him, though they would never have been able to formulate it this way, that 'space and time, and also their space-time product, fall into their places as mere mental frameworks of our own constitution.'[12] Their conviction that their status was socially constructed rather than permanently fixed was complementary to Innis's equally intense belief that time and space were the fundamental conceptions in the make-up of human consciousness. This is the basic reason that Innis chose communications as the focal point for his study of historical change. Communications technology, far more obviously than other forms of technology, directly intervenes in the structuring of time-space relations.

In order to talk about political or imperial dynamics at a profound level, it is necessary to recognize that the most significant strategies have to do with the controlling and restructuring of these central concepts of space and time by the directing groups of a society. In this regard, Innis's terms 'monopoly of knowledge' and 'monopoly of force' refer to the institutional formations within a polity that appropriate for themselves the right to construct the way in which time and space are perceived by ordinary members of the society. This vision of long-term political dynamics may seem akin to a rather cheap version of politics-as-brainwashing until we recall the profound character of time-space conceptions. In essence, the 'brainwashers' are themselves 'brainwashed' or imbued with the consciousness of time and space they are diffusing. The conceptions they espouse must remain relatively instrumental in effectively meeting the demands placed on the empire. In other words, political projects fail not so much because of external challenge (although that may be the immediate agency) but because they become caught up in the self-referring realm of their own thought.

'Monopoly' conveys this idea in a number of ways. What is 'monopolized' is the control over the structuring of space and time. The ruling group, organized institutionally (e.g., the Christian church) and backed by a particular type of communications technology (e.g., parchment), maintains its power by formulating a particular conception of time. It is a monopoly, in the case of the church, because the intricate rationalizations of this conception are mastered and actively discussed by a relatively small group of clerics. However, it is also a monopoly because these sophisticated views, diffused through the institutional framework of the church, come to dominate, in a vulgarized form, the common-sense conceptions of the era – so much so that, temporarily at least, any new and potentially competing conception of time is kept from gaining ground. As a monopoly, however, the institution tends to become more elaborate in articulating its conceptions, more out of touch with the prosaic needs of the populace, and more exposed to competition from alternative institutions and conceptions in the marginal areas of society.

Innis again and again returns to the theme that the collective consciousness, or culture, of a particular society is so fundamental that it is difficult to approach directly. There are two reasons for this. First, any society is caught up in its own conceptions so thoroughly that it is unlikely to produce a critical appraisal of them. Second, any observer, such as Innis, is so influenced by the mode of thinking of his own times that he is unable to assess an exotic culture 'objectively.' Faced with these problems, Innis opts for approaching the problem of the development of human consciousness and culture through history *indirectly*, by tracing the relationship of imperial dynamics to technological innovation in communications. Commentary on Innis generally confuses the means for the ends he had intended in the communications studies. He employs the media and imperial dynamics as a way to elucidate the bias or way of thinking of another people in another time. He believes that he has detected a general historical drift in the collective consciousness of the West that goes far beyond either the political realm (the struggles of monopolies of knowledge and force) or factors of immediate technological innovation. His view of the cumulative drift towards the tearing apart of the conception of time and towards the supreme compression of space in the consciousness of modern man is what gives the communications works their apocalyptic tone.

The strength of the communications works lies in their force of synthesis rather than in their original research. Innis's claim to intellectual profundity rests on his ability to produce new combinations of ideas rather than new ideas per se. As in other areas of his research, his dependence on secondary-source material in his treatment of the formation of conceptions of time and space is evident. The very depth and variety of sources dealing with 'time' and 'space' indicates the serious consideration he gave to the question of consciousness. We will now review a few of the most important of these theorists of time and consciousness – James T. Shotwell, Robert K. Merton and Pitirim A. Sorokin, Wyndham Lewis, and Solomon Gandz – and then examine Innis's use of their ideas.

Shotwell and the Discovery of Time

Although Shotwell is neither the most profound nor the most scholarly of the theorists Innis draw upon, he is possibly the most seminal. Innis probably talked to Shotwell about his work in this area during the 1930s when both men were active on the *The Relations of Canada and the United States* series. There is nothing in their correspondence of the period to indicate that this was so, but then Innis's correspondence in general was relegated to business or personal, rather than scholarly, letters. From the evidence of Innis's reading notes and citations, we do know that he read Shotwell. It seems reasonable, therefore, to speculate that he did so at a relatively early date given the close collaboration of the two men. Two of Shotwell's publications, his essay 'The Discovery of Time' (1915) and his book *An Introduction to the History of History* (1923), were of particular significance for Innis's communications works.

Shotwell wrote 'The Discovery of Time' in an effort to correct the space-oriented bias in approaches to the history of human 'explorations.' As he put it: 'The history of civilization ordinarily omits one half of the narrative. We live in a world of time as well as of space, in which our yesterdays, to-days, and to-morrows are as essential for us as land and sea. The process of social evolution has been one of temporal as well as of spatial mastery, for without a chart of our yesterdays we could never have mapped out the fields and built the cities or planned the empires which mark the stages of our advance.'[13]

When Shotwell turns to an explanation of why historians have ignored this crucial area, he speculates that this has to do with the nature of sensory organization. 'The sense of Time,' he explains, 'is really a sense of times, and that is not a sense at all, but the slow product of developing intelligence. A sense of time ... is to be found only where a memory has been keenly disciplined ... where the historical sense has been called into being ... It is no neglected inheritance from primitives, but the slow, and still most imperfect, acquisition of culture.'[14]

From the start, Shotwell is able to take us out of an age in which time is linear, regular, and measured externally to one in which it is dependent on the oral tradition:

> Neither calendar nor chronology was worked out in the first place to discover Time or keep track of it. They have to do with events and the problem of tracing their relationships, not with what lies between them. The one deals with recurring events, the other with those which occur but once. As the number of these increases, however, the individual events tend to lose their individuality and attention is diverted more and more to the general statement of their relationships. So, by way of mathematics we pass to the world of abstractions ... we mark Time by events rather than events by Time.[15]

Shotwell goes on to trace the increasing perfection of this process through the ages. He begins with a rhetorical question that places, as did Havelock, the topic of consciousness and mnemonic technique at the beginning of time-reckoning and of culture: 'Let us try to settle down in the world of Time and Space, as our forefathers had to do, and see how much – or rather how little – of either we could appropriate.'[16] He identifies the changeover from nomadic to settled agricultural communities as the period of fundamental revolution in time conceptions, relating the strength and limitations of these ancient paradigms to the regional milieu in which they developed: 'The floods of the Nile stimulated in Egypt its extraordinary progress in time-reckoning, as the cloudless skies of Babylonia called out its progress in astronomy.'[17] Furthermore, he identifies religious institutions as the focus of progress in early time-reckoning and associates their secular power with knowledge in this specialized field.[18]

In tracing the development of time conceptions, Shotwell outlines mechanisms that are similar to those used by Innis in connection with monopolies of knowledge and the 'Minerva's Owl' effect of progress in paradigms. For instance, he describes the speciality in the field of knowledge of the Babylonian priesthood as 'codification': 'The task of codification was forced upon the inhabitants of a country of shifting races and empires, of continual intercourse through commerce with the outside world, if the heritage of the past were to be made to fit with the contributions of the present; and the priests of Mesopotamia met the task nobly, in the spirit of scholarship. The libraries of the Assyrians bear witness to an activity such as that which codified the Roman law or harmonized the theology of the Middle Ages.'[19] Codification, then, was perfected by a monopoly of knowledge trying to meet the demands of continuity posed by an imperial project rooted in the material conditions of Mesopotamia. However, just as Innis would later do, Shotwell concludes that the very success of the monopoly of knowledge defines its own limitations at revolutionizing its paradigm or mode of thought: '... The more historical the codifier is, the less his work is likely to be of value. For when the sense of the past is strong – as is surely to be the case in all religious compilations, owing to the sacredness of origins – the result is a failure to meet the changed conditions of the present. The result is stagnation.'[20] Shotwell cites examples of this phenomenon of the stagnation at the centre of an imperial project resulting from cultural density, followed by breakthroughs to new conceptions of time in peripheral cultures. These new conceptions arise from the critical transformation of certain aspects of existing metropolitan paradigms. For instance, Babylonia made tremendous advances in astronomy and mathematics. However, it did so within a paradigm in which the moon remained ascendant both as a deity and as a time-reckoner. Shotwell explains that 'the moon [enabled] men to calculate for a given time and plan ahead so as to bring in line, for common purposes, such as hunt or war, the divergent interests of individuals. But what was a blessing to the primitive became a handicap as society developed. For the moon ... does not fit with anything else in the heavens and a month based upon it was bound to run foul of any other system of time-reckoning. We have already seen how the weeks – probably at first just fractions of the month based on a loose reckoning of the moon's phases – broke loose and ran away, as it were.'[21] Yet, according to Shotwell, this critical innovation of fixing on the same seven-day week that has come down to us was the work of a people peripheral to the Babylonian empire: the Jews.

In a similar fashion, he traces the keystone of the Egyptian priesthood's understanding of time in the geographical coincidence of the Nile flood and the heliacal rising of the star Sirius. This led to their escaping the wandering lunar cycle and basing their time conception on the more workable solar year. Yet, having revolutionized the pattern of time conception relative to Mesopotamia, the Egyptians also became inextricably caught up in their own form of knowledge in setting the year at 365 days exactly: 'The calendar year, running ahead a little over one day in four years ... gained a whole year in 1460 ... years, without ... [intervention] to stop the process. The gain was ... almost imperceptible, only ... about a month in a lifetime, – too little to bother about. Any reform would disturb business and religion even more than the retention of the old misleading cycle ... It was left for Augustus to end the agelong blunder by imposing upon Egypt that revised calendar of 365 1/4 days which Julius had himself received from an Egyptian astronomer.'[22] The deficiencies of a monopoly of knowledge in this case could be corrected only when that monopoly had become peripheral to an empire the centre of which had long since moved elsewhere (Rome).

In Shotwell's *An Introduction to the History of History*, we find a more extended treatment of themes closely resembling those of Innis's communications works. Shotwell, like Innis, is much concerned with the question of the relation of history to the historian.[23] In the epoch of the oral tradition, the 'historian' was an epic poet who, through public recitation, made the past come alive again in the present. The basis of history transmitted through this ecstatic, as opposed to didactic, educational process is that the distinction so clear in our modern minds between the past and the recording of the past has disappeared.

Shotwell grasps that the essence of history is the mobilization and presentation of events in a certain order. Since this task falls to the historian, he identifies the key element of the process as the 'attitude' of the historian towards the selection of facts. His treatment of this 'attitude' is close to Innis's later treatment of bias. It is structured by the concepts of space and time that hold sway.

> The human past, to be called history ... must be ... viewed historically, which means that the data must be viewed as part of the process of social development, not as isolated facts ... But a careful reading of our definition shows that we have already passed over into a consideration of history in the truer meaning of the word – the performance of the historian; since it is the attitude assumed toward the fact which finally determines whether it is to be considered as historical or not. Now what, in a word, is this historical attitude? It consists, as we have already intimated, in seeing things in their relation to others, both in Space and in Time.[24]

Also as in the later Innis, Shotwell goes on to stress the difficult but essential task facing the historian of assessing events outside their original cultural context. Yet he defends the validity of a carefully constructed comparative approach that always keeps in mind this problem of cultural perspective. According to Shotwell, the best examples of this type of approach can go as far as suggesting the perspective, or way of looking at the world, not only of historic but also of prehistoric peoples. He writes:

> The historian ... should not lose sight of the fact that phenomena are never quite the same outside of their environment, for the environment is part of them.
>
> However, within broad limits and used with due caution, the comparative method ... can suggest manners and customs and even, a glimpse of the mental outlook of peoples who have kept no history of their own.[25]

By relating history to the development of the fundamental 'mental outlook' of human beings, Shotwell is able to speculate that its origins can be traced back 'to the dawn of memory.' 'This means,' he says, that

> the origins of history are as old as mankind. For the dawn of memory was the dawn of consciousness. No other acquisition, except that of speech, was so fateful for humanity. We have been used to thinking of early history as a thing of poetry and romance, born of myth and embodied in epic. It demands a flight of the imagination to begin it not with rhythmic

> and glowing verse but almost with the dawn of speech ... And from that dim, far-off event until the present, its data have included all that has flashed upon the consciousness of men so as to leave its reflection or burn in its scar. Its threads have been broken, tangled and lost. Its pattern cannot be deciphered beyond a few thousand years, for, at first, the shuttle of Time tore as it wove the fabric of social life, and we can only guess by the rents and gashes what forces were at work upon it.[26]

The metaphor recalls Innis's summary of the essays in *The Bias of Communication*: 'It is assumed that history is not a seamless web but rather a web of which the warp and the woof are space and time woven in a very uneven fashion and producing distorted patterns.'[27]

Shotwell proceeds to develop a rapid survey of how the past was preserved by societies throughout history. He emphasizes the importance of the oral tradition for pre-literate peoples in accomplishing this task. He focuses on the effects of the medium of communication on the nature of history to the extent of recognizing the mnemonic techniques embedded in the structure of ancient poetry. He recognizes that the shift to print technology implied not only the diminishing of the influence of the oral tradition but also the lessening capacity of modern man in the field of organic memory.[28] He also identifies myth and epic as the principal genres that serve as precursors to what we now refer to as history. 'There are ... outside of myth and epic, two indispensable bases for history: writing and mathematics, – the one to record what time would otherwise indifferently blot out, the other to measure time itself in calendars and chronology.'[29]

Turning to the development of mathematics and the measurement of time, Shotwell articulates in an abridged form the same themes he dealt with in his essay 'The Discovery of Time.' He devotes a considerable amount of attention to what would later become a central feature of Innis's communications studies: the characteristics of the media through which history is transmitted. There are so many points in this section from which echoes rebound in the later works of Innis that it is appropriate to quote extensively from it. Shotwell begins by asking why writing did not spread more rapidly and therefore displace the oral tradition at an earlier date.

> One reason is obvious – the lack of paper. Imagine what it would be like if our libraries were stacked with chiselled slabs of stone or tablets of baked clay, if our newspapers were sun-dried bricks. When papyrus, the paper of the ancient world, came to be used in Egypt, the writing changed, lost

its slow, old pictures and became much like ours; and instead of a few walls or stelae covered with hieroglyphs, there were libraries filled with manuscripts. Stone, as a medium for writing, has a double disadvantage; it is not only hard to manipulate, it is practically immovable. One has to go to it to read. The inscription is part of a monument instead of a thing in itself, like the writing on a piece of papyrus. Babylonia never suffered from the handicap as Egypt did; owing to lack of stone it wrote on clay, inferior to papyrus and usable. It is hard to draw pictures or to write with a round hand on clay, so the Babylonian bricks and cylinders were scratched with straight little wedge-like marks.

It takes but a moment's thought to realize how the medium for preserving literature conditions its scope, and its place in society. What is written depends in a great degree upon what it is written on. It is well, therefore, before surveying the early records of history, to examine ... the manner and method of the composition.

Stone and clay, the first two media of Egypt and Babylonia, were ... definitely limited in their possibilities. There was need of a lighter, thinner substance, suitable for carrying around, yet strong enough not to break easily with general use.[30]

Shotwell continues this argument by considering civilizations chronologically. At many levels, his treatment resembles an abbreviated version of *Empire and Communications*. For instance, on the changeover from the ancient to the classical era, he writes:

It is doubtful if the antique world could have developed the classical literatures in all their variety and freedom of scope, had there been nothing better to write upon. Two substances saved the situation, papyrus and leather. Of these two, the latter played little part in the Mediterranean world during classical antiquity. In the Orient, leather was always in use, and in the fourth century of the Christian era that form of it known as parchment superseded everything else. But the *paper* of Greek and Roman times was papyrus. The Greeks were always hampered by the scarcity of papyrus, which they had to import. This partly accounts for the extent to which their literature was cast in form for oral delivery rather than private reading ... There was apparently no great library at Athens, even under Pericles ... It was in the land of the papyrus itself that the first great Greek library flourished.[31]

Shotwell's survey of civilization concludes in alarmist terms similar to Innis's: '[With] the vast and rapid increase in the output of paper in our own day comes an attendant danger to contemporary history ... For the paper made today is the most fragile stuff upon which any civilization has ever entrusted the keeping of its records ... We are writing not upon sand but upon dust-heaps ... The thought is a sobering one to any one who looks back, even in so short and superficial a survey as this, over the fate of other civilizations and the slight and fragmentary traces they have left.'[32] In other words, he identifies the 'present-minded' tendency of modern society, although in a far less profound and generalized way than Innis, as a key weakness of our mentality.

Shotwell's analysis of civilizations also leads him to conclusions similar to Innis's concerning the role of the intellectual. Consider the following passage in light of Innis's conception of the common-sense of a society being dominated by a monopoly of knowledge and his notion of the central but lonely role of the peripheral intellectual going beyond the mental framework imposed by these monopolies:

> The heretics of all ages suffer because the faith they challenge is the treasured possession of their society, a heritage in which resides the mysterious efficacy of immemorial things.
>
> Now it is a strange fact that most of our beliefs begin in prior belief. It does not sound logical, but it remains true that we get to believing a thing from believing it. Belief is the basic element in thought. It starts with consciousness itself. Once started, there develops a tendency – 'a will' – to keep on. Indeed it is almost the strongest tendency in the social mind. Only long scientific training can keep an individual alert with doubt, or, in other words, keep him from merging his own beliefs in those of his fellows.[33]

Shotwell concludes that history can become self-critical only when an approach is developed that marries 'psychic' and 'material' elements in a new synthesis:

> Pure theology or metaphysics omits or distorts the history it is supposed to explain; history is not its proper business. Materialism and economics, while more promising because more earthly, cannot be pressed beyond a certain point. Life itself escapes their analysis. The conclusion is this: that we have two main elements in our problem which must be brought together – the psychic on the one hand, the material on the other. Not until psychology and the natural and economic sciences shall have been turned upon the problem, working in cooperation as allies, not as rivals, will history be able to give an intelligent account of itself.[34]

What Shotwell reached as a conclusion, Innis took as a point of departure for his communications studies.

Merton, Sorokin, and Time

More than any other source used by Innis, Robert Merton and P.A. Sorokin, in their 'Social Time,' emphasize the extent to which 'time' is not an objective 'given' but rather a socially constructed artefact. Shotwell's primary concern is with the manner in which social structures permit an increasingly more accurate measurement of astronomical time. Merton and Sorokin carry this concern one step further. They are able to do so primarily because the discovery of relativity theory in physics had become popularized during the period that followed Shotwell's essay.[35]

Merton and Sorokin point out that, besides the far more contingent attitude towards astronomical time than had previously existed, there are a number of other 'times' in existence that require serious study:

> Concepts of time in the field of psychology are ... quite different from that of astronomy. Time is here conceived, not as 'flowing at a constant rate, unaffected by the speed or slowness of the motion of material things,' but as definitely influenced by the number and importance of concrete events occurring in the particular period under observation. As James pointed out: 'In general, a time filled with varied and interesting experiences seems short in passing, but long as we look back. On the other hand, a tract of time empty of experiences seems long in passing, but in retrospect short.'[36]

The collective expression of this psychological time is 'social time.' Merton and Sorokin describe this as follows: '"The methods which we adopt for assigning a time to events change *when the character of the events changes*, so that time may appear in various guises." Thus, social time expresses the change or movement of social phenomena in terms of other social phenomena taken as points of reference. Time reckoning is basically dependent upon the organization and functions of the group ... Even in those instances where natural phenomena are used to fix the limits of time periods, the choice of them is dependent upon the interest and utility which they have for the group.'[37] Social time is above all instrumental. It is the common-sense attitude of any particular society to the passage, monitoring, and ordering of moments. It is one of the most basic components of any project of social organization for 'all time systems may be reduced to the need of providing means for synchronizing and coordinating the activities and observations of the constituents of groups.'[38]

As such, social time becomes an effective instrument for those efforts in social-science research that are attempting to come to grips with a variety of cultural contexts: 'Social time, in contrast to the time astronomy, is qualitative and not purely quantitative ... These qualities derive from the beliefs and customs common to the group and they *serve further to reveal the rhythms, pulsations and beats of the societies in which they are found*.'[39] Merton and Sorokin are thus responsible for suggesting to Innis the possibility of using 'changing concepts of time' as one way of approaching the analysis of long-term social change involving widely different types of society. They also suggest, in general outline, the nature of the drift in time conceptions through the millennia of human history: increasingly perfect measurement of astronomical time has led to the displacement of 'local times' in a cumulative manner. This has been a prerequisite phenomenon for augmenting the size and sophistication of human societies.

> The social function of time reckoning and designation as a necessary means of coordinating social activity was the very stimulus to astronomical time systems, the introduction of which was made imperative by the inadequacy of local systems with the spread of contact and organized interaction and the resulting lack of uniformity in the rhythms of social activities. Astronomical time, as a 'time esperanto,' is a social emergent. This process was more rapidly induced by urbanization and social differentiation which involved, with the extension of multi-dimensional social space, the organization of otherwise chaotic, individually varying, activities. Local time systems are qualitative, impressed with distinctly localized meanings. A time system aimed to subsume these qualitatively different local systems must necessarily abstract from the individual qualities of these several systems. Hence, we see the important social element in the determination of the conception of a purely quantitative, uniform, homogeneous time; one-dimensional astronomical time was largely substituted for multi-dimensional social time.[40]

In fact, astronomical time has become so dominant that it has imposed a myopia on us. 'Time' has been assumed to equal the modern 'common-sense' of astronomical time as equated with mechanical measuring devices 'external' to human consciousness. 'Time' has come to be viewed as something that is fundamentally 'objective.' By pointing out the subjective, culture-bound nature of this view, Merton and Sorokin provide Innis with a powerful critical insight central to the structure of the communications works.

Lewis: Time and Western Man

Wyndham Lewis was a far more profound time theorist than Shotwell, Merton, and Sorokin, and as a result his influence on Innis was more subtle. No Innis-Lewis correspondence exists of which I am aware. However, Lewis had a more than passing appreciation of the Canadian context. His ancestry was partly Canadian. He was born, in fact, on a yacht off Nova Scotia. On the basis of this rather slim connection of nationality, he spent the First World War as a war artist attached to the Canadian Artillery Corps, the same military formation in which Innis was serving as a private. The experience of war marked him as intensely as it did Innis. During the early 1940s, Lewis was trapped by wartime shipping and banking regulations for almost five years in Canada. Most of this time was spent in Toronto during the same period in which Innis was changing over to the communications studies. It was a miserable time for Lewis, spent in a rundown hotel isolated from the university community. It seems unlikely, though possible, that he met Innis at that time. Whether or not the two men actually met, however, it is clear that Innis owed an intellectual debt to Lewis, for he cited Lewis's book *Time and Western Man* among the sources for his communications works.

Lewis's book is masterful, intimidating in its intellectual scope. In it he traces the development and disintegration of the 'time sense' of Western civilization through iconoclastic critiques of the works of numerous leading intellectuals in a wide variety of fields. Oswald Spengler, Charlie Chaplin, Albert Einstein, Gertrude Stein, Bertrand Russell, Ezra Pound, Henri Bergson, James Joyce, and A.N. Whitehead are all subjected to Lewis's extended critique. Arguments from ballet and psychology, fashion and philosophy, mass movements and physics are marshalled to support his position. It was just the sort of heretical, encyclopedic secondary-source material that Innis enjoyed mining for ideas. Indeed, at many points, Lewis's notions correspond so closely to Innis's that it is reasonable to assume some borrowing of 'philosophical approach' by Innis. At the very least, Lewis's much more philosophical treatment of certain themes must have refined and reinforced Innis's views of these areas. Furthermore, using his scanning, filing, and 'cut-

and-paste' methodology, Innis constructed parts of his text from scavenged and reorganized excerpts of Lewis's book. Let us begin by examining the former, more generalized influence.

Innis summarizes the theme of his last book, *Changing Concepts of Time*, by pointing out that 'different civilizations regard the concepts of space and time in different ways and that even the same civilization, for example that of the West since the invention of printing, differs widely in attitude at different periods and in different areas.'[41] The classicist Havelock is important to Innis because he diagnoses the terminal pathology at the heart of modern society in terms of changes in concepts of time. Lewis is equally important, for, writing twenty-five years earlier, he mounted one of the last sustained rearguard actions against this 'modern' consciousness of time. Lewis himself viewed *Time and Western Man* as

> a comprehensive study of the 'time' notions which have now, in one form or another, gained an undisputed ascendancy in the intellectual world. The main characteristic of the ['time mind,' as I have called it] from the outset has been a hostility to what it calls the 'spatializing' process of a mind not a Time-mind. It is this 'spatializing' property and instinct that it everywhere assails. In its place it would put the Time-view, the flux. It asks us to see everything *sub specie temporis*. It is the criticism of this view, the Time-view, from the position of the plastic or the visual intelligence, that I am submitting to the public in this book.[42]

At first reading this sounds diametrically opposed to Innis's approach. After all, Innis concludes that the modern world is caught up in problems having to do with space and expansion. He believes there has been a 'disappearance of an interest in time.' However, Innis was writing in an era in which the time philosophy criticized by Lewis had become hegemonic to such an extent that intellectual debate in the field had died out. Moreover, Lewis's critique contained a paradox: his insistence that the time philosophers' view of the world was becoming dominant did *not* mean, on being translated into common-sense notions, that an obsession with time (for example, in the form of widespread traditionalism) would follow.

> The profession of the 'timeless' doctrine, in any average person, always seems to involve this contradiction: that he will be much more the slave of Time than anybody not so fanatically indoctrinated. An obsession with

> the temporal scale, a feverish regard for the niceties of fashion, a sick anxiety directed to questions of time and place (that is, of *fashion* and of *milieu*), appears to be the psychological concomitant of the possession of a time-theory that denies time its normal reality. The *fashionable* mind is par excellence the *time-denying* mind – that is the paradox.
>
> This is, however, not so strange if you examine it. The less reality you attach to time as a unity, the less you are able instinctively to abstract it; the more important concrete, individual, or personal time becomes.[43]

Not only are Lewis's conclusions similar to those of Innis, his approach is as well. For instance, Lewis recognizes that the strength of the time philosophers comes from their 'specialist' technique of attack. On the one hand, the specialist approach allows for the extension of the new world-view by small increments and on narrow fronts. On the other hand, this approach defines as illegitimate any generalist attempt to define and criticize this new world-view by examining its overall advance in the many disciplines into which it has carved up the field of knowledge. Lewis is obliged, therefore, to begin his work with the same sort of defence of the generalist approach that Innis was continually making in his communications studies.

> It has been objected ... that this 'occupational' standpoint of mine should not be a starting-point for criticism of things that do not fall within the sphere of that occupation ... It has been suggested, for instance, that as an artist I have tended to imply that mathematical physics should conform to the creative requirements of the arts in which I am exclusively interested: and that I should be better advised to ignore such things, and only attend to what happens in my own field. Now that I should be delighted to do if these different worlds of physics, philosophy, politics and art were ... rigidly separated ... But my conception of the role of the creative artist is not merely to be a medium for ideas supplied him wholesale from elsewhere, which he incarnates automatically in a technique which (alone) it is his business to perfect. It is equally his business to know enough of the sources of his ideas, and ideology, to take steps to keep these ideas out, except such as he may require for his work.[44]

In other words, if one cannot escape bias, one can at least avoid automatically absorbing the biases of pre-existing paradigms in favour of establishing one's own particular point of view.

Just as Innis would do later, Lewis borrows some of the ideas from the philosophy that he was attacking and turns them against that philosophy. Innis was continually troubled by the charge that he was attributing 'causality' to large social dynamics where more 'specialist' investigators remained sceptical. His response was to abandon the direct exposition of 'causes' and to concern himself primarily with the massive cataloguing of 'effects' observed after technical innovations were introduced. As we suggested earlier, this strategy of imbedding causality in the structure rather than the content of his work makes for a far more abstruse but also more intellectually defensible position. In like manner, Lewis observes: 'I am only concerned with ... effects; and I am in that, on the principle indeed of all the most approved and most recent scientific method, thoroughly justified. For it is now quite accepted that all we need deal with in anything is the *effect* – what, for instance, can be observed *to come out of* the atom – rather than what we should commonly describe as the 'cause' of the disturbance. We are authorized, and indeed commanded, to remain sublimely indifferent to what 'causes.'[45] Lewis's philosophical background allows him to attack directly a topic that only quietly underlies Innis's communications works. 'This short historical survey will resolve itself into a history of the ego ... and its partner, consciousness.' Both men set out to examine long-term changes in the manner in which human beings thought about the world. However, while Innis feels obliged to do so through investigations of the technologies through which knowledge is advanced and disseminated, Lewis does not hesitate to tackle thought processes head on. As Lewis puts it, 'the immense influence exerted on our lives by these "discoveries" cannot leave us indifferent to the character of the instruments that are responsible for them – namely, the minds of the discoverers.'[46]

Lewis objects to the perspective propagated by the 'time philosophers' because it is fundamentally 'romantic' and '*the "romantic" is the opposite of the real*. Romance is a thing that is in some sense non-existent. For instance, romance is the reality of yesterday, or of to-morrow; or it is the reality of somewhere else. There is no real 'future' any more than there is a real "past." So according to this way of looking at the matter, the "timeless" view, "romance" would consist in apparent absence, or in a seeming coyness on the part of time.'[47] Lewis argues that what made these intellectuals so dangerous is the resonance that had been set up between their beliefs and the common-sense of liberal-democratic society. Through the application of technique, the esoteric views of a group of intellectuals is transformed into the most generally accepted values of society. The destruction of more profound traditional values is the consequence. For instance, 'advertisement ... is a pure expression of the romantic mind. Indeed, there is nothing so "romantic" as Advertisement. Advertisement also implies in a very definite sense a certain attitude to Time ... Like ... the particular time-philosophy we were considering above ... [Advertisement] is at once "timeless" in theory, and very much concerned with Time in practice ... It is the glorification of the life-of-the-moment, with no reference beyond itself and no absolute or universal value; only so much value as is observed in the famous proverb, *Time is money*.'[48]

The thinker who draws more fire than any other as a source of this new, degraded time sense is Henri Bergson. His conception of 'duration' as the telescoping 'of the past into the present' is particularly repugnant to Lewis: '"Duration" is the succession of our conscious states, but all felt at once and somehow caught in the act of generating the "new," as "free" as Rousseau's natural man released from conventional constraints but with much more élan.'[49] Duration is, to use Innis's words, 'the present-real, insistent, complex, and treated as an independent system ... foreshortening practical prevision in the field of human action.'[50] It is, for Lewis, the exact opposite of 'memory,' the chief tool of the oral tradition so important to Innis: 'Memory, again, stretched out behind us, is a sort of Space. When we cease to act, and turn to reflect or dream, immediately we "degrade duration" into a bastard Space. We make it into an old-fashioned "Time," in short ... Memory ... unorganized, with its succession of extended units, is that degraded spatial-time ... regarded with so much hostility by the inventor of "duration." "Duration" is all the past of an individual crammed into the present.'[51]

Spengler draws much criticism from Lewis for having translated the Bergsonian attitude towards time into practical history:

> Spengler ... crosses all the t's for us and dots all the i's. We see him arraigning Classical Man in his capacity of *Objective Man*, or Plastic Man, and pitting against that sensuous and 'popular' figure ... the *Subjective Man* of the Faustian or present period. We have seen the subjectivism of the 'faustian,' or modern Western Man, associated fanatically with a deep sense for the reality of Time – as against 'Space.' And the Classical Man was so shallow and 'popular,' not only because he based his conception of things upon the immediate and sensuous – the 'spatial' – but because he entirely neglected, and seemed to have no sense, indeed, of Time ... Classical Man – that inveterate 'spatializer' – was in love with *Plastic*. Modern, Western, 'Faustian' man, on the other hand, is pre-eminently interested in *Music*.[52]

Lewis objects not so much to Spengler's categorization as he does to the judgment that Faustian man is superior in all respects to his classical forerunner. To the extremism of this view Lewis opposes the 'balance' of the classical position. He counters Spengler not by choosing the Plastic over Music but by emphasizing the value of perspective that can be gained through the Plastic. Similarly, he does not denounce modern man as Faustian but insists that classical elements persist in the modern spirit and should be fostered in the interests of balance.

Lewis, like Innis, views the essence of Western civilization as the search for balance. He traces this concern with balance to the synthesis that gave rise to the Christian paradigm.

> The sacred books inherited by the Christian European were in two contradictory parts. One was a very 'realistic' account of things indeed – as barbarous and 'Pessimistic' as Darwinian theory – namely, the Old Testament. The other part was the exact opposite: it was an extremely 'idealistic' book of humane injunctions, full of counsels of perfection – namely, the New Testament: the existence of this mad contradiction at the heart of his intellectual life has probably been the undoing of the European. The habits induced by the pious necessity of assimilating two such opposed things, the irrational gymnastic of this peculiar feat, installed a squint, as it were, in his central vision of his universe.[53]

The squint Lewis refers to was the ability of European peoples to engage in the barbaric practices required by militant imperialism while still spreading the cultured moralist world-view that these practices seemed to contradict. The discoveries of science, however, made it no longer possible to maintain this uneasy balance. As Lewis puts it, 'science slowly began asserting itself in contradiction to the dogma of religion. Darwin appeared on the White horizon, and the White Conqueror began regarding himself as a kind of monkey, no longer so very little beneath the angels but wholly of the animal creation. Darwin was like another, and worse, Old Testament. The "New" lost ground daily.'[54]

Lewis criticizes the time philosophy for subverting the sense of balance that lies at the centre not only of Western culture but also of the individual consciousness of Western man:

> It is in the detaching of himself from the personal that the Western Man's greatest claim to distinction lies ... It is in non-personal modes of feeling – that is in *thought*, or in feeling that is so disassociated from the hot, immediate egoism of sensational life that it becomes automatically intellectual –

> that the non-religious Western Man has always expressed himself, at his profoundest, at his purest. That is, of course, the heritage that is being repudiated in the present 'time'-modes ... People are being taught not to reason, to cease to think ... For the essence of this living-in-the-moment and for-the-moment – of submission to a giant hyperbolic close-up of a moment – is, as we have indicated, to banish all *individual* continuity. You must, for a perfect response to this instantaneous suggestion, be *the perfect sensationalist*.[55]

Lewis is criticizing the philosophers of time for something that went far beyond typical intellectual charges of wrong-headedness or corruption of thought. He believes that, as a way of thinking, 'time philosophy' was particularly dangerous because it could dissolve traditional structures of individual personality and shatter the unity of the senses of the human organism. It is worthwhile dwelling at some length on this area, for here Lewis shows himself to be the originator of a central theme that would appear in the thought of Innis and later that of McLuhan. According to Lewis,

> the traditional belief of common-sense, embodied in the 'naif' view of the physical world, is really a *picture*. We believe that we see a certain objective reality. This contains stable and substantial objects. When we look at these objects we believe that what we are perceiving is what we are *seeing*.
>
> In reality, of course, we are conscious of much more that we immediately *see* ... Every time we open our eyes we envelop the world before us, and give it *body*, or its quality of consisting of *objects*, with our memory. It is memory that gives that depth and fullness to our present, and makes our abstract, ideal world of objects for us.[56]

In the age before the technical means existed to mechanize the 'picture,' the manner in which 'body' or 'structure' or 'bias' was given to common-sense was through memory or the technique of the oral tradition. However, Lewis goes on to relate the destruction of this worldview by that put forward by the time philosophers, based on an alternative pattern of technology exemplified by the cinema. The traditional belief in common-sense, he says,

> is, in fact, a picture. And it is this picture for which the cinematograph of the physics of 'events' is to be substituted. It is to be 'taught in schools' (according to Mr. Russell and other enthusiasts); therefore people are to be trained from infancy to regard the world as a *moving* picture ... Having

> said that all thought is 'a movement,' [the] professor-of-action will in future exact that we shall move and physically function before we can say that we have 'thought' or 'seen.' And there will, of course, be no need to think at all, or even to see. For the action will *be* the thought, or the vision: just as a thing *is* its successive 'effects' ... It is a *flat* world, it is almost also a world of looking-glass images.[57]

At the most fundamental level, 'perception' becomes replaced with 'sensation' as the central function in human consciousness.

> Perception ... has no 'date' only *sensation* had that.
>
> Thus, for accurate dating, perception has ... to be abandoned in favour of sensation ... Perception, with its element of timelessness ... smacks of contemplation ... Only *sensation* guarantees *action*. But the system of the 'percept' has been for unnumbered years the material of our life. We have overridden time to the extent of bestowing upon objects a certain timelessness. We and they have existed in a, to some extent, timeless world, in which we possessed these objects, in our fastness of memory, like gods. While we were looking at the front of a house, if we had ever seen its back we saw that back along with the front, as though we were in two places at once, and hence two times. And our infinite temporal and spatial reduplication of ourselves, this long-stretched-out chain existing all-at-once, was our *perpetual self*, which to some extent was a timeless self. It is by way of the mystery of memory, that we reached this timelessness [an] example ... to illustrate the difference between the *sensing of the sensum* and the perceiving *of a physical object*, is this. In reading a book what we notice is the *meaning of the printed words*: *not* the peculiarities of the print or paper. We 'perceive,' that is: we do not 'sense.' With all of the external world it is the same.[58]

Lewis's argument does not imply that the 'peculiarities of print or paper' are unimportant. Rather, like Innis, he is suggesting that their importance stems from the fact that they *are* generally unnoticed and, therefore, ignored.

Lewis continues by associating his argument with a particular sense. He does so not as a champion of that particular sense alone but to stress the importance of the reintegration of the senses.

> This essay was ... to provide something in the nature of a *philosophy of the eye*. That description ... it could be claimed, is the opposite of the truth. Or

rather, it would be the opposite of the truth if you wish to isolate the Eye. For it is against that isolation that we contend. We refuse (closing ourselves in with our images and sensa) to retire into the abstraction and darkness of an aural and tactile world. That sensation of overwhelming *reality* which vision alone gives is the reality of 'common-sense,' as it is the reality we inherit from pagan antiquity ... The eye is, the *private* organ: the hand the public one. The eye estranges and particularizes more than the sense of touch ...The notion of *one* Space, they say, is due to the sense of 'touch': and space is the 'timeless' idea. Space is the 'public' idea. And in order to be 'timeless,' and to be 'public,' it must be one ... It is because of the subjective disunity due to the separation, or separate treatment, of the senses principally of sight and of touch, that the external disunity has been achieved. It is but another case of the morcellement of the one personality ... Its results must be the disintegration, finally, of any, 'public' thing at all.[59]

It is primarily the ear rather than the hand, hearing rather than touch, that Innis employs to make his critique of the dominance and isolation of the visual sense in the modern era. Aside from this, his argument is essentially the same as that of Lewis, as is his conclusion that sensory imbalance and political instability are two sides of the same phenomenon. For Lewis and Innis, the restoration of sensory integration at the individual level is the key to resolving the crisis of modernity. This line of argument leads Lewis to conclusions similar to Innis's on the nature of democracy. He writes that 'democracy' has come to mean

the present so-called democratic masses, hypnotized into a sort of hysterical imbecility by the mesmeric methods of Advertisement. But whatever can be said in favour of 'democracy' of any description, it must always be charged against it, with great reason, that its political realization is invariably at the mercy of the hypnotist.

The sense of *power*, the instinct for freedom, which we all have, would cost too much to satisfy. We must be given, therefore, a dummy, sham independence in its place; that is, of course, what democracy has come to mean.[60]

Lewis suggests that the enforced lack of discussion would improve rather than weaken the general level of social intelligence, which fits Innis's conclusions in 'Discussion in the Social Sciences.' In the work of both men, 'freedom of discussion' is associated, rather than contrasted, with a lack of genuine freedom. Lewis writes that 'people should be compelled to be freer and more "individualistic" than they naturally desire to be, rather than that their native unfreedom and instinct towards slavery should be encouraged and organized. I believe they could with advantage be compelled to remain absolutely alone for several hours every day; and a week's solitary confinement, under pleasant conditions ... every two months, would be an excellent provision.'[61] Underlying these similar points of view vis-à-vis democracy is a commonly held vision of intellectual endeavour and the search for truth. This differs radically from the dominant orthodoxy concerning these matters, described by Lewis as follows:

> The mind which has truth has it so far as various minds collectively contribute their part to the whole system of true beliefs: the mind which has error is so far an outcast from the intellectual community ... Truth means the settling down of individual believings into a social whole and the condemnation of the heretical or unscientific believing ... True knowledge therefore owes its truth to the collective mind, but its reality to the proposition which is judged. The divergences of standard minds from the isolated minds of the victims of error are the mode by which we come to apprehend propositions as true.[62]

Having cited this version of the 'orthodox' scientific method, Lewis continues:

> In contradiction to this theory of a collective truth, our experience shows us that it is always an 'heretical' minority that imposes its truth upon the majority ... Truth ... is always 'heretical': and it is always the truth of a minority, or of an 'isolated mind,' that to-day is regarded as a victim of error, and is found to-morrow to have been possessed, against the general belief, of the purest truth. For, like the space-time God, the truth-bearing individual is always ahead of the rest of the world, although no one could claim that they willed him, and strained towards him, in order to reach his higher level. Rather he *drags them* up by the scruff of the neck.[63]

Their common view of the search for truth reflects the qualms of both men over the unchallenged intellectual pretensions of science. Lewis identifies the same weakness in scientific inquiry as Innis does, namely, the lack of a 'philosophical approach': 'The immense vistas opened up by science on the one hand, and this parochial progressiveness of time-philosophy on the other ... is ... disconcerting in the world of contemporary thought. The physical vastness of the scientific picture, and the emotional or intellectual narrowness of the philosophical generalization accompanying it, is very striking.'[64] The passion of both men on this score is fuelled not by abstract debate but by their first-hand experience of the war. They do not assail the philosophical poverty of modern science so much as the 'human engineering' it permits. This, in turn, is a phenomenon that takes root in the particular conditions of wartime. Both men cite with horror the naive optimism of the United States army psychologists responsible for intelligence testing. These psychologists mainly opined, 'Great will be our good fortune if the lesson in human engineering which the war has taught us is carried over directly and effectively, into our civil institutions and activities.'[65] Lewis's and Innis's objection to this sort of testing stems from the belief that it measures above all the extent to which the individual's thought processes conform to the new time sense. It measures only *presence of mind* and considers this factor to be synonymous with intelligence. In contrast, both men view 'presence of mind' as being far more synonymous with 'war' itself. Lewis writes: 'The Great War and the wars that are now threatened are the result of the historic mind. It is the time mind at work: indeed, it is peculiarly useful to the promoters of war, hence its popularity. It says, "It's time for another war." The fact that world conditions have been completely changed by the "scientific age" means nothing to it ... For the historic-mind is that of the sensationalist gallery or pit of tradition. It "expects" images of "power" and of exaggerated "passion." Then a great deal of red blood and good blue entrails. Then the curtain: a pause: then the same thing over again.'[66] This position gives rise to the same kind of Veblenian populist sentiment found in Innis. Lewis writes: 'The "captains of industry" are of one mind: – The military organization of the vast masses of people militarized during the War must be carried over into "civil life."'[67]

Lewis is a forerunner of the unpopular, 'political' vision of the role of the intellectual adopted by Innis in his paper 'Discussion in the Social Sciences.' He writes:

> In a period of such obsessing political controversy as the present, I believe that I am that strange animal, the individual without any 'politics' at all ... [My work] has been described as 'a hostile analysis of contemporary society,' which no doubt it is; but its 'hostility' had no party-label. Some of the adversaries of my recent book affected to think that I was aiming a blow at human freedom in its pages. On the contrary, I was setting in a clear

> light a group of trivial and meaningless liberties, which ... obstructed freedom ... I showed, that [the] democratic masses could be governed without a hitch by suggestion and hypnotism – Press, Wireless, Cinema.'[68]

Innis was attracted to Lewis because both men shared the vision of the marginal intellectual seeking to renew liberal-democratic principles in an era in which they had been debased by the mass media and the philosophical poverty of science. Lewis is quite clear that the old categories of politics and thought – 'right' and 'left,' 'idealist' and 'materialist' – have little critical relevance in the face of this fundamental change in human consciousness:

> What is the cause of this unanimity, so that 'realists' and 'idealists' merge in the common worship of Time and Change? Is it a political impulse ... implicit in the doctrine of Time and Change? ... Often the liaison is so apparent that it is impossible to doubt that, once engaged ... in that way of thinking, the Time-philosopher does also tend to become a 'Progress' enthusiast of the most obvious political sort. Further, no doctrine, so much as the Time-doctrine, lends itself to the purpose of the millennial politics of revolutionary human change, and endless 'Progress.' Nevertheless we believe that the impulse to this doctrine is the outcome mainly of Science: that it is really the philosophy of the instruments of research.[69]

For Lewis, the problem of breaking with this unprecedented unanimity of doctrine that is at once totalitarian and unconscious is the problem of resurrecting the thinking individual as the central figure in politics: 'As in anything else, all revolutionary impulse comes in the first place from the exceptional individual ... No collectivity ever conceives, or, having done so, would ever be able to carry through, an insurrection or a reform of any intensity, or of any magnitude. That is always the work of individuals or minorities.'[70] Sharing this purpose, Lewis and Innis should be recognized at a profound level as liberal ideologues.

Even some of the most apparent differences between the two thinkers resolve themselves on closer examination into a common concern with the disintegration of that traditional form of consciousness that had served as a basis for the achievements of Western civilization. This is clearly the case with Lewis, who was, after all, writing primarily a philosophical critique of modern thinkers. But what does this have to do with Innis's wide-ranging schema for the rise and fall of empire? We find the answer to this question in a section of Lewis that Innis cites approvingly: 'The rise and fall of Empires ... is only true of a very small stretch of historic time. It is not fundamental; not a fact of the same order as the growth and decay of individual man. With "Cultures" it is the same thing ... The clockwork rising and falling of empires, with the regular oscillations of great wars, plots and massacres seems, according to all unprejudiced information to belong to a certain limited period of history only. Had men never sought to alter things no historical, cyclic changes could ever have been got going, to start with.'[71] In other words, a study of Western empires after the Greeks is to be viewed as a *particular historical case* rather than a universal study. The peculiar nature of Western man's thinking lies behind the imperial ebb and flow of the West. Tracing that ebb and flow is but one way of examining the pathology of Western thought that has given rise to the contemporary crisis. Imperial analysis is a way of getting at the specific nature of Western consciousness. This is why Innis's imperial narrative is persistently punctuated by comments that sound much like Lewis on personal forms of consciousness. Indeed, Innis's strident warning in his final essay, 'The Menace of Absolutism in Time,'[72] quotes Lewis directly:

> What results from the isolation of the space-world of touch and that of sight, is that the pure non-tactile visual world introduces a variety of things to us, on a footing of equality as existing things, which in the world of common-sense ... do not possess that equality. Thus it is that the mirror-image draws level with the 'thing' it reflects.
>
> It may still be objected that this is only a mathematician's technical device ... This would be ignoring ... that these conceptions of the external world are intended to supersede those of the classical intelligence and of the picture of the plain-man. In other words, the 'common-sense,' of to-morrow, it is proposed – the one general *sense* of things that we all hold in *common* – is to be transformed into the terms of this highly-complex disintegrated world, of private 'times' and specific amputated 'spaces,' of serial-groups and 'events' (in conformity with the dominance of the time-factor) in place of 'things.'[73]

By the time Innis was producing his communications essays, Lewis was acknowledging that *Time and Western Man* was a 'fortress' now silent, 'a place where bats hang upside down and jackals find a musty bedchamber.'[74] The battle had been lost. The 'philosophy of time' had thoroughly overrun not only intellectual circles but the common sense of everyday life.

Yet Innis obviously found the book still germane, considering his acknowledgement of his intellectual debt to Lewis in the citations of the communications works. Once again we discover Innis using his methodology to fashion sections of his text out of Lewis's raw material. For instance, in Innis's 'A Plea for Time,' we find a passage[75] that is, sentence by sentence, a slightly abbreviated version of Lewis's original text in *Time and Western Man*.[76] To illustrate, the first three and the last two sentences are compared here:

> Innis: 'This contemporary attitude leads to the discouragement of all exercise of the will or the belief in individual power.' (Lewis: 'Discouragement of all exercise of will or belief in individual power, that is a prevalent contemporary attitude, for better or for worse' [316]).
>
> Innis: 'The sense of power and the instinct for freedom have proved too costly and been replaced by the sham independence of democracy.' (Lewis: 'The sense of power, the instinct for freedom ... would cost too much to satisfy. We must be given ... a dummy, sham independence ... that is ... what democracy has come to mean' [316]).
>
> Innis: 'The political realization of democracy invariably encourages the hypnotist.' (Lewis: 'Whatever can be said in favour of 'democracy' of any description ... Its political realization is invariably at the mercy of the hypnotist' [42]).
>
> Innis: 'In art classical man was in love with plastic whereas Faustian man is in love with music. Sculpture has been sacrificed to music.' (Lewis: 'Faustian music ... banishes the plastic of the statue' [295]).
>
> Innis: 'The separation and separate treatment of the senses of sight and touch have produced both subjective disunity and external disunity.' (Lewis: 'It is because of the subjective disunity due to the separation or separate treatment, of the senses principally of sight and of touch, that the external disunity has been achieved' [419]).

Clearly, this is a rather concentrated example of Innis's method of composition. The citations are drawn from the full range of Lewis's book. A few citations by Innis are either wrong or require supplementary page references to complete them. Finally, Innis's 'paraphrasing' of Lewis's original words is so minimal that there is virtually no new text contributed by Innis in this entire passage. Why, given the obvious parallels in the work of Lewis to major themes in the communications studies, did Innis not recognize this influence in a less mechanical way? Again, we must put forward four possible reasons: his fear of exposure to attack by scholarly specialists; his lack of ease at dealing with philosophical arguments; his state of chronic overwork; and his attempt to incorporate elements of the oral tradition into his written work.

The nature of the style itself is a tribute to how seriously Innis takes Lewis's thought. In this regard, there is one other passage from Lewis's *Time and Western Man* that seems to speak directly to the peculiarities of Innis's communications work. It refers to what Lewis termed 'the politics of style.' As Lewis writes: 'In literature it should always be recalled that what we read is the speech of some person or other, explicit or otherwise. There is a *style* and *tone* in any statement, in any collection of sentences. We can formulate this in the following way: *There is an organic norm to which every form of speech is related. A human individual, living a certain kind of life, to whom the words and style would be appropriate, is implied in all utterance.*'[77]

Gandz and the Oral Tradition

Especially during his last years, Innis developed the opinion that everything exhibiting the creative essence of human culture can be traced back to the oral tradition. Even in an age dominated by writing and print, the creative impulse lying behind the dead letters is to be found in the long periods of training during which oral methods were stressed. The oral tradition was both fecund and sterile in a fashion exactly contrary to the written tradition. The oral tradition could generate ideas but not disseminate them readily, whereas the written tradition had the ability to disseminate ideas but was obliged to depend on oral sources for creativity in thought. The oral tradition, therefore, became a central focus for Innis's analysis, and Solomon Gandz became an important source for him in understanding that tradition.

Gandz's 1939 publication 'The Dawn of Literature: Prolegomena to a History of Unwritten Literature' is an encyclopedic treatment of the contribution of the oral tradition to human civilization. As Gandz describes it: '[This is] an enquiry into the struggle of primitive society for the preservation of its traditions and for the development of its memory and consciousness.'[78] Innis is in accord with Gandz's assertion that the 'oral tradition is ... a fundamental urge, whose function it is to build up the soul of the social organism and to maintain its continuity through the ages and generations.'[79] Furthermore, Innis agrees with Gandz's tailoring of his methodology of composition to fit the subject. Gandz summarizes this writing style as follows: 'I observed, so to say, the principle of oral tradition, which is, that what was once well said shall not be changed, but repeated in the very words of the original author.'[80] As we have seen, Innis adopted a similar approach and this may be one of his greatest intellectual debts to Gandz.

Gandz begins his study much farther back than Innis. In those prehistoric times, during which man was in 'an abject state of savagery,' the oral tradition served as a positive evolutionary force in much the same way as did a better weapon or tool. The problem during this era was not a lack of belief but a profusion of contradictory personal beliefs. Innis agrees with Gandz's points of departure for his study: that the oral tradition was the cornerstone of civilization; that it addressed primarily the problem of continuity; and that it was articulated principally in conjunction with religion.[81]

Like Innis, Gandz considers the development of the oral tradition on several different levels. One of the most profound is the biological or psychological. Gandz terms this 'the theory of recapitulation which asserts ... that the *ontogeny* or the development of the individual contains an abbreviated record of the *phylogeny* or development of the race.'[82] This is the opposite of a biologically determinist view, for it assumes that 'civilization cannot be handed down as a biological trait, it can be transmitted to a new generation only by imitation or instruction. The heritage of the animal comes by instinct. The heritage of civilization must be assimilated anew by each one of us for himself.'[83] In short, the question of individual consciousness is directly related to the larger dynamics of empires and civilizations. Innis's obsession with bias is a logical extension of this recognition that a profound examination of the oral tradition necessitates an introspective approach that takes the question of the formation of consciousness seriously. For Innis, the key area for this examination is classical Greece because it was at this point, through the mediation of an indigenously perfected, efficient alphabet, that the main infusion of the oral tradition into the cultural heritage of the West occurs. Innis writes: 'The task of understanding a culture built on the oral tradition is impossible to students steeped in the written tradition ... But the similarity of the Greek alphabet to the modern alphabet and the integral relation of Greek civilization to Western civilization implies dependence on the complex art of introspection.'[84]

It is because this biologistic or psychologistic theme underlies the treatment of the oral tradition in Gandz and Innis that discussion in the works of both men persistently returns to the topic of human organs, the senses, and their extension through the application of technique. As Gandz puts it:

> The tools of civilization are designed to support, strengthen, and magnify the power of the human organ ...
>
> Civilization advances through the invention of tools which substitute for human organs and bring them to perfection. The book is an artificial memory. Before the invention of books, the memory played a much more important part than nowadays. Men carried their libraries in their memory, and people with good memories were highly esteemed as living libraries, as the carriers of the sacred traditions of the race. The chief task of education was to train, cultivate and strengthen the memory ... The development of the logical powers of the human mind weakens the memory.[85]

The issue of innovation in communications technology, which so fascinates Innis, when applied to an era in which the oral tradition was dominant, results in a discussion of mnemo-techniques. From the citations in the communications works, we can observe that Innis learned directly from Gandz in this field. For instance, he came to see the *style* of oral transmission as something of at least as much importance as its content. The dominance of poetry over prose (if prose even existed as a concept) during the era of the oral tradition was not a question of literary 'taste.' It was a fundamental component of effective communications technology. For instance, 'in carrying this immense burden of oral traditions, the memory was aided and supported by two expedients: the *verse* with its metre and rhyme, and the *proverb* with its witty and aphoristic form.'[86] The sophistication of these techniques and their overwhelming dominance as vehicles for carrying the cultural heritage of civilization made the appearance of writing seem at first like an irrelevant gimmick employed only by commercially minded individuals.

> Literature begins with poetry, which is much older than the art of writing. In the beginning, poetry was everywhere sung and recited and preserved by word of mouth. The form of poetry was adapted to the method of oral tradition; metre and rhyme aided and supported the memory. Poetry is emphatically the literature of the time of oral tradition. To write down poetry must have been considered, in the early times, as preposterous

> and absurd, like putting religion into algebraic formulae. Poetry was the oracle of the seer, the sacred word of inspiration; and what was writing in the beginning? Some nasty scratches of a money-changer or grocery-dealer, by which he would mark down how much money you owe him.[87]

In presenting an overview of human civilization from prehistorical savagery to the period of dominance of the oral tradition to the written era, Gandz suggests a change in sensory orientation: an idea picked up by Innis and elaborated to its fullest extent by McLuhan.

> The progress from oral to written literature may be compared to the change from the gesture-language to the word-language. Speech follows a language of gestures and constitutes a progress from the eye- to ear-language. With the invention of letters, the order is reversed. Writing represents a transition from the ear-language back to the eye-language. In the evolution of – gestures to words to letters – we have a successive series of visible-audible-and again visible symbols of communication and tradition. The progress consists in the inclusion of the achievements of the preceding stage in the succeeding stage.[88]

This last sentence articulates another recurring theme in Innis (and later McLuhan): that radical technological innovation often implies reaching back to a superseded technology to rejuvenate certain of its components in a new social context.

When Gandz turns to the examination of language as the basic tool of communication, he finds a paradox:

> Since language has a double function, to wit, to serve as a means of communication and as a means of tradition, we may in general distinguish two kinds of languages: a profane language, for communication, and a sacred language, for tradition; an everyday language serving the prosaic needs of everyday life, and a language of religion and poetry ...
>
> On the one hand, language serves as a means of perpetuating tradition, on the other hand, however, it is tradition which helps to preserve and perpetuate language. Dialects and primitive everyday languages undergo very rapid changes. The higher cultural languages show a certain degree of stability ... The development of religion, poetry and literature creates stability in language and causes the rise of a national, standard language.[89]

However, this phenomenon of standardization of language in no way contradicts Gandz's observation that it is not the vernacular that serves as the primary carrier of the 'social heritage' of a culture. He refers to a '*law of the heterolingual character of civilization* ... There is always a literary language as opposed to the vernacular or spoken dialect. Usually ... poetry, religion, law and science are transmitted by a foreign or a dead language, which is the language of the higher classes.'[90] This conception bears a resemblance to Innis's 'monopoly of knowledge.' The ruling class of a society is seen as maintaining its political dominance partly through controlling access to those communication technologies of central importance in the maintenance of cultural continuity.

As Innis never tires of pointing out, it is difficult to understand the nature of these ancient monopolies of knowledge based on a sacred language because our mental make-up is in large measure structured by a current 'monopoly' built around the mechanized and amplified reproduction of our vernacular. Gandz had the great gift of being able to overcome the bias of our written tradition by placing himself in the position of the old 'monopolists' of the oral technique as they surveyed the initial introduction of writing. He wrote that 'for many centuries, there was a bitter fight going on between the old guild of the minstrels and rememberers, on the one hand, and the new guild of the scribes, on the other hand. Here was the living voice, warm emotion and sweet music of the popular minstrel, and there came the new scribe with his dried skins of dead animals; here were solemn recitals, hallowed by tradition and sanctified by religion, and there came the scribe with his cold signs and scratches and symbols, intelligible to a few suspicious persons only.'[91]

Gandz had a less romantic view of the oral tradition than Innis and believed that the fundamental bias it imposed was one that minimized the role of the individual in the creative process. For instance,

> the idea of literary property is the product of two factors. (1) the rise of the individual author with his own personal inventions and creations, and (2) the invention of writing. In the old system of oral transmission and traditional literature, there can be no plagiarism, because there is no individual authorship.
>
> All knowledge and wisdom came from the past. Distinction and virtue lay in the mastery with which one was able to reproduce and preserve the old wisdom. The glory of the scholar consisted in his strong memory. He had to repeat the old traditions and poems. The great traditionists boasted that they never said anything which they did not hear from their

> teachers. The inducement was rather towards the reverse of plagiarism, which is forgery. It was more profitable to ascribe one's own ideas and books to old authorities and to present them as traditions of olden times.[92]

Gandz, like Innis, links his speculations about the psychological aspects of the oral tradition to an institutional interpretation. The monopoly of knowledge based on a special language functions only to the extent that it is supported by an effective institutional structure. In the case of the oral tradition, Gandz foreshadows Innis in viewing the primary institutional expression of the oral tradition as religion: 'Language is the physiological basis of oral traditions; religion forms the sociological mechanism through which traditions are established, preserved and accumulated. Religion ... forms the brains and will power, the soul and consciousness of the social organism. Its function is to direct and enforce cooperation of the individuals in the interest of the community, to maintain the group-life of the tribe or community.'[93] In other words, Gandz views religion based on specialized language as bearing the institutional responsibility for solving the problem of continuity in society. Like Innis, he views the fixation of religion on the problem of continuity as the undoing of the clerics. Their preservation of traditions through the use of a specialized and increasingly refined language leads them to lose touch with the vernacular of the general populace. The element of force always present in a latent form in this general populace could more readily be mobilized against the clerics by barbarians who still spoke an effective vernacular and were skilled in military, as opposed to scholarly, technique. As Gandz puts it, 'the danger to religion, in its beginnings, lies not so much in the rigidity of the traditions, as rather in the lack of stability, in the forces of change and variation ... If a civilization becomes too refined and sophisticated, then the power is usurped first by the savages out of its own midst, in the form of an organized mob or an organized army, and then by an alien race with more savagery.'[94] This portrayal of great cultures as an uneasy alliance of social groups who have mustered techniques of force (the military) and knowledge (religion) corresponds closely to Innis's conception of empire. We should not be surprised to find, therefore, that Gandz's treatment of certain empires based on the oral tradition serves as a forerunner to Innis on many crucial points. Let us consider two examples, the case of Islam and that of the Greeks.

Gandz views the birth of Islam as the development of a political movement of an illiterate people based on highly effective use of oral poetic technique disseminated through a religious institutional framework. In a passage cited by Innis, Gandz summarizes his observations: 'The development of poetry contributed to the rise of a standard, national language of Arabia long before the political unity of the Arabic nation was established. Later on the idioms of ancient poetry became also the language of the Koran and then the sacred tongue of the whole Muslim world.'[95] The Islamic empire was founded on a balanced mastery of communications technologies. Religious organization ensured continuity through sophisticated use of oral mnemo-techniques; however, this was necessarily coupled with a political administration which depended on the technology of written script. According to Gandz, 'Muhammad used to send apostles to the tribes in order to instruct them in the Koran and in the duties of Islam. These emissaries received *written orders* concerning the taxes to be collected, but no written copies of the Koran. The Koran was carried in their breasts and hearts. This is an illustration of the general rule, mentioned already ... Writing is for business only; religion and literature are transmitted orally, in the old traditional way.'[96]

Like Innis, Gandz often returns to the theme of the material basis of the media as a way of explaining the persistence of the oral tradition:

> The preservation through ... memory is for primitive people ... practically much easier and cheaper than the preservation through written records. In the Arabia of Muhammad's time writing material was very scarce, very expensive, and very voluminous. Paper was not yet known ... Moslems wrote their memoranda upon tablets of wood and slate ... For more permanent records, they made use of leather and parchment ... The writing was often washed off ... to make way for more recent compositions ... The secretary of the Prophet, when entrusted with the task of editing the Koran, collected its scattered fragments with great difficulty 'from bits of parchment, thin white stones, leafless palm-branches, and the bosoms of men' ... Other traditions add to that pieces of leather, and the shoulder and rib-bones of camels and goats.
>
> Little wonder, therefore, that under such circumstances ... to write down a poem was a very hard and laborious task. On the other hand, there were plenty of people with plenty of time, and wonderful, retentive memories and nothing in them.[97]

Innis argues that this overemphasis of the oral pole in the balance of the Islamic world had the effect described by Gandz: long-term religious continuity combined with extreme political fractiousness. Writing was introduced only in a crisis situation in which the vulnerability of oral mnemo-technique became obvious. Even then, the writing technology remained a marginal adjunct to a fundamentally 'oral' culture. In Gandz's words:

> In 682, about seven hundred 'reciters' who knew by heart the whole Koran, perished during the sack of Medina ... Soon it was observed that the Koran as read in Syria was seriously at variance with the text current in Iraq. The Caliph issued orders to prepare ... the final and standard recension of the Koran ...
>
> And yet, even after his publication of the Koran, writing was restricted to the Sacred Book only. Everything else remained as before, entrusted to memory and oral tradition.[98]

In contrast to the Arabs, the Greeks had the good fortune to experience a more 'balanced' social project in which religious sentiment played a less important role. With the advent of writing, they were able to continue to take full advantage of the creative potential of the oral tradition while exhibiting a minimum amount of its stiflingly conservative tendencies.[99] Gandz presents the Greeks as having a bias heavily in favour of the oral tradition *even after* the introduction of writing. He emphasizes that there is no automatic outcome attached to technological innovation. '[The] *introduction* must never be identified with the *spread* of writing.'[100] At a political level, the members of the traditional monopoly of knowledge remained analphabetes and resisted the introduction of a new competitive technology. At the cultural level, a population accustomed to being educated via ecstatic oral techniques found it difficult to change over to a discursive style in the search for knowledge. As with his treatment of Islam and in a manner similar to Innis, Gandz always returns to the infrastructural level or the question of the media material: 'There is also another aspect of the question ... There must be a suitable, handy and cheap writing material. This, however, was missing in Greece ... the only available materials of writing were tablets of wood, lead or stone ... The rise of a literature in book form did not begin till the sixth century and is to be ascribed to the importation of papyrus from Egypt.'[101]

In summary, in reading Gandz, Innis would have run across several themes important to his communications works: the idea of a religious elite attempting to solve problems of continuity through monopolizing a sacred language; the danger of this culturally refined group losing touch with the vernacular and falling prey to barbarian force; the sensory bias of media; the importance of the availability of media materials; and the importance of a balance of knowledge and force in a successful social project. Another area in which Gandz served as a pattern for Innis was in his belief that innovation or cultural renewal is likely to be found, not at the centre of social formations, but with the persona of marginalized man. Innis and Gandz both believe that the oral tradition was by its very nature a decentralizing force. This was partly 'as a result of the oral character of the transmission of traditions [which obliged] the student ... to travel far and wide in search of knowledge. The young scholar had to seek the teacher, who was in the possession of a tradition and often resided in a distant province of the great ... empire[s].'[102]

Aside from the call of the individual engaged in the search for truth, Gandz feels that this rule of creativity on the margin applied to more general social dynamics. He notes that 'one of the most fundamental factors in the improvement of traditions was the clash between traditions belonging to various races, tribes and countries. Migrations and wars were great stimuli for the progress of mankind. The change of the environment and of the conditions of life, the change of blood through the physical intermixture of one race with another, the clash of the contrasting ideas and traditions compelled people to think, to abandon their old traditions and to adopt new ones.'[103] As a typical example, he cites the Greeks: '[Their] old tribal traditions were discarded. "For the fugitive settlers on the shores of Ionia ... there were no tribal gods or tribal obligations left, because there were no tribes ... This fugitive had left the graves of his fathers, the kindly ghosts of his own blood ... One only concrete thing existed for him ... to supply the place of his old family-hearth, his gods, his tribal customs and sanctities. It was a circuit wall of stones, a Polis." Therefore a new system of laws, customs and traditions had to be built up, a new, higher civilization had to be created.'[104] For Gandz, this principle of intellectual creativity at the margin is a general rule rather than an observation applying only to a particular case. 'All these [epic] poems were the product of the migrations. They originated in the old places and bore no relation to the new country where the migrating tribes settled.'[105]

Yet, for all these similarities in treatment, there is a fundamental difference in attitude between Innis and Gandz concerning the oral tradition. For Gandz, the oral tradition is primarily a technique of transmission of the social heritage which, although primitive and superseded, had served a valuable purpose for a particular phase in the building of civilization. Gandz thinks of the oral tradition as something that is primarily 'in the past' and certainly inferior to current methods of cultural transmission because of its stifling conservative tendencies. For Innis, the oral tradition is the source of all genuine contributions to human culture down to the present day. He sees it as a fundamentally positive element contributing to stability *and* creativity. Far from being content with seeing the diminution of the presence of the oral tradition in modern society, Innis feels that this indicates a deep malaise at the heart of Western civilization.

Thus, Innis's 'debt' to Gandz, though significant, is also highly selective. We can see this in the diametrically opposite view held by Gandz concerning the nature of the oral tradition in contemporary society:

> Oral tradition in our times is a mere accident, a relic and survival of the past. Its contents are some trivial folksongs, fairy tales, or magic and superstition. Its bearers are the backward people of the countryside, the illiterates of some remote island. It would therefore be a great mistake to compare the oral traditions of olden times with the folklore of to-day. Once upon a time oral tradition was a sacred thing of religion, an official record of the state, entrusted to the care of professionals and high dignitaries. Nowadays it is a loose and trivial thing, a mere formless, casual hearsay, handed on by irresponsible people in an irresponsible and utterly careless way.
>
> Even in our colleges and universities an excessive and quite unnecessary amount of oral instruction is still retained as a kind of survival of archaic times.[106]

Other Influences

Innis was also influenced in his communications studies by F.M. Cornford's deliberations on space, Martin P. Nilsson's work on the oral tradition and the beginning of time, and the German philosopher Ernst Cassirer's writings on language, myth, and the development of human consciousness.[107]

Cornford's major works, *Thucydides Mythistoricus*, *From Religion to Philosophy*, and 'The Invention of Space,' are typical of the material Innis sought out to reinforce his communications themes. Although their lines of argument are quite different, Innis's understanding of the changes that have occurred in the concept of space in the twentieth century is similar to that of Cornford and the relativity theorists. Their common position is that space in the modern mind has become severely circumscribed at all levels of thought.[108]

Martin P. Nilsson's book *Primitive Time Reckoning* is another source used by Innis in the communications studies.[109] Nilsson stresses that notions of time are not automatically thrown up by the human organism but have to be constructed painstakingly over a long period. He views the growth of the oral tradition of a people as synonymous with the growth of its conception of time. As long as the conception of time depends on immediately perceivable natural phenomena which can be seen by everyone, anyone can judge time. Social development begins with the segregation of a specialized class of calendar makers (the priests). In his treatment of this class, Nilsson is close to Innis's concept of 'monopolies of knowledge.'

Of the various sources Innis consulted in his effort to trace the development of conceptions such as time and space which lie at the base of human consciousness, the German philosopher Ernst Cassirer was one of the most philosophical. Innis cited three books by Cassirer in his communications works: *Language and Myth, The Problem of Knowledge,* and *An Essay on Man*. We should not be surprised to find, therefore, that Innis's treatment of Cassirer is similar to his treatment of that other source of profound ideas, Wyndham Lewis. In Cassirer, as in Lewis, we find major themes strikingly similar to those articulated by Innis in the communications studies. If we cannot prove definitively that Innis borrowed these themes directly from a particular source, we can maintain that their presence in source material cited by Innis at the very least *reinforced* his use of similar themes in his communications manuscripts.

The direct citations of Cassirer for the most part follow the highly telegraphic, verbatim condensation of source matter used by Innis. The distinction between the limited and pedestrian nature of these direct citations, on the one hand, and the profound intellectual debt of Innis to Cassirer indicated by the similarity of major themes, on the other, is striking. I would attribute this, as in the case of Lewis, to Innis's lack of ease in dealing with source material of a philosophical nature and his fear of intellectual attack by colleagues with more specialist inclinations.

The most condensed of Cassirer's works is *Language and Myth*, and Innis seems to have relied most heavily on this book. Innis fixes on it because in this work Cassirer places 'conception' at the centre of consciousness and relates its advent to speech (language) and the primeval oral tradition (myth).[110] Innis is particularly indebted to Cassirer for his insight that an analysis of the perspective imposed by the technique of communication is more important than the analysis of the subject matter being examined through the use of that technique.

> Theoretical science remains the same in its essence no matter what object it deals with – just as the sun's light is the same no matter what wealth and variety of things it may illuminate. The same may be said of any symbolic form, of language, art, or myth, in that each of these is a particular way of seeing, and carries within itself its particular and proper source of light. The function of envisagement, the dawn of a conceptual enlightenment can never be realistically derived from things themselves or understood through the nature of its objective contents. For it is not a question of what we see in a certain perspective, but of the perspective itself.[111]

But Innis's direct citations of Cassirer are superficial and often tortured. Consider these short passages from Innis's *The Bias of Communication*[112] compared to the elements of Cassirer's thought from which they were constructed[113]:

> Innis: 'The ideal of mathematical sciences dominated the seventeenth century.' (Cassirer: 'The ideal of mathematical sciences ... dominated the seventeenth century' [170]).

> Innis: 'It was not until the Enlightenment that the historical world was conquered and until Herder and romanticism that the primacy of history over philosophy and science was established.' (Cassirer: 'The Enlightenment ... made use of it as one of the chief measures in the ... conquest of the historical world' [217]. 'Since the time of Herder and romanticism another mentality ... The primacy of history was proclaimed by both philosophy and science' [170]).

> Innis: 'Historicism was almost entirely a product of the nineteenth century.' (Cassirer: '"historicism" ... was almost entirely a product of the nineteenth century' [170]).

> Innis: 'Burckhardt and to some extent Lamprecht approached the study of civilization through fine art.' (Cassirer: 'Burckhardt ... prepared himself for insight into the historical world ... [through] fine art' [277]).

> Innis: 'The highest value of art as of all free intellectual activity was to provide release from subservience to the will and from entanglement in the world of particular aims and individual purposes.' (Cassirer: 'The highest value of art, as of all free intellectual activity, was that it gives release from strict subservience to the will, from entanglement in the world of particular aims and individual "purposes"' [277]).

Innis's methodology of composition when applied to Cassirer yields similarly dubious results as with the other sources we have discussed. Short passages are plucked seemingly at random from their context in Cassirer's tightly argued work. They are compressed and rearranged with a minimum of new words being added by Innis. They then appear in Innis's text in an order that seems largely arbitrary rather than part of a sustained argument. Finally, in the course of transposing some of the sentences, Innis inadvertently makes alterations that change Cassirer's original meaning.

Innis's use of Cassirer along with the other sources examined in this chapter indicates the importance of one level of the communications studies often overlooked: the question of consciousness. From this point of view, Innis's examination of the history of empires and the communications systems that lay at their heart can be seen as his attempt to elucidate the changes in the way human beings have viewed the external world from the time of the origins of consciousness in the language of myth to the disintegration of a critical perspective under the impact of the mechanized vernacular in contemporary society.

CHAPTER TEN

At the Edge of the Precipice: The Mechanization of the Vernacular and Cultural Collapse

Students should be taken to the edge of the precipice beyond which knowledge does not exist.

H.A. Innis[1]

The communications works should be seen as an intensely political undertaking – but in a very narrow sense. Innis's political goal was the reconstitution of a critical perspective in the social sciences that could be brought to bear on the most profound conundrums of the present era. His obsessive form of research was the opposite of antiquarianism. In order to understand the myopia of our contemporary mentality, he concentrated on past societies and how their way of thinking about their world was structured. Similarly, he supported the ivory tower concept of the university because only such an institution could support the scholar's Promethean efforts to avoid the acceptance of the common-sense of his times as a prerequisite to developing an alternative, longer-range perspective.

Innis's colleagues, because they misunderstood the essentially political thrust of his scholarly work, were shocked when at the beginning of the 1950s he began producing polemical articles. They should not have been. In a sense, Innis had succeeded in isolating himself from absorption in the common-sense of his time. It was quite natural for him, therefore, to protest as the Cold War mentality grew more pervasive and more rigid. However, the relative permeability of the university community to the Cold War's bipolar vision of world affairs determined that Innis's polemical essays would be written off as crankish anomalies attributable to ill health. Instead of recognizing an obvious link between his early 'legitimate' scholarly works and the later diatribes, his colleagues viewed the latter as a regrettable, and pitifully unscholarly, disjuncture in the overall development of his thought.

The Greeks and the Cold War

Yet, as early as 1946, Innis had focused on the Greeks as the common root of the great opposing contemporary empires: the United States and Russia.

> The significance of Greek civilization to East and West provides an approach to modern problems. Both groups in the word of Jaeger are Hellenocentric ... There could be no sharper contrast than that between the modern man's keen sense of his own individuality and the self-abnegation of the pre-Hellenistic Orient ... As against the oriental exaltation of one God-king far above all natural proportions ... and the oriental suppression of the great mass of the people ... the beginning of Greek history appears to be the beginning of a new conception of the value of the individual. 'Other nations made gods, kings, spirits; the Greeks alone made men.'[2]

Later, in his communications essays, Innis traces the history of Russia and the United States – the twentieth century's two great streams of Western civilization – back to their common Greek heritage. His litany of empires – their political and religious institutions, their ways of thinking, and their communications and military techniques – should be viewed as an extended examination of the fundamental split in Western civilization that resulted from the breakdown of the original cultural project of classical Greece.

Innis empathizes with the notion that ancient Greece provided the bedrock on which Western civilization was built. This notion accords well with his belief in the importance of the intellectual activities of marginalized groups. He returns again and again to the idea that it was the marginal position of the Greeks vis-à-vis the great ancient oriental empires that allowed them to develop an original social project. One need only think of the various levels of Innis's dialectic of empire to understand why the Greek case is so crucial to him. Until the time of the Greeks, imperial history had entailed the reformulation, under successive directing groups, of the same basic structure of empire at the level of institutions, mentality, and communications techniques. However, the Greeks successfully confronted this highly articulated tradition of oriental despotism with a fundamentally reformulated project involving new institutions, a new outlook on life, and radically different communication techniques.

Prior to the Greeks, the empires of the Near East had revolved around an uneasy alliance of absolute monarchs and scholarly theocrats. The symbolic focus of this type of alliance was found in the personage of the God-king: the king as God. Institutionally, the monarch drew his strength from a political organization that was militaristic in nature. Martial prowess, superior strategy, and innovations in military technology allowed him to conquer and weld together more localized theocracies. These theocracies drew their institutional strength from their hierarchical, cult-like organization, and they were easily integrated with a new military overlord. They simply accepted (and legitimized) him as the incarnation of God on earth.

As we discussed in the previous chapter, these institutional arrangements permitted the development of far more sophisticated conceptions of time and space than would have been possible in hunting and gathering societies. Nevertheless, both time and space conceptions remained intensely concrete and discontinuous, and therefore limited. The 'infinite,' in terms of either distance or the passage of time, does not seem to have been articulated. Similarly, the abstraction that has been thoroughly inculcated in our mentality under the term 'individual' appears to have had no meaning in these times. There was no such thing as an 'individual' in general: only beings that fit into some particular role in the cosmic hierarchy.

The most advanced forms of conceptions of space and time were principally carried by the monarch and priests respectively. In other words, the various factions within the ruling alliance exemplified differing but complementary forms of common sense that allowed them to address their particular social functions with efficiency. The monarch was typically a war leader whose grasp of the concept of space allowed him to expand his territory (and control of the subjects within them) militarily, incorporating even the most highly articulated theocracies. The priests specialized in elaborating conceptions of time and continuity. To maintain kingly rule, especially when a period of succession or new territorial conquest was involved, the new monarch was required to make an early peace with the local priests. We could view the history of the Near East as a series of bright flashes of military integration illuminating the more locally rooted cults that were the basic foundations of the societies of the area.

A similar dichotomy appeared at the level of infrastructure or technique. The original assumption of power of a particular ruling regime within an empire depended largely on developments in military technique and weapons. The element of communications technology important to this side of the imperial alliance was the means of transportation. The power of the priests, meanwhile, depended on the control of writing systems and other mnemonic techniques that served to transfer the social heritage selectively or hierarchically.

The chief problem of the priests was the imperfection of their technology. It was aimed at capturing the content of spoken language by signs. In an era prior to the true alphabet, this meant that, whatever strategy they used to reproduce the many sounds of the spoken language, priests were automatically destined to become what Innis termed a monopoly of knowledge. Leaving aside consideration of ideographic and hieroglyphic scripts, we find that they could opt for one of two strategies in formulating a script. They could develop a complex system of approximately one hundred signs – roughly one for each sound in language. This gave the advantage of precision in reading, but the long and arduous training period necessary to absorb this complex system guaranteed that it would be of little use to anyone but the members of a small elite group. Furthermore, the more effectively this group pursued its goal of freezing the oral utterance for all time, the more distant this language would be from the ever-changing vernacular. Theoretically, a message could be committed to writing by one person and transmitted over a long distance and rearticulated with accuracy by another person who was schooled in the same complex system. Practically, the very complexity of the system guaranteed that it would be restricted to relatively small groups of priests in a localized area where frequent verbalized verification of the symbols was permitted.

The other strategy for script development in the era preceding the perfection of the alphabet was to use only a limited number of symbols for the writing system – a trade-off of precision for ease of learning. Mastery of this limited number of symbols would require less time, opening the system to a greater number of individuals. However, because the system still had to carry the weight of the full range of sounds of an existing language, each written symbol was forced to represent many related and even quite disparate sounds. Reading the symbols required interpretive skill in identifying the correct sound in any particular context. This high degree of interpretation meant that messages could not be sent over significant distances with any degree of precision; maintaining common conventions of interpretation required that readers and writers remain in contact with one another. The success of this system was similarly limited in sending messages down through time. The attribution of various possible sounds to each symbol kept the system simple and flexible enough to adapt to contemporary changes in the vernacular but hamstrung scholars in interpreting the writings of past generations. This strategy, then, was likely to produce a written script that was more popular and more widespread but also more unstable.

Faced with this choice, the priesthoods of the Near East seem to have fixed on the first option. Their power was founded on the sense of history provided by their mastery of a complicated and esoteric sacred language bearing little or no relation to the contemporary vernacular. Their ability to participate in the ruling alliance of an empire was based on the complex world-view stored in this sacred language and its diffusion to the popular level through ritual and public spectacle that served as forms of mass, non-literate, ecstatic education. The subordinate groups within these oriental empires absorbed a common social vision that had its centre in the sacred writing of the priests but found its force for the structuring of society in sophisticated techniques of oral transmission. The oppressive hierarchical structure of the oriental empires was absorbed by their subject peoples through their ears via the ritualized voice of religious myth rather than through their eyes as script. The latter system, operating at one remove from the spoken word, is far more likely to create a society of discrete individuals dispassionately interpreting symbols. In contrast, pre-literate social organization is based on mass participation in emotive rituals.

I have outlined the structure of the ancient non-Greek empires, leaving out, to this point, any reference to the physical characteristics of their media of communications, to make a point about causality in Innis: namely, that he is not as much of a media determinist as he is often portrayed to be. He makes this clear when he says of the essays in *The Bias of Communication* that they 'emphasize the importance of communication in determining "things to which we attend" and suggest changes in communication will *follow* changes in "the things to which we attend."'[3] In other words, the nature of the material on which the sacred writing was imprinted is but one factor serving to reinforce the social project of the relatively localized theocracies. Using today's language, there was no need to develop a new 'hardware' of communication before a 'software' in the form of a less complex, more precise script was available. Clay tablets and stone inscriptions served the needs of the theocrats admirably. It fell to the merchants and civil administrators to undertake more radical experiments.

Why did the Greeks differ so greatly from this original pattern of empire? Innis returns again and again to the significance to their culture of the oral tradition. Yet, as we have seen above, the Eastern empires also relied heavily on oral techniques as a means of reproducing themselves. We should begin, then, by investigating the way in which the Greek oral tradition differed from that of the Eastern empires.

Innis fixes on the geographic marginality of the Greeks as a key factor in explaining their particularity. 'In contrast with the Aryans in Asia Minor,' he writes, 'the Greeks were less exposed to the influence of those whom they had conquered. Minoan civilization with its maritime empire had escaped the full impact of continental civilizations and in turn was less able to impose its culture on the immigrants of the northern mainland. The complexity of the script of Minoan civilization and its relative restriction to Crete left the Greeks free to develop their own traditions. Successive waves of Greek immigrants checked the possibility of conservative adaptation of cultural traits.'[4] The Greeks are viewed as refugee barbarians who, not having the opportunity to make peace with previously established theocracy with its attendant language and cultural features, are forced to develop elements rapidly within their own barbarian culture to ensure social continuity. The social context in which the Greek oral tradition developed was quite different from that of the Near East. Innis writes: 'The Homeric poems of the Heroic age were produced in a society in which the ties of kindred were weakening and the bond of allegiance was growing. An irresponsible kingship resting on military prestige held together kingdoms with no national basis. Tribal cults were subordinated to the worship of a number of universally recognized and anthropomorphic deities. Society was largely free of restraint. Tribal law had ceased to maintain its force and the individual was free from obligations to kindred and community.'[5]

The task of the oral tradition in this context was one of developing consensus rather than legitimizing oppression. In consequence, the oral tradition as practised by the Greeks was considerably more creative than it was in the Eastern empires. This again was largely the result of their intellectual backwardness, which forced them to adopt a more democratic effort towards accumulating and transmitting a common cultural heritage. The fact that they were less literate and had less access to writing materials forced them to develop highly sophisticated techniques of oral transmission that could be practised in a popular manner. The contrast between the ideologues of the two types of empire was extreme. On the one hand, the priests of the oriental empires guarded their sacred knowledge by committing it to an esoteric script completely unintelligible to and indeed unseen by their oppressed citizenry. On the other hand, the Homeric poets were famed for the direct popular response that greeted their recitations of the history of the Greeks. The oriental priests had developed a strategy for raising one man to the stature of God to justify the oppression of the great majority of mankind. The Greek poets reduced all the gods to the image of men and in so doing gave a new meaning to human individuality.

Innis fixes on the oral tradition of the Greeks as opposed to that of other cultures because it represents the creative epitome of the use of popular mnemo-technique as a means of building up and transmitting the cultural heritage of a people. The significance of the advanced nature of the Greek epic poetry as mnemo-technique was a favourite subject of the classicists at the University of Toronto at the time. It seems that E.T. Owen in particular was thoroughly versed in this area, and the classic study on the subject by E.A. Havelock, *Preface to Plato*, was grounded in the same intellectual milieu.

Borrowing from these classical scholars, Innis, in a scattered manner, presents an overview of how the structure of epic poetry was designed to extend popular memories to enormous capacity. It did so through the counterpoint rhythm of dance, musical (non-oral, that is, drums or strings) accompaniment, and the beat of the poetry itself; through the conventionalization of phrases and narrative structure; and through rhyme schemes.[6] The fundamental difference between the Greek method and that employed in the Near Eastern societies was the use of rhythm based on vowels rather than alliterative echoes caused by phrase clumps with similar consonants. While both traditions were at their basis ecstatic educational techniques, the Greek style appears to have been more effective in establishing the pleasurable atmosphere conducive to the effective transmission of a cultural heritage.

The backwardness of the Greeks in the employment of script and the elaboration of religion permitted them to develop a wider and more effective oral tradition. Their marginality ensured that, when the use of script finally did come to them, it would be adopted in a modified and revolutionized form. The true alphabet has been invented only once in the history of civilization, and only in the case of the Greeks was it introduced from the inside to a culture at the height of its oral tradition. That is why Greek civilization, more than any other, has been captured with extraordinary vividness.[7] As discussed above, the priesthoods tended to opt for a complex language with many signs that served adequately the needs of continuity but in so doing soon became far removed from the vernacular. The merchant groups tended to opt for a relatively simple sign system that demanded much interpretation and therefore became ambiguous with the passage of time. The latter system, however, admirably served the needs of day-to-day commerce by being flexible enough to adapt to changes in the vernacular. It was from a version of one of these types of languages that the Greek alphabet evolved: 'An alphabet of twenty-four letters which represented consonants to Semitic peoples proved exportable and adaptable to Greek demands. A different language structure and systems of sounds led the Greeks to use Semitic consonantal characters, which were useless to their language, as vowels which were indispensable to them. The Greeks with Canaanite Phoenician script "breathed into its consonantal skeleton the living flesh and movement of the vowels."'[8]

In the Near East, previous simplified systems of script based on an oral tradition stressing consonantal alliterative effects had been based on the 'reading in' of the vowels using a text composed exclusively of consonant signs. The Greeks, by appropriating consonant characters for Semitic sounds that did not exist in spoken Greek and by assigning these surplus signs to vowels, accomplished the incredible feat of abstracting the spoken word into its basic components. The profundity of this feat is lost on us because we are obliged to absorb it during long hours of study at an early age. We can gain an inkling of the tremendous leap of thought required when we recall that even now, when we 'say' aloud the series a, b, c, d, and so on, we are reduced to returning to the old system of treating each sign as a syllabic (that is, ay, bee, see, dee). Yet these sounds bear no direct relation to the interpretation of their corresponding signs in script. The abstraction of the consonant, as a non-sound in itself but something that serves to modify the basic sound of a vowel, meant that for the first time a script was available that was relatively simple, rendered accurately the spoken word, and could be readily adapted not only to changes in the vernacular but also to the reduction of all other languages to effective script form.

The Greeks were the only people in the history of mankind to become literate from 'within' their culture, adopting a true alphabet at the height of their oral culture. This process took many centuries during which script first was used strictly as a mnemo-technique for the existing oral tradition. Hence, not only was the oral tradition of the Greeks highly advanced, but their script was highly efficient, transmitting that oral culture to us in a relatively brilliant manner. On the other hand, the pre-alphabetic scripts, requiring as they do a great deal of 'reading into' them, can transmit only a shadow of the original oral culture from which they are drawn.

The introduction of the alphabet, by affording the ability to record sounds accurately, actually strengthened the Greek oral tradition – at least at first. Only with time, as the function of script changed from being a memory reference for popular recitals of epic poetry to being material for silent individual reading, did its long-term effect of dissolving oral-based culture become apparent. Alphabetic writing was biased towards rational rather then ecstatic forms of education and towards the vernacular rather than specialized (sacred) forms of language. Again the element of individuality, already present in the Greek oral culture relative to that of the Near East, became refined and exaggerated. New forms of inquiry such as dialogue and drama based on an amalgam of these two types of cultures came into being.

In summary, the coupling of the true alphabet with the rich oral tradition of pre-literate Greece brought about revolutionary social changes that ushered in the modern era dominated by Western civilization. At an institutional level,

> the strength of the oral tradition and the relative simplicity of the alphabet checked the possible development of a highly specialized profession of scribes and the growth of a monopoly of the priesthood over education. A military aristocracy restricted the influence of a priestly class and poets imposed control over public opinion. The Greeks had no Bible with a sacred literature attempting to give reasons and coherence to the scheme of things, making dogmatic assertions and strangling science in infancy. Without a sacred book and a powerful priesthood the ties of religion were weakened and rational philosophy was developed by the ablest minds to answer the demand for generalizations acceptable to everyone .... No energy was lost in learning a second language and the freshness and elasticity of an oral tradition left their stamp on thought and literature.[9]

Clearly, Innis thought of classical Greece as a society characterized by intellectual mobility par excellence. Down to the level of basic common-sense conceptions, the Greeks developed an alternate model to that of the oriental empires: 'The Greeks seized on the spatial concept as developed by Ionian philosophers and on the temporal concept emphasized by mystical religions to construct a political society which stood the test of resistance to the Persian Empire. The Greeks opposed the raising of gods and religion to an independent position dominating the state and brought to an end the threat of a theocratical and monarchical order.'[10]

Why, then, did the Greek cultural project fail on the political plane? Innis mentions three related reasons. First, 'richness of the oral tradition made for a flexible civilization but not a civilization which could be disciplined to the point of effective political unity. The city state proved inadequate in the field of international affairs.'[11] Second, 'the spread of writing contributed to the collapse of Greek civilization by widening the gap between city-states.'[12] Third, the Greek project became imbalanced in the sense of falling behind in the field of innovation in military technology. Military specialization by the less-cultured Macedonians allowed them to sweep over the Greek heartland as well as the entire territory controlled by the old Near Eastern empires.

Innis views the story of the Greeks as the fundamental intellectual tragedy of the West. The continued florescence of Western culture depends on the creative efforts of marginal peoples whose intellectual accomplishments are bound to trigger their eventual political failure. Yet the Sisyphian nature of this type of activity down through the centuries in no way diminishes, in the eyes of Innis, the accomplishments of the people who had made the first great effort. By founding a society that resisted the despotisms of the East without adopting their social mould, the Greeks had implanted the ideals of political and intellectual freedom at the basis of the cultural heritage of the West.

> The powerful oral tradition of the Greeks and the flexibility of the alphabet enabled them to resist the tendencies of empire in the East towards absolute monarchism and theocracy. They drove a wedge between the political empire concept with its emphasis on space and the ecclesiastical empire concept with its emphasis on time and reduced them to the rational proportions of the city-state. The monopoly of complex systems of writing, which had been the basis of large-scale organizations of the East, was destroyed.
>
> The adaptability of the alphabet to language weakened the possibilities of uniformity and enhanced the problems of government with fatal results to large-scale political organization. But the destruction of con-

cepts of absolutism assumed a new approach of rationalism, which was to change the concept of history in the West.[13]

Russia and Byzantium

Innis sees the failure of the Greek project at the political level as being the ultimate reason for the division of the Western cultural heritage into two main branches. This division, exemplified in modern times by the great empires of the United States and Russia, increasingly preoccupied him during his last years.

Innis's focus remained the university and its independence. As the Cold War mentality became a significant threat to that hard-won independence, he reacted vigorously against this type of thinking and its perpetrators. His initial inkling of what was to come probably arrived by way of his work on the Arctic. During the Second World War, a sudden flurry of interest in northern development resulted in a joint American-Canadian committee on the Arctic. This group included academic, government, and business representatives, and its general attitude is summarized by the following letter from one professor to another:

> Canada's greatest Polar neighbour is the USSR. For about 20 years Soviet Russia has, by all accounts, been doing remarkable things in the study and development of its own Arctic area. Even if we discount generously what we hear there seems little doubt that Soviet Russia has gone further than any other country in the world in developing knowledge of Arctic problems and in making use of Arctic areas and resources. This development in the Soviet North, set against the relative lack of development in the Canadian (North American) North, creates a situation of unbalance, which is unsatisfactory from any point of view. I need not enlarge on this point further.[14]

While Innis undoubtedly was delighted to see the advent of this interest in Arctic studies, he was equally dismayed by the biased approach that accompanied it. As he would later put it: 'There is a continuous assumption that there can only be two worlds the U.S. and the U.S.S.R. and the rest of us whether we like it or not must be stool pigeons of one or the other ... The insularity of this imperialistic self-confidence is very disturbing.'[15]

The importance of understanding Russia in this context was brought home to Innis during his attendance at the anniversary celebrations of the Soviet Academy of the Sciences at the end of the Second World War, one of the key events that drove forward his intellectual efforts in the communications field. Discovering Russia first-hand provided Innis with the revelation necessary to reinforce his earlier misgivings about specialization in the social sciences. He wrote that 'to be trained in a subject which has its roots in the west and which has suffered from the characteristic disease of specialization and to realize suddenly that a vast powerful organization built around the efforts of 180 million people has arisen with little interest in this specialization is to find oneself compelled to search for possible contacts in the broader approach of its history.'[16]

When Innis attempted such a broader approach, he cast Russia in the role of the chief balancing force to mainstream Western civilization. Innis saw Russia as a continuation of the Byzantine and later Mongol empires. As such, it was the key actor just off-stage in his presentation of the ebb and flow of Western civilization. In contrast to Cold War ideology, Innis believed that Russia, as the heir of the Eastern Roman Empire, had been a key *stabilizing* force in the history of the West. The Byzantine Empire, while not central to Western civilization, carried on several functions crucial to its survival. For instance, the continuation of the Roman Empire in the East until the fall of Constantinople in 1453 and the domination of its culture by the Greek heritage meant that much of that heritage survived to be passed along to Western Europe. Innis views this transfusion as the essential catalyst leading to the Renaissance. 'Byzantine civilization had nursed Greek civilization and turned it over to the West.'[17] At the military level, as well, the Eastern Empire made an important contribution to the West. 'Organization of the Russian empire checked the devastation of the nomads, which had threatened Western civilization over two millennia.'[18]

These specific historical tasks were complemented by the more general and long-term role that the Eastern Empire played as the negative foil to the West within the overall cultural heritage of Western civilization. In other words, in terms of the dialectic of empire, the Eastern case preserved those patterns of institutional, cultural and technological organization that, although indigenous to the Western heritage, were out of keeping with the actual social organization of Western Europe. For example, at an institutional level:

> In the East political power remained in the hands of monarchies and spiritual power based on Greek philosophical thought was in the hands of the

councils. In other words political power was monarchical and spiritual power democratic in the East and political power was democratic and spiritual power monarchical in the West. The monarchical organization of the spiritual power in the West has served as a check to the growth of totalitarianism but the democratic organization of the spiritual power in the East has left it exposed to the tyranny of the state. But the character of absolute power in the West, about which Acton spoke, in the spiritual world, contributed to the splintering of religious organizations in Protestantism and to the strengthening of power in the temporal world in the modern state.[19]

However, the element of continuity and stability that Innis assigned to the Eastern Empire was rooted in differences at the infrastructural rather than the institutional level. The overwhelming dominance of agriculture in the Russian economy contributed to Russia's stability. Agriculture imposed a regular universal and cyclical rhythm as opposed to that of machine industry.[20] At the more specific level of communications technology, the long existence of the Byzantine Empire was explained by Innis as a consequence of the nearly perfect balance of media: on the one hand, parchment used by spiritual authority, and, on the other, papyrus and later paper used by secular authority. The Eastern Empire did not face the introduction of machine industry and printing technology at nearly as early a date as the West. This meant that it did not experience the political fragmentation following an amplification of local vernaculars by their mechanized reproduction. As well, it escaped, to a great extent, the penetration of the price system.

When technological innovation was introduced, the institutional history and cultural heritage of the Eastern Empire assured that it would be used in a fundamentally different manner. Both at the end of the Czarist regime and continuing through the era of the revolution, the task of 'catching-up' to the technological infrastructure of the West (through a centralized directing autocracy) dictated concentration on producers' goods. Propaganda replaced advertising as the key function of the press.

Emphasis on producer's goods involves little need for advertising. Whereas an economy which emphasizes consumer's goods is characterized by communication industries largely dependent on advertising and by constant efforts to reach the largest number of readers or listeners, an economy emphasizing producer's goods is characterized by communication industries largely dependent on government support. As a result of

> this contrast a common public opinion in Russia and the West is difficult to achieve ... In Russia dominance of the press by the government and limited retail development which accompanies limited advertising implies unity and effective domestic propaganda, whereas in Anglo-Saxon countries extensive retail development and extensive advertising implies unwillingness to believe anything which does not get into the headlines, and not even that, including Russian news.[21]

Innis summarizes this differential application of communications technology by saying, 'The effectiveness of propaganda in Russia explains its ineffectiveness abroad.'[22] However, he notes that the two strategies for employing mechanical reproduction in efforts of social organization have a similar effect. They lessen the dependence of the directing alliance of an empire on public opinion and increase its reliance on force.

With respect to military technology as well, the Eastern Empire is fundamentally different from the West. Western technology and strategy is overwhelmingly maritime and offensive: its Eastern counterpart is continental and defensive. Innis emphasizes the inability of the West to understand the logic of the Eastern strategy by translating it into the historic context of the West. 'The strategy of defence springs forward to a position that it has not occupied since the castles of the medieval period before the invention of gunpowder.'[23] The very nature of a defensive strategy increases the potential for its being misinterpreted.

> The success of defence in secrecy, destruction or supervision of contacts with the outside world, a distinct railway gauge, a difficult language for Europeans which with Marxian indoctrination was impenetrable to propaganda, concentration on a small single party, control over the press and radio, have been evident in the underestimation of Russian strength by the Germans and by the Allies. Confusion in the Western world was in part a result of the interest in ideologies and the neglect of factors of military strength. Journalists affected by the age-old tendency of the printing industry to push to the left found themselves suddenly confronted by a powerful centralized force which was essentially right. The right which thought of Russia as left was even more confused than the left which found she was right.[24]

Innis means by this last comment that right-wing and left-wing political views are usually conceived within the mainstream intellectual tradition of the West. 'Right' or 'left' indicate the extent to which force or intelligence respectively represent the principal organizing factors of a society. Russia is an anomaly to this perspective, being a society that articulates its principles in terms of left-wing ideology but whose institutional basis lies in the mobilization of force to an unprecedented extent. Innis summarizes the paradox of Soviet 'defensive' imperialism in the words, 'The major concern [of Russia is] preparation for war against ultimate aggression from the West.'[25]

The value of the Russian case to the West as an alternative example of a society still within the Western tradition was also the reason it was misunderstood. Russia exemplified a society that emphasized stability and continuity, principles that were necessary for the development of Western civilization. Yet Russia was being presented as a totally alien civilization because the United States, with the extension of commercialism and electronic communication technology, more and more emphasized individualism and incessant change. Innis felt that it was the task of the universities to provide the longer-term perspective that would counter this tendency by presenting a more balanced picture of Russia. This task would involve implicit criticism of the extreme instability of the American Empire. Innis's 'long-term' view goes directly contrary to Cold War ideology by identifying the *United States* as the chief threat to peace and stability in the world. Bearing this in mind, we can well understand why Innis becomes increasingly sensitive to attacks on his own position and on the university as an institution. His essays oscillate between being guarded in their political implications to the point of obscurity and being shrill and propagandistic in tone.

His attitude was so out of tune with the times that in the early 1950s he came to equate maintaining communist intellectuals on the staff of the university with the final act of political independence. In reply to an attack by Gilbert Jackson to the effect that University of Toronto professors were teaching communist dogma, Innis states to university President Sidney Smith: 'I am under no illusions that we could achieve the academic freedom of the older English and French institutions where members of the communist party belong to the staff since we do not have the prestige enabling us to be unconcerned with the Mr. Jackson's. The achievement of such prestige becomes in a sense the last rampart to be taken in the struggle for responsible government.'[26]

The United States: Heir to Rome

The Russian experience was important for Innis not because it redirected his researches to an in-depth examination of the Russian Empire but because it opened his eyes to the dangers of extreme bipolarization in the world polity. As he said in one of his more optimistic letters to George Ferguson: 'Well, the world seems to be getting a little way from the dumbell arrangement with two large powers at each end of a very thin shank but it is a long way back. Our role appears to be the most difficult of all with the U.S. constantly breathing down our necks.'[27] Innis directs his critical salvoes at the United States rather than at Russia, for three major reasons. First, Innis's post-war trips to the United Kingdom brought home to him the diminution of Britain as an imperial centre.[28] This, in turn, required a change in his strategy vis-à-vis the possibility of developing an independent polity in Canada. The era in which some degree of flexibility had been possible through the countervailing pressures of Britain and the United States had given way to one in which the United States was increasingly dominant at all levels: politics, culture, the economy. The focus of Innis's articles tended to shift to meet that threat.

Second, for him to expend most of his time producing critical pieces on the Russian Empire would have been the equivalent of contributing to, rather than alleviating, imperial imbalance. Innis was convinced that the bipolar thinking that characterized the Cold War was fundamentally wrong in that it represented the triumph of institutions of force rather than knowledge within the Western pole of empire. He came to view the nuclear arms race quite literally as the extension of the technique of the Inquisition that he so abhorred. His language was anything but moderate: 'We are in grave danger of being swept off our feet by the phenomenon we are describing. We are in danger on the one hand of losing our objectivity and on the other hand of being placed under arrest. Freedom of the press has been regarded as a great bulwark of our civilization and it would be dangerous to say that it has become the great bulwark of monopolies of the press. Civilizations have their sacred cows. The Middle Ages burned its heretics and the modern age threatens them with atom bombs.'[29]

Third, Innis chose to concentrate on the United States because his perspective as a Canadian allowed him to make profound criticisms of the American Empire. There is a subtle irony in this view. On the one hand, Innis recognized the importance of the closeness of Canada to the United States both in terms of providing the security necessary for research and the easy familiarity prerequisite to access to things American. On the other hand, the marginal position of Canada vis-à-vis the United States permitted the recognition of the predatory underside of this apparently pacific behemoth. As Innis put it graphically in his *Russian Diary*, 'Canada might play [the] role of [the] small bird which picks the teeth of the crocodile and in compensation gives warning of danger.'[30]

The distinction has sometimes been drawn between Innis's more polemical and 'anti-American' essays, such as 'A Strategy of Culture' and 'The Military Implications of the American Constitution,' and the more subdued bulk of his communications works. This distinction is specious, obscuring the fact that the goal of his long-term research was to develop a critical perspective that could be brought to bear on the problems of the present as well as those of the past.

I contend that Innis's polemical essays are not an aberration but a natural outcome of the conclusions to which his long-term communications work had led. The problem is that Innis waffled on the issue of causality and made his argument more abstruse in order to avoid being attacked by specialist scholars in the case of his long-term research, and by anti-communist vested interests in the case of his polemical essays.

The genesis of his anti-American polemics in the dialectic of empire that characterizes his longer-term research is apparent in the conclusion to *Empire and Communications*. (Innis's aversion to causality and his flashing insights wedded to overall obscurity are also clearly evident in this passage.) There, he writes:

> Concentration on a medium of communication implies a bias in the cultural development of the civilization concerned either towards an emphasis on space and political organization or towards an emphasis on time and religious organization. Introduction of a second medium tends to check the bias of the first and to create conditions suited to the growth of empire ... In the United States the dominance of the newspaper led to large-scale development of monopolies of communication in terms of space and implied a neglect of problems of time. The ability to develop a system of government in which the bias of communication can be checked and an appraisal of the significance of space and time can be reached remains a problem of empire and of the Western world.[31]

To trace the development of Innis's attitude to the United States even farther back, we should turn to some of his earlier communications works such as *Political Economy in the Modern State* and 'On the Economic Significance of Cultural Factors.' Both were produced in 1944 and exhibit a much less polemical spirit towards the United States than his later works. At the same time, they demonstrate that the basis of Innis's critique of the United States is his fundamentally conservative attitude towards modern democracy. In 'Political Economy in the Modern State,' Innis approvingly quotes Jacob Burckhardt on the conundrum of modern democratic society: 'The numbers of the wise who think are little capable of increase at any time; but the numbers of the public who are influenced by opinion become yearly greater. Knowledge has less and less influence on affairs and opinion more and more.'[32] In 'On the Economic Significance of Cultural Factors,' Innis goes on to attribute the driving force of this process to the perfection of newspaper technology. This technology negated the necessity of according institutions of knowledge significant influence in the directing alliance of empires. Instead, that function fell to institutions of force directly manipulating public opinion. The soldier with his short-term tactical thought replaced the diplomat with his long-term strategic concerns as the stereotypical imperial leader. Innis views his century as one characterized by permanent mobilization for war to such an extent that he terms it the era of 'the collapse of Western civilization.' He contends that '[the newspaper as a] vast new instrument concerned with reaching large numbers of readers rendered obsolete the machinery for maintaining peace which had characterized the nineteenth century. Guizot wrote of the great evil of democracy, "It readily sacrifices the past and the future to what is supposed to be the interest of the present," and that evil was accentuated by the reign of the newspaper and its obsession with the immediate.'[33]

Innis's long-term research, therefore, had led him to progress from a concern with the problems posed by the exhaustion of natural resources on the periphery to the problem of exhaustion of economic as well as political and cultural resources at the centre of empire. Newspaper technology was like an addictive drug that accelerated this destructive process while at the same time obliterating those long-term critical faculties that could identify the basis of the problem.

> The circulation of printed matter cheapened thought and destroyed the prestige of the great works of the past which were collected and garnered before the introduction of movable type ... European civilization lived off the intellectual capital of Greek civilization, the spiritual capital provided by the Hebrew civilization, the material capital acquired by looting the specie reserves of Central American civilizations, and the natural resources of the New World.
>
> ... The enormous capacity of Western European civilization to loot has

> left little opportunity for consideration of the problems which follow the exhaustion of material to be looted.[34]

Particularly after the Second World War, it was clear to Innis that the United States represented this general malaise in its most exaggerated form. In a series of essays, he sets out to probe the dynamics of this disintegrative process, concentrating on the various levels of the dialectic of empire that he had been developing. 'Technology and Public Opinion in the United States' (1949), 'Military Implications of the American Constitution' (1951), and 'Roman Law and the British Empire' (1950) dwell respectively on technological, political, and cultural aspects of his critique of the United States.

Innis begins 'Technology and Public Opinion' by describing a typical situation of technical innovation on the periphery of empire occasioned by monopolistic forces operating at the imperial centre to prevent such innovation. Newspaper technology had been introduced on the periphery as an agent of imperial control (through the printing of legislative decisions). It had been strengthened indirectly by the banning of the political press at the imperial centre. This action had stimulated the rise of the social novel and the book trade at the centre, and the effective depression of this type of production on the imperial margin. Other attempts at the centre to curb political journalism through restriction on paper production had released rags for export as raw materials to the presses of the New World.[35] In short, the development of the press in the United States is viewed as a counterpoint to its active suppression in Britain. The role of the American press in providing a key technological infrastructure supporting the revolution was recognized for the first time in the incorporation of the principle of 'freedom of the press' in the constitution of the new republic. The influence of the newspapers was likewise reflected in their preferential treatment under United States postal legislation. Far from seeing them as acts affirming individual liberty, Innis sees these pieces of legislation as guarantees of a monopoly of communication central to the ruling alliance of the United States.

The political influence of these technological developments carried a significance out of all proportion to communications innovations in the past. Previously, technological innovation, by its very nature, had led the class primarily concerned with it to become more and more cut off from everyday society and to become less and less innovative. This phenomenon of increasingly inefficient monopolies of communication less and less able to understand and direct the masses had led eventually to imperial stultification accompanied by technological innovation carried out by marginal sectors. The result was either internal revolution or external 'barbarian' invasion to replace the traditional ruling class.

However, newspaper technology, unlike other technologies, did not carry with it the seeds of a revolution against its directing class. Because it represented the mechanization of the vernacular, it could not become removed from the language of the masses. Moreover, because it carried the ability to amplify a centrally produced message in the vernacular, it had the capacity to mobilize and transform, with hitherto unheard of rapidity, the overall psychology of a society. It not only could deliver the message, it also could change the nature of the mind-frame interpreting the message. It was a peculiar technology that intervened in the popular psyche in a manner that went far beyond the immediate ends of those who used it to manipulate day-to-day issues.

Innis traces in minute detail the advances in the technology and its gradual movement away from being a political instrument towards becoming an advertising tool. This disturbed Innis greatly, for the underlying logic of the former is a political project that by definition involves a concern for duration and stability. In contrast, advertising, aimed as it is at the expansion of markets, emphasizes the permanent excitement of the largest number of people to the prospect of change. The same basic mechanisms apply in such widely separated fields as fashion and politics. It is as if we have gone full circle and returned to the ecstatic basis of social education characterizing pre-literate societies. Unfortunately, the new ecstasy created by the newspaper is oriented towards fuelling the most banal forms of consumerism rather than towards the conservation of a cultural heritage.

Furthermore, newspaper technology is viewed as pernicious because it is so fundamentally linked to the vernacular. As a result, it can be challenged successfully only by technological innovations that themselves surpass the newspapers' ability to penetrate the senses (particularly the visual and aural) of the citizenry through the mechanization, amplification, and transmission of the vernacular. These technologies may permit the reorganization of political coalitions and relatively short-term trends towards increased centralization or decentralization. However, their overall effects are similar to those of the newspaper. They increase social hysteria, irrationalism, and the appeal to force at the expense of rational, contemplative rule based on an appreciation of the strengths and limitations of the social heritage of the West.

Innis views communications innovations based on the increasingly effective mechanization of the vernacular as predatory. He feels that they are self-limiting only to the extent that they destroy the cultural heritage that had given rise to them. In this sense, the United States represents to Innis something akin to a cultural apocalypse. He concludes that 'dependence [of the United States] on organized power and a traditional antipathy to coloured peoples weakens political sensitivity, and lack of experience with problems of continuity and empire threatens the Western world with uncertainty and war.'[36]

In 'Technology and Public Opinion' Innis is principally concerned with tracing the relationship of innovations in mass communications to the end-game principle in the American imperial project. In 'Military Implications' he attempts to outline the preponderant position of strategies of force in American political history. Innis begins his essay by observing that

> revolutions leave unalterable scars and nations which have been burned over by them have exhibited the most chauvinistic brand of nationalism and crowd patriotism ... The founders of the American Constitution appear to have recognized the danger [of this context] by framing an instrument which put limits on the number of things concerning which a majority could encroach on the position of the individual ... This Conservativism and an emphasis on the theory of divided powers led to provisions strengthening the executive power [by] making the President Commander-in-Chief of the Army and Navy and giving him control over patronage[37] [and thus creating a countervailing force to the radical state legislatures].

Freedom of the press in the Bill of Rights had a similar end in mind (that is, the fostering of countervailing powers) in an era in which communication technologies were relatively inefficient.

At this point in the essay, Innis embarks on an extended political history of the United States designed to show that the untrammelled development of mass-communication techniques subverted the ideal of the rational, individual citizen through the perfection of techniques for mobilizing and manipulating public opinion. These developments strengthened the executive, which could utilize the bias of the mass media by activating public sentiment through tactics emphasizing the deployment of force. The citizenry could be inoculated against absorbing different perspectives on the world – perspectives developed on the periphery of the American Empire – by the executive's describing, these as artefacts of an 'enemy' rather than by challenging them in rational debate. The overall result was a highly inconsistent, force-oriented foreign policy that was subservient to short-term domestic politics:

> The disequilibrium created by a press protected by the Bill of Rights had its effects in the Spanish American War, in the development of trial by newspaper, and in the hysteria after the First World War. Holmes wrote 'when twenty years ago a vague tremor went over the earth and the word socialism began to be heard, I thought and I still think that fear was translated into doctrines that had no proper place in the Constitution or the common law' ... As a result power shifted increasingly to the executive and involved reliance of the executive on force.
>
> In the conduct of foreign affairs, a lack of continuity ... was in strong contrast with the continuity evident in Great Britain and in Russia.[38]

The specificity of this situation to the United States is evident in the regularity with which military men serve in its executive arm. Innis feels strongly that a '1984'-style state of permanent war was being reached as a result of this dynamic at the heart of the United States. 'We must have "war to solve unemployment in order to ensure against internal anarchy, instead of war solely to protect employment (ordered life) against external aggression."'[39]

In 'Roman Law and the British Empire,' Innis deals with the pathology of the American Empire primarily at a cultural level. He sees the rule of law as one of the highest achievements of Western civilization. To him, it represents, in institutional form, the coalition of force and knowledge directing the social project. It is the main convention linking the technological infrastructure of society to its political organization. On a cultural plane, it exemplifies archetypically a rational, non-arbitrary way of thinking that permeates even the common-sense of individuals in a society.

Innis's article was a response to study by James Bryce that compared the contribution of Roman and common law to the ancient Roman and British Indian empires respectively. He maintains that both Roman law and common law were great achievements of Western civilization, and in any great empire of that civilization we could assume that a mix of the two traditions would exist even though one might at first appear dominant. 'Roman Law and the British Empire' attempts to trace the role of the subordinate mode of thought within the Anglo-Saxon imperial tradition. Roman law functioned by arguing from the general to the particular. It involves assessing particular actions on the basis of a universal code. Common law, in contrast, proceeds from the particular to the general. An act is assessed as just or unjust in being compared in all its details with the precedent of past acts brought before the court. The former stressed the written code; the latter, the oral record. Innis summarizes some of the consequences of these traditions for societies dominated by one or the other:

> Common law implied concern with local customs and facilitated the development of the British Commonwealth by peaceful means or by minor rebellions.
>
> With the increasing importance of legislation ... lawyers continued to play an important role in Parliament ... Common law countries favour the election of lawyers as legislators to the exclusion, for example, of journalists, in contrast with Roman law countries which seem to favour the latter. In common law countries the state became a part of customs and traditions and the revolutionary tradition was weakened. Marx's withering of the state had reference to Roman law and not common law countries. Common law traditions which made politics a part of law and emphasized the relation of the state to law implied an absorption of energies in politics and a neglect of the cultural development which has characterized Roman law countries.[40]

Innis views imperialism based on common law as more balanced, less brutal, and more stable. This is significant because it has implications for his negative assessment of American imperialism, for he views the United States as the Anglo-Saxon fragment most permeated by a strain of Roman law. As Innis puts it:

> The British Empire emerged as a result of a balance between the oral tradition and the written tradition, between common law and Roman law. The element of Roman law, especially precipitated the American Revolution, and the written constitution of the United States was designed to restore it and to protect its position. Emergence of a federal government with a constitution which gave enormous powers to the courts involved protection of fundamental law but, in protest against the divine right of Parliament, assumed the divine right of the United States. Without a written constitution Great Britain was able eventually to master the problem of empire and to digest the element of Roman law or rather to cast it out into regions which left the empire, for example the United States.[41]

> The reaction of the United States and members of the Commonwealth to protect fundamental law has left them in their attempts more imperialistic than the mother country ... The reassertion of common law in Great Britain and the decline of imperialism [is matched by] its decline in the other Anglo-Saxon regions and the rise of imperialism.[42]

In Innis's view, because of its degree of certainty concerning the 'truths' of its ethnocentric principles, this imperialism is particularly pernicious. Unlike other European imperialisms founded on Roman law traditions, the United States is not characterized by a strong cultural tradition of its own. It is, therefore, blind to the nature of its own imperial impositions and insensitive to the resistance 'satellite' cultures put up on the basis of their own traditions. It not only dominates the regions coming within its sway politically, it also annihilates their cultural heritage and replaces them with its own inferior culture. In post-Innisian terms, this was a devastating critique of 'consumerism' and 'globalization.'

In his conclusion to the article, Innis makes clear his belief that this process of cultural homogenization is the key threat facing Canada. The principal strategy at the heart this threat is a generalized mobilization in the name of anti-communism.

> Common law traditions assume that the state is part of the law and the subject has greater difficulty in separating himself from the state. Change is consequently more gradual and less subject to revolution ... But Roman law tradition favoured by [the] written constitution in the United States ... leans toward imperialism, and threatens the beneficial effects of common law in western civilization. As a result of a firm belief in the impossibility of the spread of Communism in common law countries and in the danger of American imperialism in exploiting us through its propaganda about Communism I have felt compelled to seize this opportunity to describe our difficulties.[43]

Canada, the United States, and the Realization of Innis's 'Project'

From the above analysis it should be evident that Innis's diatribes against the United States are not anomalies but the logical result of the direction in which his long-term research was steadily carrying him. As we suggested in the introduction to this book, Innis conceived of his life's work as a highly significant, wide-ranging project whose highest attainment would be to make a contribution to universal knowledge based on the particular perspective developed by Canadian scholars. He remained opposed to political statements on the part of scholars because this distracted their efforts from the fundamentally important area: basic research. However, his own efforts to make a contribution to knowledge through the development of a theory of imperial dynamics and communications systems led him in the end to believe that his own project was threatened at every level by overwhelming American influence. Only when faced with this realization did he unleash a series of articles unremittingly critical of the United States.

Yet even these strident essays should not be regarded as simplistically anti-American. Innis's brand of 'nationalism' was often highly critical of Canadian nationalism, especially when it appeared to replicate the most inconsistent and chauvinistic aspects of American imperialism. The tortured nature of his writing in the last years – torn between dispassionate scholarly analysis and bitter diatribes – reflects his realization that the worst aspects of American society were confronting Canadians not only as an external threat but also internally. The process involved the systematic truncation of a budding national psyche. Innis continually identified the double-talk involved when this process was hastened in the name of Canadian 'nationalism.' The anti-British, isolationist nationalism of the left in the 1930s fell into this category. Four decades later, Eric Havelock would concede:

> In a curious way, both the imperialists and the isolationists shared in the twenties and thirties a common illusion, namely that Great Britain was still the dominant world power. The former accepted this as a concept they wished to embrace; the latter saw it as a legacy which they wished to escape. The result was an overestimate on both sides of opinion of Britain's capacity to wage and win a second war. The radicals were also unimpressed ... by a Hitler who they mistakenly thought would inevitably be undone by the contradictions of his own system ... For such illusions, nourished by Marxist theories, Innis had no use.[44]

Not surprisingly, in the post-war era, Innis's left-wing nemesis of the 1930s, Frank Underhill, 'became the major academic defender in Canada of the American view.'[45] At the same time, Innis found himself articulating an anti-cold war position with Underhill-like vitriol.

Innis's nationalism was a pessimistic, conservative one similar to the position that George Grant would articulate two decades later in *Lament for a Nation*. Innis writes:

> Canada has had no alternative but to serve as an instrument of British imperialism and then of American imperialism. With British imperialism, she had the advantage of understanding a foreign policy which was consistent over long periods and of guidance in relation to that policy. As she has come increasingly under the influence of the United States, she has become increasingly autonomous in relation to the British Empire. Paradoxically, the stoutest defenders of the Canadian tariff against the United States were the representatives of American capital investors. Canadian nationalism was systematically encouraged and exploited by American capital. Canada moved from colony to nation to colony.[46]

Innis's intensely critical attitude towards the Liberal Party was also similar to the later position of Grant. In fact, Innis believed that the totalitarian spirit being nurtured in the United States in the name of anti-communism had found ready proselytizers in all three federal parties in Canada. However, the Liberal Party, as the party in power, was the one that reaped the benefits of this trend. The diminished valorization of 'balance' undercut popular acceptance of the role of the political opposition. Within all legislative jurisdictions, the resurgence of 'Roman law' concepts tended to support the view that the authority of the legislature was legitimately extendable to a totalitarian degree. Provincial legislatures followed the federal in this pattern. Innis expected that a prolonged crisis of federalism would be the result of the increased pretensions of assemblies with regard to the limits of their power. This trend would manifest itself in the decline of opposition parties within each legislature and their corresponding strengthening in legislatures at another level or in another region. In short, by the late 1940s, Innis was playing the role of a scholarly Cassandra. He was not taken seriously as he accurately predicted a profound and long-term international crisis of stability, as well as recurring federal-provincial crises and the diminishing role of the political opposition. Furthermore, he was alone in situating the latter within the framework of the Cold War mentality of the former. Consider how outrageous the following comments would have seemed to the world of 1948:

> The size of the Liberal majority in the last election ... [indicates] the electorate felt that a strong opposition was not important. It may be suggested that militarism played its role in that emphasis was given to it by all parties and that such emphasis could have no other effect than strengthening the party in power.

> As a result the political shape of Canada began to assume characteristics similar to those of Russia. A politburo in Canada comparable to and paralleling that of Russia effectively diverts attention from its character by pointing to the dangers of the politburo in Russia ... In Canada a powerful bureaucracy, in part a product of bilingualism, built up in the depression and during the war, continued to exercise a powerful influence in a period of peace to an important extent by insisting that war had never ceased ... The results of an overwhelming majority for the Liberal Party in the federal government and of their control of the Senate and the bench have been evident in various directions. This domination has left individual provinces as the only opposition ... and accentuated the problem of federal government. Parties other than the Liberal party tend to dominate the provinces. Consequently dominion-provincial relations occupy a more important role in Canadian politics ... General disequilibrium and instability have necessitated enhancement of the power of the Dominion. The tendency towards centralization has accentuated an interest in defence and created an impasse strengthening the influence of the United States. The sense of omnipotence derived from an emphasis on the theory of the divine right of legislatures developed in the federal government compels a sense of omnipotence in provincial governments ... The basis of federalism in which the provinces maintained or acquired control over natural resources has been largely destroyed as a result of an increasing emphasis on monetary policy and particularly of large-scale resort to income taxes.[47]

For some, this passage gives rise to a strange sense of déjà vu.

Only when Innis begins to recognize that his hopes for an authentically Canadian cultural contribution to Western civilization were in jeopardy does he openly, publicly, and stridently defend his goal: 'We can point to the dangers of exploitation through nationalism, our own and that of others. To be destructive under these circumstances is to be constructive. Not to be British or American but Canadian is not necessarily to be parochial ... "A cultural heritage is a more enduring foundation for national prestige than political power or commercial gain."'[48] He further recognizes that the French–English question was also exacerbated by the influence of American dominated media: 'The effects of these developments on Canadian culture have been disastrous. Indeed they threaten Canadian national life. The cultural life of English-speaking Canadians subjected to constant hammering from American commercialism is increasingly separated from the cultural life of French-speaking Canadians. American influence on the latter is checked by the barrier of the French language but is much less hampered by visual media.'[49]

We can judge the gravity the situation held for Innis by the fact that he felt obliged to abandon his long-held scholarly principles in favour of adopting an impassioned political stance in support of public broadcasting.

> We are indeed fighting for our lives. The pernicious influence of American advertising reflected especially in the periodical press and the powerful persistent impact of commercialism have been evident in all the ramifications of Canadian life. The jackals of communication systems are constantly on the alert to destroy every vestige of sentiment toward Great Britain holding it of no advantage if it threatens the omnipotence of American commercialism. This is to strike at the heart of cultural life in Canada. By attempting constructive efforts to explore the cultural possibilities of various media of communication and to develop them along lines free from commercialism, Canadians might make a contribution to the cultural life of the United States by releasing it from dependence on the sale of tobacco and other commodities.[50]

The almost paranoid tone of this passage underscores, not only the extent to which Innis was a politically engaged individual looking on dark times, but also the extent to which this gathering darkness, so apparent to Innis, was invisible to the majority of his colleagues.

CHAPTER ELEVEN

Cassandra's Curse

This is the most cruel pang that man can bear – to have much insight and power over nothing.

H.A. Innis[1]

At the beginning of this book, I made the case for viewing Innis's life-long work as a political project with multiple levels and distinct phases. From the start it was a precocious project, involving not only the educational research path of an individual scholar but also a commitment to building scholarly institutions and having a policy impact on the country's development from colony to nation. In the normal scheme of things, a scholarly career plan would become more focused and specialized as time went on. Not so for Innis. His research continually led him to broaden his interests to new areas over longer periods of time; his institution building carried with it ever greater responsibilities and new challenges; and his advice was more sought after outside the university as his reputation grew.

On top of these pressures, Innis recognized that the waning influence of Great Britain, which, circumstantially, had supported his project during the 1920s and 1930s, was being replaced and undercut by the increasing influence of the United States. Again, a normal scholar faced with these realities might have chosen to trim back what he was seeking to accomplish. But Innis was not a normal scholar. He chose to redouble his efforts. In the end it was cancer that claimed him. But in fact Innis was set on a course of working himself to death.

A Life Cut Short

It is probable that the disease that killed him announced itself in a breakdown in his health that sent him to bed for nearly a week in late 1948. Although he recuperated from this bout, his intensive schedule indicates that he was aware that his overall project was far from complete and that his time remaining was limited. In 1949 he launched into the travel and work of the Royal Commission on Transportation. For much of the year, he was essentially commuting between Ottawa and Toronto, working on royal commission affairs during the week and pursuing teaching and academic duties over weekends.

In late July 1950 his health broke down again, and it is likely that he was diagnosed with prostate cancer during hospitalization in September.[2] This was followed by a long period of recuperation at home where he worked on the commission report and departmental correspondence. In June 1951 he embarked on a tour of academic institutions in Europe with his son Donald, visiting Dublin, Glasgow, Paris, Oxford, Cambridge, and London. This was followed in the fall of the same year by a series of academic conferences at Kingston, Princeton, Yale, Philadelphia, and Boston. He probably knew in his heart that these were farewell tours. In January 1952 he was forced to take to his bed again and only briefly returned to the university thereafter.

In an age before radiation treatment and chemotherapy, cancer was a phenomenon that everyone feared but no one spoke about. This was true of even close friends and loved ones. Because of this, Innis's death struggle from the perspective of our more open times takes on a surreal air. It was widely known, though not acknowledged, that he was dying. Irene Biss Spry, with whom Innis had never communicated after her resignation from the University of Toronto and marriage in 1938, heard rumours in England. She sent a note of concern, not to Innis himself but to Mary Quayle Innis, and not directly to her but via her old colleague Vincent Bladen. Bladen replied to Irene in March 1952: 'Harold talked of being fattened up for a minor operation. No one really knows whether this is an "act" – or whether he really is ignorant of what is happening. I gather the doctors look on 3 weeks as about the limit – but I think they don't realize the toughness of the man.'[3] Innis was in hospital again from 17 April to 31 May for a major operation.[4] Mary Quayle Innis wrote to Irene Spry on 25 June:

> Harold has several times been desperately ill and then rallied with a vigour that surprised the doctors. He was in hospital 6 or 7 weeks and is now at home again. I am nursing him and at the moment have no domestic help so I'm kept pretty busy.
>
> At the moment also – we just live from day to day – he is getting along

> pretty well. Since he came home his appetite has been better and he has been able to walk about a little. The outlook is extremely grave but Harold doesn't know how sick he is. He is very cheerful and busy dictating by turns to three secretaries.[5]

During the summer of 1952, on his deathbed, Innis met with numerous friends and engaged in two last projects: his autobiography and the publication of his last book, *Changing Concepts of Time*.

Anyone expecting a 'tell all' autobiography would be sadly disappointed with Innis's version. It is just the opposite. One gets the impression of a man reviewing his extensive personal archives and producing a dry summary of facts from the contemporary documents. Any sense of emotion or the emotional impact of events is absent. There is a curious distancing that takes place; it is as if a stranger is sorting through the material. He would repeat what he wrote about people or events in the language of the contemporary documents but offer no retrospective interpretation. It is clear that he wanted his personal thoughts and feelings to remain private.

I suspect that this was especially true of his relationship with Mary Quayle, who, despite having been his lifelong partner, travel companion, academic supporter, keeper of the home front, and a talented professional in her own right, is all but absent from the 'Autobiography.' It is true this document comes to an abrupt end in 1922, but nevertheless it is odd that her role as Innis's student, their love affair, her psychological support during the post-war period, her extensive work on his thesis, and her participation in their early trip to Europe including the visit to Innis's old battlefields are all absent. The only reference to her at all is in two sentences – 'My wife and I were married on May 10th, 1921, and the summer was spent chiefly in finding a temporary apartment ... and continuing work on my thesis'[6] and 'My wife had spent endless hours typing it'[7] – and the use of the plural 'we' in describing the post-war trip to Europe. Innis has no hesitation in naming Mary Quayle's relatives whom they meet along the way, but nowhere in the 127-page 'Autobiography' is she ever referred to by name.

This makes sense, however, when the 'Autobiography' is viewed as representing Innis's exercise – in the knowledge that he would die soon – of putting his papers in order. It would fall to Mary Quayle to complete the task after his death. She would ensure that the papers that were eventually deposited in the university archives would be subjected to the same filtering out of the emotional and personal.

Innis's decision on his deathbed to devote himself to bringing out *Changing Concepts of Time* indicates its significance to him. The book hints at the perspective that his devotion to long-term scholarship could bring to current affairs. He attempts 'in this volume to elaborate the thesis developed in *The Bias of Communication* (Toronto, 1951) and *Empire and Communications* (London, 1950) in relation to immediate problems.'[8] In it Innis deals with the key theme of his communications phase: 'The problems of understanding others have become exceedingly complex partly as a result of improved communications.'[9] The essays are characterized by extreme pessimism regarding the rising influence of the United States in world affairs. Innis is uncharacteristically clear and vitriolic in these essays. For example, he writes: 'American foreign policy has been a disgraceful illustration of the irresponsibility of a powerful nation which promises little for the future stability of the Western World.'[10] The analysis is radically out of sync with the mentality of the Cold War, the McCarthy era, and the Korean War. It is as if Innis knew that he would be attacked for his thinking and therefore had held off releasing it in book form until he knew his death was imminent.

Late summer saw the final downturn in his health. He was no longer able to receive friends. He was in extreme pain, alternating between agitation, delirium, and semi-consciousness; morphine injections provided the only relief. Finally, Mary Quayle Innis notes in her diary on 8 November 1952: 'H. died peacefully about 8.'[11]

In the days that followed, masses of friends and colleagues visited to express their condolences; the Innis home was packed with flowers, and obituaries appeared in all the major Canadian papers, a wide range of professional journals, and as far afield as the *Times* of London. A large contingent of friends and family came down from Otterville. On 10 November, classes were cancelled at the University of Toronto, and President Sidney Smith delivered a memorial eulogy at Convocation Hall.

Innis after Innis

The many obituaries published after Innis's death were characterized by a striking unanimity in assessing his life. His colleague W.A. Mackintosh expressed the general view: 'The work which Innis did from 1920 to 1940 will stand as his most lasting work, or so it seems to me. His later work is suggestive and will stimulate further enquiries in the field, some of which will bear fruit. But it differs from his earlier work, in that there is not beneath it the massive base of research which will make *The Fur Trade in Canada* and *The Cod Fisheries* works to be consulted in the next century and beyond.'[12] There was much praise for Innis as the model of the quintessential scholar; great admiration for his dedication to the ideal of the university as ivory tower; and recognition that he had produced the definitive studies on fur and cod that provided the basis for understanding the development of Canada 'from the inside.'

Another recurrent theme in the obituaries was Innis's character as a shy, kind, diligent man who had many friends. In assessing his life, his colleagues stress that the only time Innis was hard on others was when he applied his austere and monk-like standard of scholarship to their careers. This was a trait that was universally forgiven even by those who suffered its application directly because they knew that Innis was subjecting them to the same superhuman standards to which he subjected his own scholarly life. Aside from this, there is general recognition that Innis was a good man: unfailingly kind, supportive, and attentive. With his wry sense of humour and his lifelong ability to strike up conversations with common people, this scholar/statesman was the opposite of the intellectual snob.

Let one comment speak for all on the issue of Innis's personality. It is a short poem written for him by a scholar who visited the University of Toronto shortly before his death.

They say he was a good soldier and a good man,
A good scholar and even a good dean.
I do not know the curve of his high span
Nor over what waters its long flight could lean.
His beginnings and ends and the great all
That lay between elude me. I remember
Only he dragged his lame foot down the hall
To the reading room all through a dark December
In the library. I remember, though long gone,
That he moved across a crowded common room
To ask me, a bird of passage from the West,
How I did. His eyes with simple kindness shone,
With light that quite dissolved the Hart House gloom.
For such things he may be remembered best.[13]

Given this general assessment of his character, it is not surprising that, after his death, a hagiographical portrayal of Innis as a kind of scholarly saint set in. In 1957 Creighton's biography appeared, stressing all those elements brought out by the obituaries and sealing the process of canonization. Forty years later, writing, interestingly enough, from the Innisian margin of Australia, Brian Shoesmith notes the detrimental effects of this process: 'Innis has been canonized in Canada. Everybody knows Innis, and most have an opinion about him. I would argue that because of his canonization he is largely misunderstood and frequently misinterpreted.'[14]

Innis hagiography resulted in his being explained rather than understood. A.R.M. Lower describes a common sentiment even among Innis's closest friends: 'For me, the intriguing and unanswered query is, to use a colloquial phrase, what made Innis "tick"? What was there inside the man that kept driving him on? What were his beliefs? What was his ambition? Why did he have this mania for publication? What God was he serving? Except superficially, I cannot answer these questions. Very few of us can know other men well enough to penetrate into the deeper recesses of their psyches, and certainly I could not have done so with Innis. I can recall no long heart-to-heart talks with him, such as I have had occasionally with other men.'[15]

Nowhere is this puzzlement more apparent than in the explaining away of the communications works in the years after Innis's death. Again Lower provides a typical interpretation: 'Persons like Innis and myself have had to start too far behind the line in the scholarly race to be able to tackle these wide-ranging themes; we have had to begin by exploring our own backyard, and by the time we have done that life has been drawing to its close for us ... Innis ... was getting beyond the narrow confines of his discipline, economics, in his last years and, if he had had a cultural background of sufficient depth or had had his training in a more liberal discipline, he might have gone on to philosophy, on the verge of which he was hovering. That was not to be.'[16]

As we have seen, there is an element of truth in what Lower is saying in essentially dismissing Innis's communications works. Innis *did* run out of time. But Lower misses the point in many other respects. First, he overlooks Innis's insight that he, Lower, and others like them had the unique ability to develop a critical insight because of their position on the margin of empire. On the periphery, they had to relate their research more directly to the world in which they lived, unfiltered by the perspective of overdeveloped metropolitan paradigms passed on through a fully established educational system. Second, he fails to grasp that Innis viewed his work as part of a multigenerational project based on the oral tradition. The oral tradition is a living one in which knowledge and culture are handed down through time by the face-to-face interaction of teachers and students. The extent to which Innis contributed to this living tradition, the extent to which he perfected a philosophical approach, would be confirmed not by the acceptance of a 'big book' by the most sophisticated metropolitan scholars but by the critical work done by the next generation of scholars who had been mentored by Innis. Most important, Lower completely misses Innis's success at developing an approach so fundamental in its understanding of the contemporary impasse of the West that it could not be readily absorbed before the cyclonic impact of changes in communications technology made themselves fully felt in the last half of the twentieth century. Innis was simply too far ahead of his time.

The intellectual tragedy of Innis is that the work on communications essentially came to an abrupt halt with his death. Neither his colleagues nor his students took up where he left off. The solitary nature of his research, his fear of attack by specialist scholars, the obscurity of his methodology and style, and the unpopular views to which his research led provide only partial explanations of this intellectual disjuncture.

Innis's colleagues in economic history did not continue his communications work because, to be blunt, they neither understood it nor appreciated its significance. Moreover, these economic historians themselves were becoming more and more out of fashion within the discipline of economics, which was moving inexorably towards a preponderant concern with equilibrium theory and mathematical modelling. By the time I arrived at the department of political economy as a student in the late 1960s, the human capital directly linked to the Innisian tradition had been reduced to a few old men viewed condescendingly by the rising stars of econometrics. The revival of an interest in Innis's take on the economy, in the late 1960s, came not from them but from a new left-wing oriented generation of economic nationalists. By 1982, the department of political economy itself was dismantled and the economists went their own way into a new department of economics, with no discernible difference in orientation or interest from their colleagues south of the border.

The classicists continued to teach at the college level, but their discipline was out of fashion and quickly shrinking. In any case, Innis had not been involved in a direct dialogue with them that might have led to some of their number taking up his research path. Some went on to make major contributions 'in the gathering dusk' of classical studies that Innis would have found fascinating had he lived. But their effect on Innis was one-way only. Since they were quite oblivious of their influence on his work, it is not surprising that he had no influence on theirs.

The development of communications departments and Canadian studies programs was two decades in the future. They would eventually lead to a new interest in the communications works, but only through the rediscovery of the equivalent of ancient canonic texts.[17] The living chain of the oral tradition had been cut with Innis's death.

This is not to say that Innis did not have an influence on other scholars. He served as a model for scholarship, motivating his colleagues to pursue their own research with diligence and enthusiasm. For instance, Creighton, shortly before Innis's death, wrote to announce that he was dedicating his new biography of John A. Macdonald to Innis:

> You are the first person who really showed an interest in what I was doing. You got me the opportunity of giving my first paper on 'The Commercial Class in Canadian Politics' to the Canadian Political Science Association away back in 1933. When all my colleagues were fooling around with contemporary politics and the C.I.I.A. [Canadian Institute of International Affairs], you told me that the only real thing was research and writing. And so it is. But you helped me to see its real importance. And, when I look back, all my scholarships, all my opportunities to do research and writing, seem to have come to me through your help and support. So that the dedication is only just: Macdonald is really your book anyway.[18]

Innis deserves credit for spurring on the research of a number of his colleagues in this way. But it was their research, not his, that they pursued after his death.

And what of his students? Why did not some among them take up what Innis was pursuing? As we have mentioned, Innis was delivering in his fourth-year economic-history class in the late 1940s a running narrative of his communications research as it unfolded. This course was required for all students in economics, commerce and finance, and political science. In the late 1940s, the class numbered from two hundred to three hundred students. Many of them were veterans making up for lost time on the way to a business career. Forty years later, one of them, George S. Blodgett, who went on to a successful career as a business executive, would present to the Harold Innis Foundation at

Innis College the one hundred typed pages of lecture notes he had prepared as an undergraduate in Innis's 1948 class. He would recall fondly:

> Professor Innis gave his lectures ... surrounded by almost complete silence – without the shuffling of feet or the incessant coughing which might have evidenced boredom, but the kind of silence which comes with concentration. For the Innis lectures – make no mistake about it – were hard work. Most listeners, like myself, were fully occupied taking notes, or straining our mental faculties to understand why 'the most striking effect of the introduction of the horse was the rise of Mohammedanism ...' or tracking the logic behind throwaway lines such as 'the Russian Revolution led shortly to the founding of Watrous, Saskatchewan ...' You should realize that in 1948 the majority of the university student body consisted of ex-servicemen. As a group we were in a hurry to complete our education, get our degrees, and get on to whatever careers we could make for ourselves in the real world. In some respects, we regarded our four year detour as simply a necessary evil, something to be endured so that we could improve our chances for success. Certainly for me, and I know for many of my compatriots, that last course from Harold Innis made the four year university experience worthwhile. It was perhaps the only contact some of us would ever have had with an original mind, and we were grateful for having had the opportunity. If you brought nothing else away from university at least you could hardly have come away from that course without a deepened respect for scholarship.[19]

This exposure of future Canadian business leaders to the long-term perspective provided by scholarship was an important element in Innis's project, but it did little for the prospect of the next generation of scholars following up on his research work. But not all Innis's students were destined for business; some would go on to graduate school. Why did they not continue Innis's work? His young colleague and former student, S.D. Clark, recollects:

> Only within the last six or seven years of his life were there any great number of graduate students in the Department of Political Economy at Toronto and few of those did their Ph.D's under his supervision. He had very little contact with graduate students in history where his influence might have been expected to be greatest. I have often felt it was unfortunate that because of the departmental structure of the university Innis spent his

> years teaching students [in his fourth-year course], the vast majority of whom, in economics and commerce and finance, had little interest in economic history. The result was that no great number of students came out of Toronto over these years as followers in his line of work. Most of them, indeed, went off into the business world. Of those who pursued further study, and entered upon on an academic career, very many of them engaged in work in the area of economic theory or political science ... In the years after the Second World War, however, the number of students going on to graduate study in the Department of Political Economy vastly increased, and most of those students enrolled in the Ph.D program took Innis's graduate seminar. The seminar sessions seldom varied in format. Papers were prepared and presented by various students in turn. Vigorous discussion ensued. Yet throughout Innis said very little. There was no expounding on his part; no effort to present, certainly in any systematic manner, his ideas. A pointed remark now and then constituted what appeared to be his total contribution to the discussion.[20]

This reticence of Innis to be more directive with his students during the period when he was pursuing his communications studies (and incorporating the results in his lectures) to a large extent explains why his line of research was not picked up by the next generation of scholars. Innis's reserve is attributed by Clark and others to his shy character and his dislike of authoritarianism. And yet this analysis does not account for the very different style of scholarly supervision that he exhibited during the 1930s. In his correspondence with Irene Biss, for instance, he was far more controlling than Clark's comments would lead one to expect.

To understand why this is so, we must return again to the methodology of research and composition that Innis developed during the 1940s to produce the communications works. Innis could not expect those he taught to carry on his work in communications by simply presenting them with the final product of his labours. He would have had to teach them the process of research and presentation that led him to produce these enigmatic texts. This he could not do without opening himself to attack on charges of plagiarism. He was trapped in a contradiction. The methodology he developed to deal with a new field and increase his scholarly productivity offered a defence against specialist criticism but could not be taught to his students for fear of being considered suspect.

It is likely, then, that Innis's communications works would have been consigned to the scholarly dustbin had not a young intellectual peripheral to Innis's world come on the scene in the last few years of his life. That young scholar was Marshall McLuhan.

McLuhan on Innis – The Oral Tradition and the Aural Tradition[21]

The literature builds up around the name of Keynes or Marx, or someone else, and everything else is dropped.

– H.A. Innis[22]

Marshall McLuhan begins his 1953 essay 'The Later Innis' with this quote from his subject: 'The polished essay was introduced as a clever contrivance adopted by a former dynasty to prevent the literate from thinking too much.' McLuhan then embarks on an analysis of what he views as the most important aspect of Innis's scholarly contribution: the enigmatic style of presentation so at odds with the typical narrative style of scholarship. What McLuhan intuitively recognized, and all of Innis's other closer colleagues missed, was that Innis's communications studies were based on an experiment in a methodology of scholarly research and presentation. It was this realization that would imbue McLuhan's life's work. Moreover, McLuhan would consistently cite Innis as the source of many of the ideas he would popularize.

Without McLuhan's rising fame as an intellectual star in the 1960s, Innis's communications works, largely unread, probably would not have been reprinted. As we have seen, contemporary reviewers generally viewed them as a failed effort when compared to his earlier scholarship in economic history. McLuhan played a role similar to that of the Byzantine monasteries and the Arabs in preserving ancient thought through the Dark Ages – he transferred Innis's ideas on communications through a generation that had dismissed them to a later group of scholars who would value their insights anew.

It is no exaggeration to say that the majority of scholars who approach the communications work of Harold Innis, even today, do so by way of McLuhan. In the main they are searching for clues to the development of McLuhan's thought through an examination of his mentors. But, even if they are seeking out Innis on communications without reference to McLuhan, they will find the latter, in the role of a critical reviewer, present as the author of the introductions to Innis's two major volumes on communications. (The current edition of *The Bias of Communication* contains a long overdue replacement of McLuhan's introduction. However, this underscores my point. *Bias* was not reprinted until McLuhan added his introduction in 1964, a full thirteen years after it originally appeared in print. The McLuhan-introduced version of *Bias* was then reprinted in 1968, 1971, 1973, 1977, 1982, and 1984. Even the 1991 edition, with its new introduction by Paul Heyer and David Crowley, retains the name McLuhan prominently on the cover, presumably in an effort to increase sales.)

McLuhan came to a consideration of communications from an intellectual trajectory that was quite distinct from Innis's. He was an English professor, not a social scientist, and developed an interest in the world-view transmitted through American mass-advertising based on insights provided by movements in art (the surrealists) and literature (the New Criticism).[23] As was the case with Havelock and his book *Prometheus Bound*, McLuhan first became aware of Innis's communications work when Innis put his book *The Mechanical Bride* on the reading list of the fourth-year economics course.[24]

W.T. Easterbrook, who was present at the first meeting of Innis and McLuhan, reports that they did not take well to each other. The lapsed Baptist, Innis, and the convert to Catholicism, McLuhan, had a heated discussion over the Spanish Inquisition.[25] In short, they did not have complementary personalities. One wonders, for example, what the tone-deaf Innis made of the conclusion of a letter addressed to him by McLuchan: 'the actual techniques of economic study today seem to me to be of general relevance to anybody who wishes to grasp the best in current poetry and music. And vice versa. There is a real, living unity in our time, as in any other, but it lies submerged under a superficial hubbub of sensation. Using Frequency Modulation techniques, one can slice accurately through such interference, whereas Amplitude Modulation leaves you bouncing on all the currents.'[26] Despite this, McLuhan is sometimes cast as a disciple whose thought has outgrown the profundity of his teacher. It is not my intention to present a critique of McLuhan's work here. Rather, I am concerned with the way in which McLuhan interprets Innis. I will not deal with the validity of McLuhan's studies themselves but with the accuracy of his perception of Innis's efforts.

McLuhan did attempt to identify a common intellectual background to his and Innis's work. He is justified in identifying Wyndham Lewis as having a significant carry-over into Innis.[27] He also posited a link between Innis on communications (since Innis did his PhD at the University of Chicago) and the work of G.H. Mead and Robert Park. However, James Carey has pointed out, quite correctly in my view, that this supposition is an 'absurdity.'[28] There is no biographical documentation to support McLuhan's assertion. Moreover, Innis's work in thematic terms can be seen as a life-long critique of the basic elements of the frontier thesis of the Chicago scholars. Innis's perspective was more historically rooted, political, and sceptical than this scholarly stream. It led to a dark vision of a fundamental imbalance in the American Empire rather than McLuhan's celebration of the 'electrical sublime.'[29]

Not surprisingly, given this misreading of the genesis of Innis's thinking, McLuhan is the most effective proponent of interpreting Innis's communications work in terms of a unitary dialectic that mechanistically links media, their durability, their sensual bias, and the characteristics of empire.[30] 'Once Innis had ascertained the dominant technology of a culture he could be sure that this was the cause and shaping force of the entire structure.'[31]

But Innis's approach is far more complex than McLuhan would have it. Innis returns again and again to the importance of the 'living tradition.' For him, this was in essence the oral tradition in the sense of the spoken word, rather than the aural (or heard) tradition as in McLuhan. Innis contrasts the living tradition not with the written tradition but with centralized mechanized communication:

> My bias is with the oral tradition, particularly as reflected in Greek civilization, and with the necessity of recapturing something of its spirit ... We should try to understand something of the importance of life or of the living tradition, which is peculiar to the oral as against the mechanized tradition, and of the contributions of Greek civilization ... Creative thought was dependent on the oral tradition and ... the conditions favourable to it were gradually disappearing with the increasing mechanization of knowledge ... The oral dialectic is overwhelmingly significant where the subject matter is human action and feeling and it is important in the discovery of new truth but of very little value in disseminating it.[32]

McLuhan is correct when he states that 'it would be too much to say that Innis had anywhere fully explored the dramatic interplay of the written and oral, the visual and auditory forms of human organization,'[33] but only in an unexpected sense. For McLuhan, the oral tradition equals the auditory, that is, communication by sound waves, and is in complete opposition to the written tradition, which is visual. In Innis, this simple equation is not possible. In his work, all other dichotomies must be viewed in terms of the overall interplay of the living and mechanized tradition. The written tradition may at times reinforce the living tradition (for example, manuscript study), just as oral communications (for example, the radio) may dissolve the oral or living tradition.

McLuhan bases his analysis on a different dialectic without realizing it.[34] As a result, his critique of Innis is characterized by a 'straw man' mechanism. All McLuhan has to do is point out 'inconsistencies' in Innis's thought that, if corrected, will lead to the adoption of the same optimistic conclusions as his. McLuhan's one-dimensional dialectic of communications leads him to criticize Innis for inconsistency where, in fact, there are only different conclusions. His comments on Innis's treatment of the introduction of the alphabet in Greece and the effects of electronic media are two cases of this. The unitary dialectic focusing on the senses that is at the centre of McLuhan's thought also allows him to posit that Innis maintained an apolitical stand when the entire point of Innis's work was political, in the sense of being a demonstration of the dangers to Western civilization inherent in the mechanization of knowledge.

Having presented Innis's method as essentially a form of laboratory testing of how media that are ear- or eye-oriented affect society, McLuhan faults Innis for inaccurate observation in the statement, 'The Greeks took over the alphabet and made it a flexible instrument suited to the demands of a flexible oral tradition by the creation of words.'[35] McLuhan immediately points out that the alphabet 'is a technology of visual fragmentation' and cannot strengthen the oral ('heard,' in McLuhan's sense) tradition. Innis makes the point that initially the alphabet strengthened the living (oral) tradition of the Greeks in a positive sense, namely, that 'the improved alphabet made possible the expression of fine distinctions and light shades of meaning.' The reason why the written tradition played such a positive role in strengthening the oral or living tradition is that it maintained a balance with the spoken forms of communications in the absence of mechanized transmission. The statement that McLuhan makes to criticize Innis is, ironically, precisely the Innisian critique of McLuhan's one-dimensional dialectic: 'As long as the oral culture was not overpowered by the technological extension of the visual power in the alphabet, there was a very rich cultural result from the interplay of the oral and written forms.'[36] In other words, mechanized reproduction, and not writing per se, is responsible for the destruction of the oral (living) tradition. In McLuhan's work, writing, the eye, and the written tradition are one, and their effect is the same. He is, therefore, incapable of seeing anything but inconsistency in Innis's statement that, with the introduction and adaptation of the Phoenician alphabet by the Greeks, 'the ear replaced the eye. With the spread of writing the oral tradition developed fresh powers of resistance evident in the flowering of Greek culture in the sixth and fifth centuries.'[37]

But the problem is with McLuhan, not Innis. McLuhan is looking for an absolute and deterministic set of principles whereas Innis is dealing with the complexity of a real political situation. If we look at the passage McLuhan quotes or at the more extended treatment in *Empire and Communications*, we find a description of a people using their resources of force and intelligence to withstand a political threat. Of course, writing *in general* emphasizes the eye rather than the ear. But in this situation, Innis is *not* dealing with writing in general but with the nature of the Greek alphabet relative to the Phoenician during a time in which the Greeks faced the threat of conquest. It is the very scholar Havelock, cited by McLuhan in contradiction to Innis, who supports the Innisian position by examining in detail how the Greeks developed the technical innovation that allowed them to make their oral tradition *politically* more effective in the face of the challenge from the empires of the East. Because Innis's treatment involves a political dialectic, there is no contradiction in statements such as 'the carriers of the oral tradition, the rhapsodes and minstrels, resisted the threats of the written tradition by the use of writing as a means of guaranteeing accuracy.'[38]

For Innis, the oral tradition is thought of not in the abstract but only as it actually appears in history and survives. We are still living with the heritage of the Greek alphabet because it provided the technological innovation that even today tends to corrode monopolies of knowledge that develop around imperial projects. The crucial element in the alphabet of the Greeks was its effective linking of the spoken and written word. This had global implications for the history of empires, for 'the problem of political organization was in part that of efficiency incidental to the mobility with which ability was attracted to administrative positions. In part such efficiency was dependent on the success with which writing linked the written to the spoken word. A breach between the written and the spoken word accompanied the growth of monopoly incidental to complexity of writing and invited invasion from regions in which such breaches were not in evidence and in which technological advance was unchecked.'[39]

When we turn to an examination of modern empires, we find that McLuhan is once again critical of Innis's treatment of the media. Let us review the case of radio, first in terms of the textual criticism developed by McLuhan and then with regard to the methodological assumptions underlying his criticism.

McLuhan fixes on the following passage in his critique of Innis: 'Shifts to new media of communication have been characterized by profound disturbances and the shift to radio has been no exception. An emphasis on continuity and time in contrast with an emphasis on space demands a concern with bureaucracy, planning and collectivism. Without experience in meeting these demands an appeal is made to organized force as an instrument of continuity.'[40] Innis is here referring to the United States, but in a footnote to the first sentence he points to similar dynamics in Germany in which the telephone, teleprinter, and wireless led to the production of the Eichmann personality type that served as a basis for fascism. Innis ends by saying: 'Dependence on organized power ... and lack of experience with the problems of continuity and empire threatens the Western World with uncertainty and war.'[41] I take this as a clear statement of the consequences of the triumph of the mechanized tradition as opposed to the living tradition. There are several points worth noting. First, for Innis, it is a general occurrence in Western civilization. (This is why the example of Germany is raised suddenly in an article that deals with the United States.) Second, the new media represent the increasing domination of power in the power/knowledge dialectic. Third, the problems of continuity demand 'a concern with bureaucracy, planning and collectivism,' but the present-mindedness of mechanized knowledge is such that it cannot provide solutions, and so organized force becomes the only instrument of continuity. An obsession with sophisticated models of strategic 'planning' is present only in contexts that have failed to solve the problems of social continuity. Fourth, the situation is obviously not presented in a positive or even neutral manner. Finally, the basis for the current context is elaborated in the last part of the footnote: 'Former dictatorships needed collaborators of high quality even in the lower levels of leadership, men who could think and act independently. In the era of modern technique an authoritarian system can do without this. The means of communication alone permit it to mechanize the work of subordinate leadership. As a consequence a new type develops: the uncritical recipient of orders.'[42]

McLuhan consistently misconstrues these passages. He quotes the footnoted section without mentioning its use in Innis, overlooks the last three sentences quoted above, berates the moral indignation of the author as unworthy, and then goes on to suggest that Innis's basic conclusion is that 'new technical media for managing information, when used for older ends established by older media, result in utter confusion and disorganization.'[43]

This is a case of McLuhan's being blinded by his own ideas, which link electronic media to decentralization and a new sense of humanism.[44] In Innis, and in the original author (Albert Speer, in his Nuremberg trial speech), the meaning of the message is clear: that the new media result in a *frightfully efficient* pursuit of ends that are inherent in the structures of those same new media. In Innis we find that the effect of the newspaper and the electronic media is cumulative. It allows for the mobilization of public opinion behind monopolies of force and at the same time undercuts the possibility of a solution to the problems of continuity.

McLuhan overlooks Innis's political stance against both fascism and American imperialism explicit in this statement. At the same time, he takes one sentence out of context and suggests that this is a prescription by Innis: 'An emphasis on continuity and time in contrast with an emphasis on space demands a concern with bureaucracy, planning and collectivism.' McLuhan can then accuse Innis of making a 'simple prescription of more spatial control as a remedy for the excesses of spatial monopoly of knowledge and communication.' If we look at the statement in context, however, we find that Innis is describing a specific historical case: the shift of modern society from the domination of the newspaper to that the radio. The radio era ushers in 'a concern with bureaucracy, planning and collectivism' because the problems of continuity are not being solved through traditional agencies. The point Innis is making is that these *new efforts* fail and an appeal to force is made precisely because radio is characterized by the same form of 'present mindedness' that characterizes the newspaper. Once again, the mechanization of knowledge, not the relative sensual bias of media, is the key in Innis's work. This also underlies the politicization of Innis's position vis-à-vis that of McLuhan. For Innis, the yellow press of the United States and the Nazi loudspeaker had the same form of negative effect: they reduced men from thinking beings to mere automatons in a chain of command.[45]

At the level of theory, McLuhan has based his critique on the assumption that Innis's work should be consistent with a unitary dialectic that stresses the sensory bias of the medium. In light of this, McLuhan finds that Innis's treatment of the advent of radio 'is an example of Innis failing to be true to his own method. After many historical demonstrations of the space-binding power of the eye and the time-bending power of the ear, Innis refrains from applying these structural principles to the action of radio. Suddenly, he shifts the ear world of radio into the visual orbit, attributing to radio all the centralizing powers of the eye and of visual culture.[46]

Let us clarify the methodology behind this criticism. In McLuhan's dialectic of communication, the dominant dimension is the sensual – be it auditory or visual. The chief effect to be traced is how new media change the ratio of perception by the ear and eye of each individual and, in so doing, determine how they transform the psychological and structural make-up of any given society. McLuhan's elaboration of this dialectic is a surprisingly rigid one that equates the eye, the visual, written tradition, centralization, space-binding media, and secular political organization with one pole of the communications dialectic; and the ear, the auditory, oral tradition, decentralization, time-binding media, and quasi-religious organization with the other. When he criticizes Innis, he does so by pointing out that Innis has broken the logic of this dialectic.

But I have tried to show that Innis's later work was concerned mainly with an examination of how a different dialectic, a dialectic of power and knowledge, has played out in human history, and that he uses communications systems as a focus for an analysis of this process. The main poles of the dialectic of history in Innis are the living tradition and the mechanized tradition or the concern with knowledge and the concern with power. The dialectic of communications in Innis is not rigid. Therefore, the ear and the eye, the auditory and the visual, the oral and written tradition, decentralization and centralization, time-binding and space-binding media, and religious and secular organization are not necessarily opposite poles of various levels of the same continuum. Their interrelationship does not follow any rule but is radically indeterminate, or, more correctly put, it depends on the various circumstances present in a particular historical context. What is important in Innis is how each of these aspects relates to the overall dynamic of power and knowledge at a given time. Innis saw clearly that the outcome of this dynamic was the triumph of the mechanized tradition and organized force. This triumph encompassed both poles of the communications dialectic.

In criticizing Innis, McLuhan applies his own central dialectic of ear and eye to Innis's treatment of radio. In fact, Innis is concerned with the effect that the advent of radio would have from the point of view of the dialectic of power and knowledge. From this perspective, he sees in the new media an unprecedented potential for the dissolution of the living or conversational or spoken tradition since the new media made use of the same type of transmitting wave (sound) but subjected it to a previously unheard of degree of mechanical transmission. As never before, the spoken communications impinging on people's lives were similar and were controlled by a central source. The effect that Innis was predicting was a tendency away from critical thinking and towards following orders on a mass scale. If we start from Innis's presuppositions, and not McLuhan's, there is no internal inconsistency in this argument. While McLuhan happily proclaims the switch from seeing to hearing as the dominant form of perception, Innis is less optimistic. He sees no new cycle but only an endgame called the collapse of Western civilization. We are speaking and writing less and hearing and seeing more material that is now mass-produced by a central source. Innis's vision thus becomes an increasingly dark one.[47]

In terms of his general approach, Innis assumed that modern empires demonstrate the same sort of mutually complementing relationship in their media structure that characterized their ancient antecedents. Thus, the newspaper, which emphasizes space, is counterpointed by the radio, which emphasizes continuity of time. However, the general tendency is for new media to be increasingly characterized by lightness or, more precisely, by instantaneous transmission. The result is an incredible ability to solve problems of space (for example, to depose, militarily, hostile regimes in Afghanistan and Iraq) but an increasingly apparent inability to deal with problems of time (for example, to reconstruct these societies in the image of liberal democracies with market economies). The historical drift is away from living (or oral) communication and massively towards passive acceptance of mechanical messages. As we suggested in chapter 8, Innis was extremely interested in the perceptual level of analysis that dominates McLuhan. However, Innis's view of changes that were taking place in the everyday 'common sense' of people was far less optimistic than McLuhan's.

From the point of view of the balance between power and intellect, Innis projects a tremendous shift towards the former. The excellence of modern media, in terms of control through space (or, in more common terminology, the orchestration of public opinion), on the one hand, and the application of military force, on the other, has led to an unprecedented present-mindedness. The control of the media and their use in the formation of public opinion has negated the necessity for the politically dominant group to foster a climate suitable to the development of the type of scholarship that could provide answers to the problems of continuity. The bias of the present is such that even the field of knowledge becomes imbued with a present-mindedness that emphasizes administrative technique and technical specialization. Information replaced intelligence.

Innis's point revolves around the concept of power more than the bias of media. What is critical is not the fact that messages are reaching us in a changed proportion of sight and sound but that they are centrally produced and mechanically distributed. This induces passivity rather than thought, and manipulation rather than personal interaction. The essence of the oral tradition is not that it is heard but that it is dialogue. 'The limitation of mechanization of *the printed and the spoken word* must be emphasized and determined efforts to recapture the vitality of the oral tradition must be made.'[48]

Innis's vision during the last years of his life, while dark, was not as bleak as it might have been. There was a great deal of positive thinking in his approach. Empires were thought to continue in existence *only* if they solved the major problem of balance that Innis described. For this reason, there was an objective reciprocity of interests between the political administrators and the intellectuals of an empire. Innis appealed to this factor of complementary long-term self-interest even in his most pronounced diatribes against the current situation. His position was not an anti-imperialist one per se but a demand for stabilization of empire through its reformulation along balanced lines.[49] He hoped that rational analysis could have an impact on history.

To understand the basis of Innis's ambivalent attitude to power and imperial systems, we would do well to remember an incisive comment on Innis by McLuhan:

> His hostility to power may well be a clue to his essential Canadianism, because power for Canadians has at all times been absentee and irresponsible power. At first Europe and England and now the United States – these are the centres where the decisions concerning us were and are made. The effects of this situation on national psychology have been according to basic temperaments. In the man of 'inner direction' it has bred an extremely independent attitude and a distaste for every kind of authority and social hierarchy. In the 'outer directed' man it has fostered an inclination to rigid bureaucratic structures. So that Canadians, on one hand, distrust Ottawa or have no interest in it, and on the other hand look to centralized government action to achieve the most ordinary local results. In neither case is there any vestige of American acceptance of power as essentially local and amenable to personable intervention.[50]

Thus, it is Innis's colonial background that provides an explanation for his intellectual tragedy. It offered him the orientation and subject matter that eventually led, at the height of the Cold War, to his incisive critique of American imperialism. And yet, the same background dictated that his thought, though lauded, would not be fully appreciated or pursued.

The most serious consequence of this colonial myopia has been an inability to recognize the essentially political dimension of his thought. If we assume the viewpoint of either the staples-theorist or communication-theorist commentators on the later work of Innis, then we must take the media and their characteristics as the determining factor. From this perspective, the intellectual may be seen as in or out of tune with the times but, given the determinacy of the dominant medium, involved in activities that are irrelevant to the course of history. This contrasts sharply with Innis's championing of the role of the intellectual and the university in the modern world.

From the point of view of the classicist, however, there is no contradiction in the work of Innis, except that which reflects the historical motor force of history, the conflict of intellect and power. For this reason, it is Havelock, rather than Easterbrook or McLuhan, who ends by drawing the same conclusions as Innis:

> The intellectual must by definition be pushed to the wall, because his science cannot be competitive [with the powers that be]. To compete for power would destroy his premises and his mental processes ...
>
> So by that virtue which is his [foresight], he is called on to bear an emotional burden which his rival does not have to shoulder. Every time he attempts a fresh effort of foresight he risks offence to the established chain of command in society. If he extends this to the science of man, he sets up an automatic malaise in the machine of society, and provokes active distrust and antagonism from those who enjoy operating it.
>
> ... This brings to intellectual man a certain loneliness.[51]

The heartbreak of Innis's work is that he did not find a solution to the conundrum of linking his critique of imperialism to a social actor who could rebalance the imperial project of the West if given the time and freedom to do so by those exercising contemporary power. His resolute faith in scholarship and the university is no less admirable because of this.

In fact, his most powerful statements on the human condition were made in full recognition of the impasses that had been reached in modern times. The scholar and his institution, the university, were simply not the social agents equal to the task set by Innis.

> The Industrial Revolution and mechanized knowledge have all but destroyed the scholar's influence. Force is no longer concerned with his protection and is actively engaged in schemes for his destruction. Enormous improvements in communication have made understanding more difficult. Even science, mathematics, and music as the last refuge of the Western mind have come under the spell of the mechanized vernacular. Commercialism has required the creation of new monopolies in language and new difficulties in understanding. Even the class struggle, the struggle between language groups, has been made a monopoly of language. When the Communist Manifesto proclaims, 'Workers of the world unite, you have nothing to lose but your chains!' in those words it forged new chains.[52]

Borrowing from the classicists, rather than the surrealists, Innis had brought into his thinking the basic problem traced by his friend and colleague C.N. Cochrane in classical thought. This was the inadequate handling of the concept of fortune. Cochrane traces its passage from the pre-classical epoch into classical culture by way of the contemporary poets: 'From poetry it passes into science, there to become a stumbling-block to historians and philosophers alike.' Thus, the attempt of the classical mind to extend scientific materialism to an understanding of all aspects of human society led to a resurrection of irrationalism at the time of the later republic in the form of a deification of chance. The reintroduction of irrationalism following the domination of the intellectual world by an uncritical materialism 'illustrates in a most serious form the artistic and philosophical vice of *fantastica fornicatio*': the prostitution of the mind to a fantasy world of its own making.[53]

As I have indicated, Innis felt that a similar process was occurring in modern society. Social scientists were increasingly called on to express opinions on the basis of which programs of social engineering were carried out. Innis viewed this as an example of *fantastica fornicatio* to the highest degree. His work was a political project that sought to correct this perversion of the social sciences and to re-establish them as indicators of the long-term limits to human activity. This strategy was 'political' because it is clear that Innis viewed his work as an effort to dissolve specific vested interests in the form of what he termed a 'monopoly of knowledge.' His conception of the destruction and reconstruction of monopolies of knowledge in history lay behind his treatment of imperialism. The essential difference between Innis's work and classical thought is that his ideas were imbued with a Christian concept of progress. Specifically, he viewed the current monopoly of knowledge as so total that its destruction would not necessarily entail the re-establishing of balance but an apocalypse that he referred to as the collapse of Western civilization, or the coming of the new Dark Ages. Modern technology led to conditions in which peripheral spaces where critical thought could be developed were rapidly disappearing. It is this conception that leads to the strange paradox in tone of the communications studies. On the one hand, we find a measured, apolitical, and scholarly treatment of the history of civilization; on the other, a politicized, Cassandra-like tone of urgency in his articles referring to present problems.

In the end, the application of classical themes to history by the 'modern' Innis had a similar outcome to their application by the ancients. Both Innis and the classical thinkers end by concluding, with Cochrane, that 'the process to which mankind is subject is therefore self-defeating; it is like the oscillation of a pendulum ... The role of mind in the historical process ... is simply that of a passive spectator ... Self-consciousness thus resolves itself into a consciousness of impotence in the grip of material necessity ... The acceptance of this conclusion must necessarily breed a profound and ineradicable pessimism ... "Of all the sorrows which afflict mankind, the bitterest is this, that one should have consciousness of much, but control of nothing."'[54]

Precocious to the end, Harold Innis was attempting to provide in his communications works a grand new synthesis that would combine a theory of politics or imperialism (based on the classicists) with a theory of consciousness (based on the theorists of time and space) and a theory of technology (based on his interpretation of the bias of the media). The methodology he employed to develop this synthesis allowed him to develop great insights but undercut his ability to articulate the synthesis in a philosophical manner. Had he been given another ten years, would he have successfully completed the synthesis? While we will never know, posing such a question is worthwhile for it makes two things clear. First, Innis would have had to develop an underlying philosophical framework not dependent on his dubious research methodology to articulate his grand synthesis. Secondly, in doing so, he would have removed many of the pitfalls into which later Innisian scholarship has fallen.[55]

Epilogue

The United States, with systems of mechanized communication and organized force, has sponsored a new type of imperialism imposed on common law in which sovereignty is preserved *de jure* and used to expand imperialism *de facto*.

H.A. Innis, 1950[1]

The new imperialism is creating a new form of sub-sovereignty, in which states exercise independence in a name, without real independence in fact, as formal or informal protectorates of the great powers.

Michael Ignatieff, 2003[2]

After finishing the initial research on this project, I had the opportunity of managing a major development project in newly independent Zimbabwe. This was the beginning of a career in international development: I never returned to the university. Instead, for the last seventeen years, I have been leading one of Canada's major overseas charities, CARE Canada. This has provided me with the opportunity to travel widely in the Third World and the Balkans, often working in countries wracked by war and famine. Along the way, I have also had the good fortune to go from being a bachelor to being a family man.

When colleagues become aware that I have done extensive doctoral work on Innis (an intellectual who is invariably unknown to them), they often ask me if my research has been relevant to my work. It is a question that always gives me pause. On the one hand, I have no doubt that my perspective on the world is largely influenced by Innis. On the other hand, Innis never visited any of the places that have been central to my life and wrote little about them.

In the end, my answer is always 'Yes; Innis remains highly relevant to a critical understanding of the world in which we live.' A complete answer would form the script of another book. I will end this one with some small epiphanies I have had which seem to indicate that this is so.

Good Night Stories

Having seen the light under my son's door at midnight one night, I knocked to give him a fatherly 'lights out' directive. He was playing a computer game. His reply astonished me, 'O.K., George has to go back to school now anyway.' He was playing the game over the Internet with George in Perth, Australia, and George did indeed need to go back to school while Max was going to bed.

Thinking about the incident afterwards led me to appreciate Innis anew. It seemed to me to confirm what Innis, over five decades ago, was identifying as a trend of modern times long before the key technologies mainly responsible for producing it were even in existence. If we think through what this story means in terms of the way my son will view the world, this should become apparent. He is able to 'play with' and 'talk to' friends on the other side of the world in real time. Clearly, his concept of space is radically diminished, as is that of all his contemporaries who are making use of this new technology in the course of their childhood play.

But there is another side to this story. Computer games have a short shelf life. Today's 'hit' game becomes obsolete and unplayed in a matter of a few years. In short, the perception of time in the consciousness of this game-playing generation is much more cut-up and short-term compared to previous generations.

The unprecedented rate of change in basic consciousness with the introduction of these new technologies is apparent if we think back to how a similar incident would have played out a generation earlier. In my case, growing up in the final years of pre-television times, my father would have found me reading a classic children's story (Walter Scott and Robert Louis Stevenson were favourites), or more likely he would have been reading a story to me that he had enjoyed in his youth. In the case of my family (immigrants from the north of Scotland), I was also told the stories of the 'old country' – Macpherson, the Robin Hood of the North; the 'green lady' ghost of Crathies Castle; the stone eagles at the gates of Duff House, the local manor house that came to life once a year, and so on. These stories complemented the full range of Bible stories we absorbed over many years of Presbyterian Sunday school. When we think through what these changes mean in terms of individual consciousness, Innis's insights are borne out. If we concentrate on our sense of 'belonging' or 'identity,' one of the most personal elements of our consciousness, the trend seems clear.

When I was a boy, my friendships were local, neighbourhood ones. Unless one's father's work required frequent family moves, obliging one to establish a new set of friends, this association of friendships and play within a local neighbourhood was true for everyone. Space was so real and 'large' that relations at a distance were utterly exotic: for example, a foreign 'pen pal'; or 'Stephen,' the orphan in Bihar supported by our Sunday school; or international Scout 'Jamborees.'

In contrast, both my sons' sense of belonging to a peer group extends over vast distances every day without impediments. By the same token, their sense of belonging to a group in time has been radically truncated and foreshortened. They do not 'know' the stories of the north of Scotland nor the Bible stories I was exposed to (nor the equivalent stories from my wife's Caribbean/South American tradition). They have few connections through time with a peer group. To put it another way, when either son attains my age and gets together with an international group of the same age, they will be able to talk fondly of the great games they played in their youth. But they will not be able to do so with people, even from their own national context, who are fifteen years older or younger than they. They will certainly not have the same sense of belonging to a religious tradition that extends centuries into the past.[3]

I am not suggesting that video games per se are leading to a massive change in individual human consciousness. I am suggesting that the general trend in innovations, leading to ever 'lighter' and ephemeral communications technology which is highly efficient over distances but of short duration, is leading to unprecedented and accelerating rates of change in human consciousness.[4] It seems to me that Innis was right both in identifying this trend and in positing that it was difficult to become aware of these developments because they had a fundamental impact on human thinking itself.[5]

The Afghan Farmer

When I was visiting one of CARE's projects in the Shomali Plain, east of Kabul, shortly after the fall of the Taliban, I was approached by one of the farmers with whom we were working. There was still widespread bombing in the rural areas as the United States tried to hunt down remnants of al-Qaeda and the Taliban and there was significant 'collateral damage' (civilian deaths) as a result of this activity. The farmer was concerned and wanted to know when the bombing was going to stop. U.S. actions puzzled him, so much so that he was uncertain as to what would happen next. He was approaching me to see if I, as a North American, could better explain what motivated the U.S. war effort.

He started by summarizing the situation from his perspective. He explained how the Afghans had fought against the Russian occupation and how much they had appreciated the financial support of the United States and the Saudis for their resistance. He said that this support had brought many 'Arabs' (by this he meant any foreigner fighting with the Mujaheddin, whether Arab or not) to fight alongside the Mujaheddin. The Afghans particularly appreciated the Arabs because of their bravery. He characterized the Russian occupiers as cruel people who killed many Afghans arbitrarily. The Arabs, he said, were always ready to lead dangerous attacks against the Russians. The Afghans appreciated this because, as he said, 'we Afghans are brave but we do like to live.' So the Arabs often led the attacks. The course of the war changed when the United States supplied Stinger missiles to the Afghans and soon the Russians were gone. What surprised him was that the Americans left too: with the end of the Russian war, U.S. financial and material support ended.

He said that the Afghans let the Arabs stay after the war because they were war heroes. The Afghans also felt sorry for them because many had no place to go after being declared *persona non grata* in their own countries. Besides, he said, many of them had married Afghan women.

With the end of the war and the cut-off in American support, the Mujaheddin factions fell to fighting among themselves, more effectively destroying their country than the Russians did. The only outsiders who remained interested in the country – the Pakistanis and Saudis – then started supporting the movement of religious students – the Taliban. Most Afghans welcomed the establishment of the Taliban regime because they were tired of fighting and they understood the 'Islamic' principles that were the basis of that rule.

Then the Americans became interested again. All of a sudden, the war heroes who had originally arrived in Afghanistan had become evil in the eyes of the United States. They had to be turned over, 'but these were our heroes and our guests so we could not do this.' So the Taliban was attacked. With bombs and money, the United States worked with the regional warlords and routed the Taliban, but what followed was a new period of fighting and instability.

Nothing impresses an Afghan more than military know-how, and the farmer was obviously *very* impressed with how the Americans could dominate the sky and call down fire which could not be countered and from which there was no escape. What puzzled him was the unpredictability of what the United States would do next. One minute they were supporting the Mujaheddin against the Russians. Then their minds seemed to wander and they lost interest. They returned with their own soldiers to fight the Arabs and the Taliban whom they had helped to come into power. Now the Russians were their friends. How could this be? How could they bring the Afghans Stinger missiles and then ten years later insist on taking them all back? What really scared him was that some Afghans had been given telephones by the Americans and were using these phones to call down fire from the sky on their local enemies who often had nothing to do with the Taliban.

I tried my best to explain the trauma of September 11th and al-Qaeda's role, but it was clearly beyond him. He viewed the lack of predictability in U.S. strategy as a kind of pathology – a mental deficiency of an otherwise remarkably sophisticated people. Moreover, in his view, it was a syndrome that was communicable to close allies of the United States. Otherwise, he said, how do we explain the flip-flops of the Pakistani support for the Taliban?

Again, it was one of those epiphanies that called Innis to mind again. He always stressed that Russia was an essential counter-balancing factor to the West and this balance in the Cold War provided a predictability that underwrote forty years of relative peace and stability. With the end of the Cold War, we are faced with an unbalanced situation in which the world faces a new imperialism characterized by a total lack of self-awareness, and a conviction that force alone can solve most problems.

After the last great human blood-letting in the Second World War, the United States led the establishment of an international system whose goal was to replace war as an instrument of national policy with structured mechanisms of diplomacy, negotiations, containment, and deterrence.[6] War was to be a legitimate option only under a United Nations mandate designed to roll back the illegal aggression of one state against another.

Now the country that largely was responsible for putting in place this international system has gone over to an official policy of 'preventative' war. This means a national policy of aggression when the United States decides on aggression for defence, since it has made clear that only the leaders of the United States will be able to determine where and when such a 'preventative' war will be unleashed.

It appears to me that the critical commentary on these extraordinary changes has been less than profound. Returning to where Innis left off may provide a new point of departure for developing an understanding of the United States and its present-mindedness and an answer to the Afghan farmer's query. Consider the following commentary applied to current events although it was written by Innis in 1948:

> American foreign policy has been to a large extent determined by domestic politics ... Under these circumstances a consistent foreign policy becomes impossible and military domination of foreign affairs inevitable ...
>
> Formerly it required time to influence public opinion in favour of war. We have now reached the position in which opinion is systematically aroused and kept near boiling point ... [The] attitude [of the U.S.] reminds one of the stories of the fanatic fear of mice shown by elephants.[7]

What the Hutu Thought

While I would argue that Innis is relevant and deserves reconsideration for his insights into the relation between individual consciousness and innovation in communications technology, and also into the present-mindedness dominating current United States foreign policy, I would also sound a note of caution over Innis's high hopes for the 'oral tradition.' In Innis's writings, the 'oral tradition' is invariably brought in as a positive element which allows for the development of a critical perspective that serves as an alternative and a balance to the 'present-mindedness' instilled by the trend towards increasing levels of 'lighter' media.

In 1994 in Zaïre and Burundi, we at CARE found ourselves dealing with the consequences of the Rwandan genocide. We established camps originally for those Tutsis and Hutu moderates who had managed to escape the genocide and then, in July, larger camps for the hundreds of thousands of Hutu (among whom were many implicated in the genocide) who fled Rwanda before Kagame's advancing Rwanda Patriotic Front (RPF) forces.

One of our key staff had married in Africa and knew several African languages, including Kirundi, and I took advantage of his skills to carry out an informal survey, wandering first around the camp in Kitale, Zaire, and later among the long line of Hutu refugees walking to Zaire from Gikongoro, Rwanda, as the French troops withdrew from their so-called 'Zone Tourquoise' protectorate. I was trying to determine how the refugees were thinking in order to plan for their eventual return to Rwanda. The results astonished me. What was revealed was a kind of lunatic alternative reality which was consistent within its own framework but totally at odds with what was going on in the real world. In this self-referring realm of the Hutu militants' world-view, it was the Tutsi who (with the active support of the Belgians) were massacring the Hutu. The Hutu genuinely believed they were the victims and they lived in a state of total fear.

The crazy world-view had a frightening consistency. The same bizarre answers were given by a wide variety of individuals in a wide range of locations. The genius of it was that, even if I asked follow-up questions which were crafted to raise doubts about the accuracy of this world-view, the interviewees had already been provided with follow-up answers which kept them convinced they were the victims. When a few refugees (usually women) began to show some scepticism towards the world-view, the young men who gathered at the edge of the interviews would make threatening comments and we would be obliged to move on for fear of exposing our interviewees to violence.

From an Innisian perspective, we can see from these incidents that some of his insights were valid. The Cold War did provide a kind of balance in which local political elites jockeyed for power by aligning themselves with the West or the Soviets or by seeking support from both sides as the price for remaining 'non-aligned.' When this sort of support dried up with the end of the Cold War, local politicians cast around for other strategies to build a political constituency and they invariably found them in the old techniques of hate politics which had so decisively been defeated in 1945.

In Rwanda, this political project had been entirely constructed in the local vernacular. Consistent with Innis's expectations, the vectors for spreading the world-view depended on communications technologies which stress the ear: the genocidal radio station, 'Milles Collines,' and the loudspeakers of countless rallies that spurred the hatred. The spread of the hatred and eventually the bulk of the killings were done through civil institutions such as soccer clubs. In short, an understanding of the local oral tradition and its technologies was used to develop a genocidal project. Moreover, this project was largely misunderstood by the international community because it was 'off the radar screen' of developed societies dominated by imperial language and television and newspaper news. The results were horrific. In short, the oral tradition, amplified by loudspeakers and radio, had been used successfully to mount a genocide.

This case would be bad enough if it were a single anomaly, but many post–Cold War conflicts exhibit the same mechanism of political projects mounted through these old techniques of hatred. We have seen these first hand in many places: Bosnia, Kosovo, East Timor, and Zimbabwe, to mention a few.

This is not to argue that Innis is irrelevant to understanding these situations. His grasp of the importance of the oral tradition and its mechanisms, and his understanding that it represents an alternative to the current domination of the present-mindedness fostered by development of electronic communication systems, is valuable. But his expectation that the oral tradition could have a positive balancing effect has proved illusory.

I also think that he underestimated the extent to which elements of the politics of hatred can infect metropolitan contexts with sophisticated communication infrastructure. It is difficult to understand the generalized fear that pervades the United States today without recognizing the ongoing impact of oral elements such as fundamentalist Christianity, conservative talk shows, and the concentration of anger and fear that takes place when fanatics, previously isolated in society, converse with each other and reinforce their paranoia through Internet technology.

Innis Today

The ruin of the concrete silo stands alone on the hillside – the old barn has long since disappeared. The dilapidated old farmhouse is hardly visible behind the row of unruly evergreens. No farm family lives here anymore, only a non-farming bachelor down on his luck.

The railway lines that criss-crossed the township are long gone; the only trace of them is the occasional line of trees at an odd angle to the lot lines and the relocated station now used as the Otterville Museum. The farms are much bigger than in Innis's day, and the farm families smaller and fewer. The schools are bigger too, and the one-room elementary school is a thing of the past. The poor grandmother of a local family now occupies what used to be S.S. #1 South Norwich.

Innis's uncle's country store in Hawtrey, where he worked as a boy, is a still impressive but derelict building. Hawtrey itself is slowly being occupied again after a period as a ghost town, but the new people have jobs in the city and commute to their homes in the country for lifestyle reasons. No industry, stores, or services have returned with them.[8]

Otterville itself has fared better. It has a regional high school which should be named after Innis but is not because 'Innis' to most people in Otterville reminds them not of Harold, the scholar who left, but Sam, the younger brother who stayed, told tall tales, and made many enemies. The mill has long since closed but is maintained as a historical site. But the mill pond gives Otterville a centre and a beauty that Hawtrey lacks. The old farmer's bank is no longer an independent operation, but a branch of a national institution especially set up to serve the needs of its largely aged clientele.

Otterville Baptist Church still exists and remains a fundamentalist 'fellowship' church, having sided with the nemesis of the progressive McMaster professors, the Reverend T.T. Shields, in the late 1920s. The congregation, however, is now small, with many old people and few young.

There is still industry in the town but the many independent operations that served the local market with goods and services are gone. So, too, are the cheese factories that shipped millions of pounds of cheddar to England in Innis's youth. Yet the county successfully produces for the world market a crop never heard of in his early years – ginseng. The landscape is dotted with the small-scale ruins of agricultural production – derelict tobacco sheds. The tobacco companies now demand that the exhaust from the heaters used in drying the leaves be kept away from the leaves and many farmers have abandoned tobacco rather than make the required investment to comply with the new regulations.

The roads are far better than in Innis's day, yet this has resulted in a net outflow of population in search of jobs, education, and the excitement of the city, as well as the replacement of locally produced goods and services by those from outside.

There is little to indicate that this place produced one of Canada's greatest thinkers. One still finds his name on the brass plaque of the First World War memorial on a rise above the mill pond, and also on the faded government 'historic sites' plaque located beside the road in front of the rundown farmhouse that was his birthplace. Innis would probably be quite sad to see what has become of the farm today. And he would undoubtedly be angry at what has happened to his beloved university, too, where the tide of specialization and concentration on immediate problems has swept away the type of university he had worked diligently to build. I believe that he would also be disappointed that Canada has not managed to achieve the degree of sovereignty that he had hoped.

Innis was not alone. He was one of a group of profound figures produced by Canada whose gestation period was between the end of the First and Second World Wars. Innis, Macpherson, Grant, Frye, McLuhan, Havelock, and the other great intellects are gone now – leaving us with a rich tradition on which to build. And yet Canada does not seem to be producing a new generation of thinkers that will attain their stature.

Innis seemed intuitively to understand what had made him an exceptional scholar. After all, he could just as well have viewed his rural background as a drag on his progress. In that case, he would have cut himself off from his roots, been embarrassed by his family, and denied his provenance. This almost certainly would have been his reaction if his story had played out in a more class-bound society. Success would have implied a clear change in class by an individual in a society where this was rare.

Yet Innis never lost his feeling of being indebted to his roots, of loving 'his own.' This is apparent when we look at the key factors that he described as underlying great scholarship. First, there was his sense that breakthroughs came when intellectuals recognized hidden patterns in the great masses of facts they had mined from their real world; second, his contention that great scholarship must have, as one of its ingredients, the gathering of the great wisdom available from the perspectives of ordinary people. In the end, I believe that this was why he placed the oral tradition at the centre of his thinking. For him, the only way to get at that wisdom was through one-on-one conversations with the characters who came from his time and place.

Indeed, it is difficult to imagine how Innis could have valued his background more highly. After all, his contention that great intellectual contributions that re-energize Western civilization are likely to come from the periphery of empire was, in effect, the ultimate compliment that could be paid to the milieu of his youth. To put the matter in a negative sense, his alarm (towards the end of his career) at the direction Western civilization was taking was an acknowledgement that the conditions that had made him were passing. An Otterville in an era of letter post, kerosene lamps, newspapers, and railways was fundamentally different from Otterville in an era of electricity, telephone (and now the Internet and e-mail), television, and good roads. The people are fewer, older, and more prosperous, but something has been lost along the way. There are no more opportunities for conversations with interesting strangers on the Woodstock-bound train. There are only rare, if any, discussions within large farm families on how they can best generate more cash income by adjusting the complicated mix of current and potential crops, the division of labour among family members, and the identification of unused capacities. Nor would these discussions take on the same concentration and focus as they did in an era where their outcome could imply going hungry before spring arrived or leaving school at adolescence.

Otterville (and thousands of places like it) has stopped being part of the margin or periphery. It is now 'connected' in ways that would have seemed fantastic to Innis as a boy. Innis well understood that this was happening. This realization is the basis of his often repeated paradox: better communications lead to a decrease in understanding. He might have put it another way: increased entertainment leads to decreased contemplation.

We can learn various things from the legacy of Harold Innis, it seems to me. The first is that the task of understanding our world in a profound manner is hard work and demands a lifetime of effort, as well as modesty in the conclusions that one arrives at. To put it another way, Innis would view punditry as the antithesis of scholarship. Second, great intellectual contributions are founded on a thorough knowledge of one's own – a knowledge that is not only passively absorbed (culture, if you like) but actively sought after through scholarly 'dirt' research. Finally, the really great contributions to knowledge are likely to take place on the margins of civilization where people are still wrestling with real problems of livelihood.

Innis's vision was a dark one because the margin, as he knew it, was disappearing. New technologies were integrating marginal areas in an unprecedented manner. The quiet that had stimulated thought on the margin has been replaced by a cacophony of noise and data available everywhere. Similarly, real local problems are not now stimulating collective intellectual efforts to find their solution, for the solution (worked out by others to be sure) is available elsewhere, usually 'off-the-shelf' at no cost. Little wonder that Innis's conclusions were so bleak in the end.

And yet, when we look at how things are unfolding in the world today, might not he be considered prescient? Historically, has there ever been an imperial imbalance as great as that which now exists? Militarily, there has never been as overwhelming an imbalance of power, reflected in the technological sophistication of American weaponry and the ideological doctrine of preventative war. Neither has there been an empire so caught up in its own self-referring realm of thought that it sees itself as non-imperialist: an 'empire lite,' in Michael Ignatieff's phrase.

Where are we likely to find the intellectual resources to develop a profound understanding of this imbalance? Where is the 'margin' now?

Innis's hope that scholarly excellence would develop in Canadian universities, underwritten by the economic and cultural independence of their nation-state, has not been realized. Canada is more thoroughly integrated with an imperial economy than ever before, and our universities are far too focused on specialist solutions to practical problems to fulfil the role Innis assigned to them.[9]

Perhaps if we look at another image that Innis was fond of using – Minerva's Owl – we will be able to mitigate the bleakness of his vision. Innis did not expect intellectual fervour to arise on just any margin of empire. It had to be a margin to which refugee intellectuals were fleeing, bringing with them imperfect knowledge of the traditions they had left behind. If we focus on this element of recent arrival and the challenge of making a new environment one's own, perhaps there is better hope for the future.

If so, in the national context in which this is written, it will depend upon Canada remaining an open, pluralistic, and socially tolerant society. This is necessary, for the new intellectual contributions are likely to come from the geographic margin (the first generations of newly arrived immigrants and refugees), the social margin (women and gay intellectuals), the cultural margin (our First Nations), and the linguistic margin (the francophone and Hispanic communities in North America). (In other contexts, the dynamic may play itself out in ways that are specific to those contexts.) The best way for the Canadian context to make a contribution to rebalancing the current imbalance of empire is to build and maintain a tolerant and inclusive society. Because we, too, are 'American,' perhaps we can offer an alternative to the current dominant belief of many of our southern neighbours that fear, force, and good intentions are enough to ensure the future of the West.

And there is one final margin to which we should look for inspiration and that is the margin of time. In the Canadian context, that would mean the exercise of ignoring current academic fashion and returning to the great thinkers of our past to see if they are still speaking to us across the years. Surely, this is one of the best ways to value the oral tradition in our time and place? I am certain that Innis will still be among the clearest voices if we choose to listen.

Notes

Acknowledgments

1 A note is in order here on the imbalance that has been observed by one of my readers between Part I and Part II of the manuscript, with Part I emphasizing predominantly biographical matters and Part II predominantly textual influences. The book focuses on the 'dark vision' Innis developed in the last ten years of his life. It seeks to understand and explain that vision by exhaustively delving into primary-resource material. Innis's writings up to 1940 are far better understood than those of the communications period, which is why the manuscript concentrates on them only inasmuch as they have carry-over into the communications period. On the other hand, I believe that there are formative biographical elements that have been overlooked or underemphasized during this early phase. Similarly, post-1940, Innis was working alone in fields in which he had little familiarity. Hence the direct biographical connections from his life to his work during the period are far less present. In short, the 'imbalance' in the book replicates the imbalance in Innis's scholarly career.

Introduction: The Innisian Puzzle

1 Quote from Innis recorded in the handwriting of Mary Quayle Innis, Innis Papers, University of Toronto Archives.
2 Riendeau, 'On the Critical Edge,' chapter 2. The composition of this Innis College theme song has been attributed to Bob Bossin, an Innis College alumnus and noted Canadian folk musician. The lyrics were, of course, an ironic composition, highlighting the typical Canadian trait of applauding

home-grown success stories only after they have been recognized outside Canada.

(Sung to the tune of the 'Battle Hymn of the Republic')

Who the hell was Harold Innis?
Will someone tell me who the hell was he?
We'll cheer his name with all that's in us
If only you will reveal his identity.
But don't say that he's a politician
Or that he'll reach sainthood any day,
But if he lectured students, we'll forgive his imprudence,
And cheer and sing his praises anyway.

Chorus
Glory, glory, Harold Innis
We'll cheer his name in all that's in us
If you aren't with us, you're agin' us
For we're people that you don't meet every day.

Did he invent a better mousetrap?
Or discover something new?
We'll gladly follow in his footsteps
If you only will inform us as to what size shoe.
He's famed in academic circles,
But you can't get in without a PhD
We'll follow his example, if you'll just give us a sample,
Oh, Harold Innis, who the hell was he?

3 See Watson, 'Marginal Man.'
4 See, for instance: Katz, Peters, Liebes, and Orloff, ed., *Canonic Texts in Media Research*.
5 Watkins, 'A Staple Theory of Economic Growth.'
6 I am indebted to one of my anonymous readers who pointed this out.
7 Examples include: Evenden, 'Harold Innis, the Arctic Survey, and the Politics of Social Science during the Second World War'; Usher, 'Staple Production and Ideology in Northern Canada'; Bickerton, 'Too Long in Exile: Innis and Maritime Political Economy'; Gagnon and Fortin, 'Innis and Québec: Conjectures and Conjunctures'; Salée, 'Innis and Québec: The Paradigm That Would Not Be'; Jenson, 'From Silence to Communication? What Innisians Might Learn by Analysing Gender Relations'; Black, '"Both of Us Can

Move Mountains": Mary Quayle Innis and Her Relationship to Harold Innis' Legacy'; Whitaker, '"To Have Insight into Much and power over Nothing": The Political Ideas of Harold Innis'; Noble, 'Innis's Conception of Freedom'; Buxton, 'The Bias against Communication: On the Neglect and Non-Publication of the "Incomplete and Unrevised Manuscript" of Harold Adams Innis'; Campbell, 'From Romantic History to Communications Theory: Lorne Pierce as Publisher of C.W. Jeffreys and Harold Innis'; Salter and Dahl, 'The Public Role of the Intellectual'; Ferguson and Owram, 'Social Scientists and Public Policy from the 1920's through World War II'; Gauvreau, 'Baptist Religion and the Social Science of Harold Innis'; Angus, 'Orality in the Twilight of Humanism: A Critique of the Communications Theory of Harold Innis'; Carey, 'Innis "in" Chicago: Hope as the Sire of Discovery'; Wernick, 'No Future: Innis, Time Sense and Postmodernity'; Stamps, *Unthinking Modernity: Innis, McLuhan and the Frankfurt School*; Bonnett, 'Communication, Complexity and Empire: The Systemic Thought of Harold Adams Innis'; Dudley, 'Space, Time, Number: Harold A. Innis as Evolutionary Theorist'; Angus and Shoesmith, 'Dependency / space / policy: An Introduction to a Dialogue with Harold Innis'; 'The Staple Theory Revisited'; Drache, 'Introduction' to *Staples, Markets, and Cultural Change*; Parker, 'Innis, Marx and the Economics of Communication: A Theoretical Aspect of Canadian Political Economy'; Jhally, 'Communications and the Materialist Conception of History: Marx, Innis and Technology'; Blondheim, 'Harold Adams Innis and His Bias of Communication,' in *Canonic Texts in Media Research*; Fisher, 'Harold Innis and the Canadian Social Science Research Council: An Experiment in Boundary Work'; and Beale, 'Harold Innis and Canadian Cultural Policy in the 1940's.'

These examples represent a cross-section only. Over the last twenty five years, many theses and papers have been produced on Innis. For a wider selection, I would direct the reader to three collections of essays: Melody, Slater, and Heyer, eds., *Culture, Communication and Dependency: The Tradition of H.A. Innis*; Shoesmith and Angus, eds., *Dependency/Space/Policy: Continuum*; and Acland and Buxton, eds., *Harold Innis in the New Century*.

8 Among these I would highlight in particular the work of: Evenden, Noble, Campbell, Salter and Dahl, Ferguson and Owram, Carey, and Fisher.

9 An outstanding example of this approach is found in John Bonnett's thesis, 'Communication, Complexity and Empire: The Systemic Thought of Harold Adams Innis.' By carefully re-reading Innis's work, in particular those elements that deal with social change in specific societies over the long term, Bonnett has significantly advanced our knowledge of Innis. He

has demonstrated continuity rather than disjuncture between the 'early' and 'late' Innis. He has also convincingly countered the portrayal of Innis as a technological determinist. His thesis is especially good in its presentation of Innis's view of the importance of the oral tradition.

Bonnett brings in the ecological theory of 'complex adaptive systems' and the economics of 'increasing returns' as frameworks on which to hang his understanding of Innis. I find this the least convincing part of his argument. There is little or no evidence that these schools of thought had any influence on Innis's research or writing. If they helped Bonnett gain an advanced understanding of Innis, so much the better. However, to go beyond this and suggest that they provide us with the hidden architectonics underlying Innis's work is not credible.

10 I do not wish to convey a sense of 'politics' so general as to destroy the usefulness of the term. All personal activities are not political in the older sense. Here, only those that involve civic matters shall be considered so.

1. The 'Herald' of Otterville, 1894–1913

1 Grant, *George Grant in Process*, 63.

2 The tributes following his death preserve much of the character of the man. Creighton's memoriam sketch, *Harold Adams Innis: Portrait of a Scholar*, is still the best overall work. Innis himself provides us with an autobiography until 1922. Other authors such as Robin Neill, Carl Berger, and Robert Babe have included biographical sections in their important writings on Innis. The 1972, 1979, and 1994 CBC radio documentaries on Innis, prepared by Elspeth Chisholm, Paul Kennedy, and Donald Cayley, respectively, present Innis through the eyes of those who knew him or studied his life. Tom Cooper's thesis on Innis and McLuhan carries this personal approach one step further. Taken together, this material gives us a creditably thorough version of Innis's lifetime itinerary.

I would maintain that Creighton's work is still the best summary biography of Innis partly because, as a contemporary, he had a first-hand knowledge of Innis and his context. The most recent biography of Innis, by Paul Heyer, seeks to correct Creighton's bias against Innis's later works by devoting more space to Innis's communications thinking. Unfortunately, while Heyer's approach offers a simplified summary of Innis's thought, which may be useful for undergraduate students in communications studies, he introduces new lines of argument that are faulty and devalue his overall presentation.

I am referring to the arguments put forward in essays by Buxton and Black and included as appendices to the Heyer biography. Buxton argues that Innis's 'A History of Communications' remains unpublished for no good reason, given that it could contain hidden insights into Innis's communications thought. Similarly, Black portrays Mary Quayle Innis as a previously ignored intellectual influence on Innis's work. Both of these notions are off the mark and are not borne out by an examination of primary-source material. Moreover, any close contemporary of Innis's, like Creighton, would have recognized that this was so. I will deal with these arguments below (see chapter 7 for Buxton on 'A History of Communications' and chapter 3, n.23 for comments on Mary Quayle).

My point here is that, while Creighton and other early biographical treatments of Innis suffer from a hagiographic bias, later treatments, in seeking to correct this, can suffer equally from factual mistakes that would have been evident to Innis' contemporaries. Aside from the speculations of Black and Buxton, there are an annoying number of faulty details which Heyer and his pre-publication reviewers missed. Let me cite a few examples. No 'old growth forests' (page 2) exist in the Otterville area – timber was a valuable resource and the forests were cut down by the first generation of pioneer farmers. The Otterville church did not mellow so that it 'now bears a United Church of Canada affiliation' (page 6). It remains a 'fellowship' Baptist church, meaning an institution so fundamentalist in its Christianity that it broke with the mainstream Baptist Convention. Innis was not wounded while on a general 'reconnaissance patrol' (page 4), but while serving as an artillery 'spotter.' He did not teach 'in an adult continuing education program in nearby Hamilton' (page 5) but in a Workers' Education Association night course. The difference is more than semantic for Innis believed that he was under police observation as a result of this activity. The first draft of *The Fur Trade in Canada* was not completed in 1927 (page 11). What was completed that year was an earlier study, 'The Fur Trade of Canada.' It is not accurate that, after the crisis over his promotion as associate professor in the 1920s, 'later promotions went smoothly' (page 14). As we shall see, virtually every career advancement made by Innis met with resistance which was overcome by the threat of resignation.

Heyer's insights into Innis as a 'Red Tory' and 'intellectual guerrilla' have merit but, given the factual errors in his biography, coupled with the speculative fantasies of Buxton and Black, I prefer the shorter, clearer treatment of Innis in Babe, *Canadian Communications Thought*. The best recent biographical research on Innis has focused on narrower fronts and the same kind of primary-source 'dirt' research that Innis espoused (see for example, Pierce, Salter and Dahl, Fisher, Evenden, and Ferguson and Owram).

3 Mary Quayle Innis is virtually an enigma behind the enigma of Innis. She helped him prepare material for publication and provided domestic support for him during his lifetime. She oversaw the publication of revised editions of Innis's books and essays in Canadian economic history after his death. Finally, she worked on his personal papers over a twenty-year period following his death before they were deposited in the University of Toronto Archives in 1972. Yet she has left us with virtually no commentary on her husband. She has noted only a half-dozen comments made by Innis when she and no one else was present. Not one of her letters to Innis has been deposited in the archives, and her own papers, including several completed but unpublished manuscripts, have been, for the most part, destroyed. The more recently deposited 'Harold A. Innis Family Records' (B1991–0029) at the University of Toronto Archives contain her personal diaries, but these are so spare and unemotional as to reveal little. Similarly, her personal correspondence includes mostly that having to do with matters bearing on Innis's intellectual estate or letters to her immediate family in the United States.

4 She never allowed herself to be recorded and these remarks were made to Elspeth Chisholm while the latter was preparing the 1972 CBC Radio documentary on Innis. See Accession B74.007, University of Toronto Archives. Mary Quayle Innis did not like the Innis farm or feel at home there and the remarks might have been related to this personal aversion. Interview with Hugh Innis, Friday, 21 September 1979.

5 See Innis Family Records in the Norwich County Archives.

6 See *The Illustrated Historical Atlas of Oxford and Brant Counties 1875–76* (Toronto: Walker and Miles/Page and Smith 1969. Available online at http://digital.library.mcgill.ca.

7 Griffith Taylor, 'Towns and Townships in Southern Ontario,' *Economic Geography*, 88–96. See Accession B91–0029/053 file 7, University of Toronto Archives.

8 Gail Lewis, interview by author, Otterville, 7 July 2003.

9 Eacott, *Of Other Times*, 53.

10 See 'The Grand Trunk, Georgian Bay and Lake Erie Railways,' accessed online, 25 June 2003, at http://www.globalserve.net/~robkath/railgte.htm, pg. 2 of 4, 'The Port Dover and Lake Huron Railway.'

11 Eacott, *Of Other Times*, 54.

12 Lewis, interview, 7 July 2003.

13 Harold Innis, 'Autobiography,' Innis Papers, University of Toronto Archives, 12.

14 Ibid., 3 and 8.
15 Mary Adams Innis's sketches were preserved by Harold and eventually deposited with his papers in the University of Toronto Archives.
16 The current website for the school can be accessed online at http://www.castle-ed.com.
17 Innis, 'Autobiography,' 1–6.
18 Ibid., 24.
19 Another example of how childhood circumstances mark adult traits is found in Innis's legendary frugality. During Innis's youth, a changeover was taking place between individual erasable chalk slates and the more expensive pencil-and-paper alternative. The sense of the preciousness of paper instilled in him lasted a lifetime. Even when he had long lost the need to do so, he would make his reading notes by crowding in two lines of writing to each ruled line of stationery. He left no space for the margins. Both sides of the page were crammed full of notes.
20 'Autobiography,' 3.
21 'Autobiography,' 39.
22 Innis Family Records, Norwich Township Archives.
23 Innis, 'Autobiography,' 23.
24 Woodstock Sentinel, September 1889 and January 1896, Local History Files, Woodstock Public Library.
25 By 'titling technology' I mean the constellation of technology (surveying and mapmaking) and social organization (land registries, deeds, courts) that allowed property to be held by individuals through written documents rather than communally through conventions based on local oral tradition.
26 *Explorations*, June 1969, 86, in Biography File, H.A. Innis, Canadian Baptist Archives.
27 Innis Family Records, 'The Vision Which Became a Reality, Involving the Norwich Quakers' (signed 'Sam Innis, "Sunnyside"'), Norwich Township Archives.
28 Donald Innis, interview by author, 18 July 1979.
29 'Minute Book' – School Section #1,' 27 December 1911, Otterville Museum, Otterville.
30 See Entry Birth Registration, 024074 no. 16, 20 November 1894, Registrar Alex McFarlane, 'When Born' – Nov. 5th 1894, Name – Harold Adams Innis.' Enlistment: See 'Attestation Paper' in Innis's War Service File, Library and Archives of Canada. Curiously, Innis enlisted as 'Harold Adams Innis' but his birth date on this document is incorrectly entered as 'Nov. 21. 1894.'
31 Innis, 'Autobiography,' 24.

32 Ibid., 7.
33 Ibid., 24.
34 Other non-traditional first names do occur in the family but their introduction is 'explained' in the family oral tradition in ways similar to the explanation of 'Harold' as 'Herald' misspelled. Harold's sister 'Hughena' is an example: 'The name Hughena which has passed down through three generations was acquired in a unique way. Mary Adams' grandmother Christina McDonald was about to give birth in 1840 when she and her husband were on board ship on their way to Canada. A fellow passenger offered them a substantial sum of money to name their child after him. As his name was Hugh, and the baby a girl, the name became Hughena. This gentleman in the end had succeeded in passing down his name in its original form. After three generations of Hughenas there are now two of Hughs: Harold's son Hugh Innis and his grandson Hugh Dagg.' Innis Family Records, 'The Name Hughena.' Norwich Township Archives.
35 Proverbs 29:15–18.
36 Innis, *The Fur Trade in Canada*, 383.
37 In the days in which the vast majority of the population was illiterate, the church attempted to transmit its normative sanctions by presenting 'morality plays.' The central character in these plays was often called 'Everyman.' The general populace was invited to associate with this figure as he faced, in an allegorical fashion, the trials and tribulations that the audience, as individuals, faced in their day-to-day existence. An everyman character, then, is one which at some very profound level represents the character of the community as a whole.
38 This is not to say that such a character can never develop in a peasant context. I am only suggesting that, when it does so, it is in spite of the dominant communal attitudes of the childhood milieu.
39 This phenomenon has been dealt with in historical fiction, by Hugh Hood in *A New Athens* (Montreal: Oberon 1977).
40 As Reid puts it in *The Scottish Tradition in Canada*:

> there was little or no class distinction in the influence which the Scottish Protestant tradition exerted ... The Protestant tradition seems to have fostered considerable social mobility. 'The lad o'pairts,' whatever his background, social and economic origins, felt that he had the right and duty to make the best of himself and to rise in the world. One can think of farming families which have produced ministers, doctors, lawyers, nurses, and school teachers ... Scottish Protestants did not accept any rigid class structure, but stressed the importance of every

man developing his God-given gifts to the best of his ability in this life ... One thing that stands out very clearly in the Protestant tradition is the desire for intellectual and technical training.

41 Innis, 'Autobiography,' 9.
42 Interview with Hugh Innis, 21 September 1979.
43 Ibid.
44 In the white-settler colony, one of the most problematic psychopathologies is the rather ugly personality traits that can be aroused through the interface of white-settler and indigenous cultures. In the short term, this may involve the total denial of the humanity of autochthonous peoples and their slaughter in the face of an imperial project for which their cultural traditions serve no purpose. Yet, ironically, the most serious pathology (institutionalized racism) seems most prevalent in those white-settler colonies in which a relatively large group of indigenous but subordinate peoples continues as an essential part of the white settler state (South Africa, Israel, South America).
45 Innis Papers, University of Toronto Archives.
46 Ibid.
47 CBC Radio documentary on the *Family Herald*, Saturday, 13 June 1979. CBC Radio Archives, Toronto.
48 Innis, 'Autobiography.'
49 *Cotton's Weekly/The Observer*, Microfilm Periodicals Holdings, September, 1908–, Toronto Reference Library.
50 Ibid.
51 Ibid., 11 March 1909.
52 This newspaper, begun as a weekly small-town paper called the 'Cowansville Observer,' changed its name to 'Cotton's Weekly' in 1908 and later became the 'Canadian Forward' after moving to Toronto in 1914. It was the voice of the Socialist Party of Canada (1904) and later the Social Democratic Party of Canada (1911).
53 Clark, *Church and Sect in Canada*, 346.
54 Ibid., 166.
55 As Hotson writes in *Pioneer Baptist Work for Oxford County*, 16: 'As soon as a new church was started its ledger began the listing of persons excluded: 'for falsehood ... for intoxication; for fighting with one of his neighbours on the Lord's Day; for accusing another with guilt and failing to give any proof and by likewise having joined the freemasons; for lying; for refusing to travel with the church; for railing against the church and minister; and for joining finity with the world in plays and sham marrying.'

56 Clark, *Church and Sect in Canada*, 346.
57 Innis, 'Autobiography,' 23.
58 Innis, 'A Tragedy of the Bluffs.'
59 Johnston, *McMaster University*. 11.
60 Ibid., 13.
61 Ibid., 14.
62 Ibid., 16.
63 Ibid, 7.
64 Ibid., 10.
65 Ibid., 33.
66 Ibid., 50–51.
67 Ibid., 74. Johnston writes: 'Over the four years of instruction that led to the B.A. degree, students were required to take annually – English, a modern language, Latin or Greek, natural science (biology or chemistry), and mathematics or physics. Beyond the freshman year these staples were augmented by mental science (psychology) in the second year, by a combined course in constitutional history and political economy in the third year, and by church history, education or metaphysics, evidences of Christianity, and Hebrew in graduating year. Provision, however, was made for those students who obtained high standings in their first-year examinations to do a limited amount of work in the departments in which they excelled. This led to the creation of special courses designed in large part for prospective high school teachers.'
68 Honours courses were not instituted at McMaster until its move to Hamilton in 1930. Innis took his degree with *two* special courses: political economy and philosophy.

2. The Great War, 1914–1918

1 Marc Bloch, French First World War veteran, cited by Fussell, *The Great War and Modern Memory*, 115.
2 This section is heavily indebted to Johnston, *McMaster University*, and Innis's 'Autobiography."
3 Harold Innis, Otterville correspondence (11 September 1915), University of Toronto Archives.
4 Johnston, *McMaster University*, 90–113.
5 As James Ten Broeke preached before the Baptist Convention of Ontario and Quebec, on 31 May 1896 ('Educational Sermon: Education a Means to Personal Development,' Toronto: Baptist Convention of Ontario and Quebec 1898), for Baptists, the Christian ideal of life implied 'the unfolding of

the life [of the individual] into the perfect likeness of Christ.' Formal education representing part of that ideal had to be universal with regard both to access and to the scope of inquiry. Ten Broeke, who had possibly more influence on Innis than any other professor at McMaster, articulated the Baptists' position as follows: 'Education should be universal, because all knowledge is the birthright of all men. Upon this broad foundation rest the public school system and the final argument against sectarian exclusion ... The young should be encouraged to go farther and see more clearly than their fathers, if the spirit is really leading the race forward into all truth, indeed, if we have any faith in the possibility of man's attaining "into the measure of the stature and the fullness of Christ"; for who has compassed the length and breadth of that life of perfect wisdom, beautiful in holiness, faithful in love even into death?' While the mature Innis would reply in the negative to Ten Broeke's 'if' clauses, he would just as passionately hold to the principle of searching out the intellectual talent of society so that, no matter where it was located, the individual would be permitted to pursue his gifts to the maximum possible extent.

In his presentation, Ten Broeke identifies what Innis will later term 'monopolies of knowledge' as the key impediment to the advancement of civilization: 'Civilization advances in proportion as knowledge increases and becomes universal. So true is this that in consequence of the inequality in the distribution of knowledge we find barbarians in the midst of our boasted civilization.'

6 Johnston, *McMaster University*, 83

7 Ibid., 122

8 Innis, 'Autobiography,' 41. 'I won the D.E. Thomson Scholarship in Economics and the Teetzel Prize in Philosophy and first class standing in history in the examinations of 1915.'

9 Michael Gauvreau, in 'Baptist Religion and the Sociology of Harold Innis,' outlines in some detail the lifelong influence on Innis of the 'Christian pragmatism' he encountered at McMaster and especially in Ten Broeke's classes. Nevertheless, this essay, which is admirable in its 'dirt' research – that is, its rooting in primary material – is highly flawed and gives a false impression of the primacy of religion in Innis's thought. It is always a danger in dealing with a complex character like Innis to concentrate on one element of influence (in this case the Baptist Church) and to seek to explain his later writings in terms of this hidden influence.

Gauvreau contends that 'military service overseas ... challenged, but did not fundamentally alter, Innis' Christian faith and concepts of service.' He goes on to say that '[Innis] continued to regard himself as a Christian both in terms of belief and in strict adherence to an evangelical moral code.' He sees it as significant that Innis 'lectured at least once ... to the Otterville Baptist Young People's Union on the subject of his travels in northern Canada,' and that 'Innis was an instructor in economic geography in the School of Missions established by the United Church in the 1920's.' As for the debates of the 1930s, Gauvreau reaches the conclusion that 'this, in turn, compelled him in his scholarly writing to accentuate the role of religion in human society and to a public reassertion of his evangelical convictions.' As far as politics is concerned, Gauvreau says, 'Not only did Innis believe that all political movements should be undergirded by a religious spirit, but he welcomed a particular kind of religious revival, one based on the outbreak of spontaneous evangelical preaching.' He concludes his essay by presenting Innis as a believer and insider speaking to the United Church's Board of Evangelism and Social Service, opining that 'what was needed was for the churches to stress not the letter of Christ's moral teaching, but the very essence of the New Testament, the worship of Christ, incarnation of God in man, divinization of man in God.'

This is folly. As I will show, Innis abandoned organized religion shortly after the First World War because of his belief that it had provided the ideological underpinning that had sent him and thousands of others to senseless, organized slaughter. There is no evidence that he ever regained his faith. He also spoke to the Workers' Educational Association, the Bankers' Educational Association, and political party summer schools, and he participated in the original meetings of the group that would become the League for Social Reconstruction in Toronto, but this should not lead one to conclude that he was a labour leader, banker, card-carrying party man, or socialist. He never reversed his decision to decline a profession of faith and baptism, and he actively lobbied against the influence of the evangelical fundamentalists at McMaster in the 1920s.

We can see how far off the mark Gauvreau's analysis takes him by looking at his concluding quotation from Innis in the context from which it was taken: 'Christianity would have remained a Jewish sect had it not been made at once speculative, universal, and ideal by the infusion of Greek thought, and at the same time plastic and devotional by the adoption of pagan habits. The incarnation of God in man, and the divinization of man in God are pagan conceptions, expressions of pagan religious sentiment and philosophy. Yet what would Christianity be without them?' Instead of Innis, the believing Baptist trumpeting the orthodox Christ, we find Innis, the intellectual agnostic, introducing his audience to a 'Catholic atheist' (George Santayana), the better to acquaint them to the influence of the classics and paganism in the worship of Christ!

10 F.W. Waters, 'A Century of Philosophy at McMaster,' chapter 2, 'The Ten Broeke Years,' n.p., draft manuscript held by author.

11 Innis, 'Autobiography,' 40.

12 This was particularly so since later developments in science often proved the anti-agnostic side in the scientific debate to be in error. As an example, we could refer to discussions from course readings noted by Innis concerning George W. Crile's 'A mechanistic view of psychology' (*Science*, 29 August 1913, 285). This perspective reduced human consciousness, social decision making, and so on to a pinball-machine concept of atoms, chemicals, and nerve endings requiring no reference to divine inspiration. As an opposing viewpoint, Ten Broeke suggested Sir Oliver Lodge's defence of the more conservative scientific theories of the time. The defence of more traditional currents of science against the radical implications of evolution, 'mechanistic psychology,' and relativity meant that Innis was exposed to a wide variety of some of the more advanced elements of contemporary thought. He was surely to be led towards an agnostic position by the soon-to-be-proved-absurd notions of the traditionalist scientific beliefs discussed in Ten Broeke's class. Since science tends to remember only its successes, it would be appropriate to recall an example that indicates both how initially radical beliefs have become the common-sense of today and how absurd were some of the scientific 'truths' of the time. In the pre-relativity era, science had relied on an emperor's clothes concept – the universal ether – to make sense of the universe. Relativity was incorrect, argued Lodge (pages 391–419 of the above-mentioned issue of *science*), because

> I cannot imagine the exertion of mechanical force across empty space, no matter how minute a continuous medium seems to me essential ...
>
> It is an ancient and discarded fable that complications introduced by the motion of an observer are real complications belonging to the outer universe ...
>
> The ether of space ... is the writing and binding medium without which, if matter could exist at all, it could exist only as chaotic and isolated fragments: and it is the universal medium of communication science between worlds and particles.

13 Innis, 'Autobiography,' 40.

14 Johnston, *McMaster University*, 105.

15 Ibid., 72

16 Innis, Debating Note, Innis Papers, University of Toronto Archives.

17 Ibid. It is a short step from this formulation to McLuhan's 'Global Village.'

18 Johnston, *McMaster University*, 125.

19 Box 385, McMaster University WWI File, 'National Service,' Canadian Baptist Archives.
20 Ibid.
21 Innis, 'Autobiography,' 50.
22 Johnston, *McMaster University*, 131.
23 Ibid., 132.
24 COTC files, Canadian Baptist Archives.
25 Harold Innis, letter to Otterville (*c.* July 1916), Innis College, University of Toronto.
26 Ibid., letter from Edinburgh, 10 November 1916.
27 Innis, letter to Otterville, 4 April 1916, Innis Papers, University of Toronto Archives.
28 For example, as Tom Cooper writes in 'The Unknown Innis' (111): 'Although Innis' World War I letters to his parents indicate a continuing trust in Christian theology, it is likely that these letters conceal his real thoughts on religion ... Innis' later statements about the cataclysmic effect of World War I upon his attitudes call into question a literal interpretation of his letters to his devout Baptist relatives in complacent, conservative rural Ontario.' The problem here, I would contend, is our present-day tendency to favour an agnostic bias (the later Innis) over a Christian one (the young Innis). The words 'Innis' later statements about the cataclysmic effect of World War I' are simply a recognition of the *intense sincerity* (rather than the reverse) of the Christian-beliefs that he abandoned in the post-war era. Innis was quite straightforward in his correspondence. He would avoid talking about the horrors of war in his letters home by commenting on the weather; he would not, however, adopt a policy of conscious deception.
29 Innis Papers, University of Toronto Archives.
30 Ibid.
31 Donald Innis, interview by author, 10 July 1979.
32 Tuchman, *The Guns of August*, 50.
33 See Leed, *No Man's Land*, on the First World War.
34 Innis, 'Autobiography,' 56.
35 Leed, *No Man's Land*, 106.
36 Ibid., 185.
37 Innis wrote in his 'Autobiography,' 57: 'Occasionally German batteries would send their shells to stir up the enemy and we in turn attempted to pay back. On the whole it was a peaceful area with the troops on both sides wondering why we should not let each other alone and enjoy the winter's rest.'

38 For veterans' comments, see the interviews for 'Flanders Fields,' CBC Radio, sixteen one-hour programs beginning 11 November 1964, CBC Radio Archives, Toronto, and Robertson, *A Terrible Beauty*.
39 CBC Radio, 'Flanders Fields,' program on Vimy Ridge.
40 Leed, *No Man's Land*, 23.
41 Ibid.
42 Ibid., 4.
43 Innis, letter to Otterville, 31 March 1917, Innis Papers, University of Toronto Archives.
44 For instance, his letter of 31 March 1917 is seventy lines long. Four lines refer in a general manner to the war. Four mention his general good health. Twenty-eight are commentary on letters from home. Five contain apologies for his writing. Twenty-nine are concerned with the weather.
45 He often refused to do so point-blank, changing the subject to the weather. 'As to increasing Uncle Will's intelligence with regard to the war, I could scarcely imagine myself of service. Every letter could probably be summarized up in three words – rain, cold and mud.' 31 March 1917.
46 Innis, 'Autobiography,' 60.
47 Will Bird, 'Going Home,' in Robertson, *A Terrible Beauty*, 100.
48 Innis, *Political Economy in the Modern State*.
49 Leed, *No Man's Land*, 185.
50 George Ferguson, letter to Robin Neill, 25 September 1967, and interview by Elspeth Chisholm, University of Toronto Archives.
51 Notes in the Innis Papers, University of Toronto Archives.
52 Gwyn's treatment of Innis in *Tapestry of War* is largely based on the following material.
53 Waters, 'The Ten Broeke Years.'
54 Leed, *No Man's Land*, 81.
55 Pay certificate, War Service Records, Library and Archives of Canada.
56 Innis, 'Autobiography,' 84–5.
57 From letter to Innis from Invalided Soldiers Commission, 10 May 1918, University of Toronto Archives:

> In cases such as yours, arrangements have usually been made to pay tuition fees for one term at the University and place the student on Vocational allowances during this time. I would suggest that this would be a good deal better in your case as if you were placed in a Lawyer's office you would only receive your Vocational pay and allowances for say six months, and it would be necessary for you to pay your own fees at the University you would receive the benefit of

the Commission's paying of fees at the University besides placing you on pay and allowances during this term and for one month after.

It might be quite possible for you to get in touch with a lawyer through one of our Vocational Officers in the West, in order to have you placed with him for the summer. However I would recommend very strongly that you would adopt the scheme as I have laid down in the above paragraph.

58 Innis Papers, University of Toronto Archives.
59 Innis, 'Autobiography,' 62.
60 Innis Papers, University of Toronto Archives.
61 Leed, *No Man's Land*, 90.
62 Innis, letter to mother, London, 22 July 1917, Innis Papers, University of Toronto Archives.
63 Innis, letter to mother, 16 July 1918, Innis Papers, University of Toronto Archives.
64 Innis, 'Autobiography,' 50. In a more humorous but also more bitter anecdote, Innis told his son of a sadistic British NCO who would insist that the untrained men attempt to gallop their horse immediately upon commencing their cavalry training. When they fell off, risking serious injury, the NCO would laughingly refer to them as falling 'maple leaves.'
65 Innis in his 'Autobiography,' 126: 'I remember the resentment which many of us felt when we were publicly thanked in England for coming over to help the Mother Country. We had felt that we were concerned with fighting for Canada and Canada alone. It was this feeling which strengthened my determination to work in the general field of Canadian Economics.'
66 Grant, 'Canadian Fate and Imperialism,' *Technology and Empire*.
67 The toll of the First World War for Canada was as follows: enlisted, 595,000; served overseas, 418,000; total dead, 60,383. The figures in the text project losses proportionate to the more highly populated Canada of today. Canadian Military Heritage Project, accessed online at www.rootsweb.com/~WW1can/stat on 15 September 2003.
68 A famous example of this phenomenon is the riot at Kimmel Camp among Canadian soldiers waiting to return home in March 1919.
69 Leed, *No Man's Land*, 28–9.
70 Innis, 'Autobiography,' 57.
71 Leed, *No Man's Land*, 28–9.
72 Innis's position as an artillery signaller gave him an ideal perspective for gaining an overall understanding of the war. As already noted, operating as close to the front as possible, Innis had the task of calling down fire on potential targets visible from his post. The horrific nature of the front lines,

the technological innovation of over-the-horizon artillery, and the power potential of improved communications systems were all abundantly clear from this position in the labyrinth of the war.

73 Each shell had a characteristic 'whine' that could indicate what calibre it was and give some idea of where and when it would hit. During his first night back at the farm, Innis gave his brother Sam a lecture in the dark, complete with the various imitated 'whines' of the incoming shells. It was a subject he would reproduce for his son in later years. (Interview with Sam Innis by Elspeth Chisholm, University of Toronto Archives, and interview with Donald Innis, 18 July 1979.)
74 Innis, letter to President Cody, n.d. [1943], President's Papers, University of Toronto Archives.
75 Creighton, *Harold Adams Innis*, 107, quoting letter to Arthur Cole.
76 Ibid., 108.
77 Innis, 'The Returned Soldier,' 1918 MA thesis, McMaster University Library, 7.
78 Ibid., 7, 10.
79 Ibid., 4.
80 Ibid., 7.
81 Ibid.
82 Ibid., 6.
83 Innis, War Service Records, Library and Archives of Canada.
84 *McMaster Monthly*, March 1919, Canadian Baptist Archives.
85 Innis, 'The Problems of Rehabilitation,' *Political Economy in the Modern State*, 56–7.

3. One of the Veterans, 1919–1923

1 Innis, Foreword to *The Commerce Journal* (1945), xi.
2 Robertson, *A Terrible Beauty*, 80.
3 A good account of the discipline at the front is recorded by Canon F.G. Scott in his account of an execution of a young Canadian soldier. See Robertson, 'The Iron Hand,' in *A Terrible Beauty*, 82.
4 Leed, *No Man's Land*, 173.
5 Ibid., 181.
6 The description of neurasthenia is extracted from *The Comprehensive Textbook of Psychiatry* – II, vol. 1, 2nd ed. (Baltimore, Md.: Williams and Wilkins 1967), 1264–7.
7 Innis, letter to his mother, 6 July 1917, Innis Papers, University of Toronto Archives.

8 The most often quoted passage in this regard is the following: 'Since Christmas I have run across a lot of fellows in McMaster who tend towards materialism or who believe there is no God which was an astonishing fact to me.' The significance of the passage is usually missed. Surely it is the presence of 'a lot' of agnostics in the student body of a pre–First World War Baptist school rather than Innis's naivety that is the surprising fact.
9 Innis, 'Autobiography,' 32.
10 Ibid., 26.
11 Ibid., 41.
12 George Ferguson, letter to Robin Neill, 25 September 1967, and interview by Elspeth Chisholm for CBC Radio, University of Toronto Archives.
13 The occasion was a new campaign by the Reverend T.T. Shields and the fundamentalists within the Baptist Church attacking the 'modernist' faculty at McMaster. Innis leapt to the defence of his old professors, writing to the *McMaster Graduate* in 1925:

> I am not a Baptist. Perhaps I should be the last one to write you about the subject.
>
> Nevertheless I do feel very strongly as a member of another University faculty that one of the most precious assets of our civilization is in very serious danger, namely, academic freedom. I sympathize very much with a new professor of the Baptist denomination for which I have always held the highest regard because of its splendid record in religious history in the struggle for tolerance and freedom of thought – a new professor from England where the traditions of academic freedom have been built up more solidly than in probably any country.

Innis also kept in his family in Otterville informed about the campaign, which was playing out at the level of individual congregations. His mother replied: 'Our church is kind of split now most all old members are for McMaster + young like A. Hall, E. Graham, S. Davies are Shields.'

By the end of the 1920s, the Baptist Church had split into 'modernist' (Baptist Convention) and 'Fundamentalist' (Union of Regular Baptists) entities. The Otterville Church went with the fundamentalists.

14 Jim Doze, letter to Innis, 14 December 1919, Innis Papers, University of Toronto Archives.
15 Innis, 'Autobiography,' 25.
16 Ibid., 26.
17 Ibid., 41.
18 Ibid., 83.
19 Ibid., 87.

20 Ibid., 82.
21 Ibid., 125.
22 Innis never attended any formal courses in conjunction with his MA studies. He completed his degree through self-education as he recuperated in several hospitals in England during the period July 1917–March 1918. Yet his University of Chicago transcript shows that he was credited with nine graduate-level courses by Professor Wright for this work. For this reason, he was able to complete his PhD. course work at Chicago in only six quarters. He did not start research on his CPR thesis until the summer of 1918, completing it by July 1919. Finally, he accomplished all of this while carrying on teaching duties, which was the only way he could support himself during his studies. During the summer of 1919, he received a break by being appointed the head of Snell Hall men's residence, a job less onerous than his teaching duties. During the winter and spring of 1918–19, he had received a number of job offers before the completion of his thesis. All in all, given the general shortage of faculty at the time and Innis's stature as a returning veteran, it appears that he was given quite a few breaks that would not be open to a doctoral candidate during a more normal historical period.
23 'Self-Assessment,' completed 24 July 1920, Innis Papers, University of Toronto Archives.
24 In his essay, '"Both of Us Can Move Mountains": Mary Quayle Innis and Her Relationship to Harold Innis' Legacy,' and in his appendix B to Paul Heyer's *Harold Innis*, J. David Black speculates on 'her possible influence upon Innis' political economy and media research' and suggests that 'in understanding Mary Quayle Innis better, puzzling features of Harold Innis' own work are illuminated.'

I do not believe there is any merit to such speculation. Indeed, Black's entire argument seems founded on the retrojection of the feminism of Harold's and Mary's youngest daughter, Anne Dagg, onto the lives of her parents. Black cites two areas of intellectual influence. He credits Mary Quayle with stimulating the interest of Innis in biographical studies such as *Peter Pond* as opposed to the macro-historical works he usually produced. He also suggests that Mary Quayle, particularly in her textbook *Economic History of Canada*, foreshadows Harold's 'bias of communications' concept.

Neither of these assertions withstands scrutiny. Popular biography was an effective way of disseminating the new way of looking at the Canadian past that Innis was developing. He always viewed this as an important element of his project. In any case, he wrote *Peter Pond* in response to the popular biography his friend C.N. Cochrane had written on David Thomson.

Innis felt that Pond had not been given due credit for the role he played in opening up the Canadian interior because, as opposed to Thomson, he was American and semi-literate.

Similarly, the *Economic History of Canada* was written by Mary Quayle undoubtedly at the urging and with the support of Harold because no textbook yet existed incorporating the staples approach to Canadian history. (Innis indicates as much in his 'Editorial Preface.') This, too, was an important work of popularization of the research being carried out by Innis, Mackintosh, MacGibbon, and others on staples. However, as with all works of this nature, the primary intellectual influence was in one direction only.

There is no merit in the assertion that the communications works are foreshadowed in this book. Contrary to Black's assertion, communications references make only a 'cameo appearance' in the book, always subordinate to developments in transportation and the staples trades and never in reference to their effects outside the economy of the periphery.

If Mary Quale had had been a significant intellectual influence on Harold, one would have expected to find it acknowledged in Harold's prefaces to his published works or seen evidence of it in their correspondence or in discussions noted in Mary's extensive dairies. It simply is not there.

What is there is the semi-autobiographical novel, *Stand on the Rainbow*, published by Mary Quayle in the mid-1940s. The image it portrays is one of a harried mother taking care of the homefront and a habitually absent father pursuing his career with little reference to his wife's and family's day-to-day activities.

Finally we have the comments, cited by Black, of Innis's older daughter, Mary Cates, who says of her father and mother's relationship: '[I] do not believe there was any influence of one upon the other. My mother was a very remarkable woman, who had a very interesting life, and I think she can be respected in those terms' (Black, in Heyer, *Harold Innis*, 121). This rings true in a way Black's thesis does not.

25 See Mary Quayle Innis, *Stand on a Rainbow*. Margaret Brown, the wife of Professor E.K. Brown, was pregnant with her first child when the novel was published. She was shocked by the general tenor of the novel and telephoned Mary Quayle to ask if being a housewife was actually as miserable an experience as had been portrayed in the novel. Mary Quayle was surprised to find that the novel was being taken that way. She said in puzzlement that her own son had asked her, 'Mom, are we really that much trouble for you?' (Margaret Brown, interview by author, 20 October 1979.)

26 Innis writes in a draft letter to F.E. Quayle, Liverpool, n.d., Innis Papers,

University of Toronto Archives: 'She had just returned from war work in Washington, and knowing she would find it difficult as I had not long since to get back into the daily grind, I became interested in her career ... Our interest in each others troubles steadily grew ... Mentally, I have managed to get back to a civilian basis and morally, I have left alone totally, wine and women though to have been a soldier is not a good recommendation. I have kept myself straight under conditions of army life. You may have some conception of army life and if you have it is needless to go further or if you have not, it is even less necessary.

27 Innis, letter to Mary Quayle, n.d., Innis Papers, University of Toronto Archives.
28 Ibid., n.d. [1920?].
29 Ibid., n.d. [1920?].
30 Ibid., 10 December 1921.
31 Ibid., 8 October 1922.
32 Ibid., 17 April 1923.
33 Ibid., 12 July 1925.
34 Ibid., 28 July 1925.
35 Ibid., 30 September 1922.
36 The depression was long term. On 23 November 1928 Innis's mother wrote to him: 'From the way you wrote in your last letter I expect to hear of you being down in 999 Queen St. What is the use in killing yourself like that.' Innis College Archives. The letter to which his mother refers is not found in the archival holdings.
37 From family reminiscences included in the Innis Papers, University of Toronto Archives.
38 Blanche Graham, interview by author, 22 October 1979.
39 Innis, 'Autobiography,' 104.
40 Innis, *Essays in Canadian Economic History*, 387. My emphasis.
41 Donald Innis, interview by author, 18 July 1978.
42 Creighton mentions this bout in his short biography, even though in later interviews he seems to have forgotten the incident. Why Creighton would have known this when other close acquaintances of Innis seem to have had no knowledge may be related to the early access Creighton had to the Innis Papers.
43 The only written documents that mention the incident are the travel schedules prepared by Mary Quayle after Innis's death. There, March 1937 is labelled as Harold's 'breakdown.' In another list she refers to nervous exhaustion. That files of a confidential nature were removed before the Innis material was placed in the archives is (further) apparent from

the old file labels on file folders reused for classifying non-controversial material.

44 S.D. Clark, interview by author, 28 June 1979.

45 Her cooperation on the radio series was partly the result of her friendship with its producer, Elspeth Chisholm, a friendship that had been formed when Chisholm was directing the research team that was set up by the CRTC to prepare the 'History of Communications' manuscript for publication. It is ironic that both had a meeting of minds that the manuscript should not be published: Chisholm, because she was convinced that it was nothing more than a collection of reading notes; and Mary Quayle, because she thought its publication would damage her husband's reputation. In chapter 7, I argue that both are partly correct. 'A History of Communications' is largely a collection of summarized reading notes, but, rather than being atypical of Innis's published work, it represents the central methodology of work that underlay all Innis's communications publications.

46 Innis, letter to Mary Quayle, 5 September 1919, Innis Papers, University of Toronto Archives.

47 Innis, 'Autobiography,' 98.

48 Neill, *A New Theory of Value*, 27.

49 James Carey, presentation to the Innis Conference at Innisfree farm, 1972. Tape originally available at the Innis College Archives, later transferred to University of Toronto Archives. See also Carey, 'Space, Time and Communications: A Tribute to Harold Innis,' in *Communications as Culture*, 142–72.

50 Ibid.

51 The most prominent intellectuals at Chicago concerned with communications were George Herbert Mead and Robert Park. Indeed, McLuhan has argued that Innis was 'the most eminent member of the Chicago group headed by Robert Park.' In fact, as Carey points out, there is no biographical evidence of a direct influence of Park or Mead on Innis when he was in Chicago. Paul Heyer discusses this paradox in a balanced way in *Harold Innis*, 4.

52 Carey, *Communication as Culture*, 145.

53 Innis, 'Autobiography,' 126.

54 Berger, *The Writing of Canadian History*, 113.

55 Innis, 'Autobiography,' 124.

4. The Search for a New Paradigm, 1920–1929

1 Thomson, *The Poverty of Theory*, iv.

2 Medical Records, H.A. Innis War Service Records, Library and Archives of Canada.

3 The reason Innis gave for accepting the University of Toronto offer over the Brandon one was that Brandon did not control the awarding of its degrees (that is, it came under McMaster's degree-granting powers) whereas the University of Toronto has degree-granting power in its own right.
4 Itinerary compiled from notes made by Mary Quayle Innis, Innis Papers, University of Toronto Archives.
5 All three sets of notes are available in the Innis Papers, University of Toronto Archives.
6 In fact, much of the work done by Innis students on this project was incorporated into the text of *The Fur Trade in Canada*.
7 Comments are excerpted from Paul Kennedy's interview with A.R.M. Lower, CBC Radio *Ideas* program tapes, University of Toronto Archives.
8 Urwick, letter to Falconer, 23 January 1929, Accession A68.006, President's Papers, University of Toronto Archives.
9 Urwick, letter to Falconer, 29 November 1929, Accession A68.006, ibid.
10 Toronto Star, 5 January 1926.
11 Bladen, *Bladen on Bladen*, 32.
12 Innis, letter to Falconer, 15 November 1929, Accession A68.006, Box 120, President's Papers, University of Toronto Archives.
13 Innis, letter to Falconer, 31 May 1929, President's Papers, ibid. A copy of Innis's identical letter to Urwick is included in the Easterbrook Papers, University of Toronto Archives.
14 Berger, *The Writing of Canadian History*, 109.
15 Brebner, 'Innis as Historian,' 9.
16 Gilbert Jackson, letter to C.R. Fay, 21 January 1959, Accession B72.017, University of Toronto Archives.
17 Jackson remained senior to Innis in the department despite having attained only a BA degree.
18 Bladen, *Bladen on Bladen*, 40.
19 Vincent Bladen, interview by Robin Harris, 'Oral History of the University of Toronto,' University of Toronto Archives.
20 The department was called the department of political economy and constitutional history from 1888 to 1892, the department of political science from 1892 to 1924, and the department of political economy from 1924 to 1982. Department of Political Economy Inventory, Introduction, University of Toronto Archives.
21 Innis, letter to Falconer, n.d., Innis Papers, University of Toronto Archives.
22 Innis, confidential memorandum to president, n.d., [1929?], Innis Papers, University of Toronto Archives.
23 Ibid.

24 Falconer, letter to Innis, 14 November 1929, President's Papers, University of Toronto Archives.
25 Innis, letter to Falconer, n.d., ibid.
26 Ibid.
27 Innis, letter to Falconer, 29 October 1931, ibid.
28 Innis, letter to Cody, 6 April 1938, ibid.
29 Claude Bissell, interview by author, 5 December 1979.
30 Jackson, letter to Falconer, 7 January 1929, President's Papers, University of Toronto Archives.
31 Innis, confidential memorandum on economics course, 15 April 1929, 1–2, ibid.
32 Ibid., 4.
33 Ibid., 3.
34 Ibid.
35 Ibid.
36 Urwick, letter to Falconer, 17 February 1932, President's Papers, University of Toronto Archives.
37 Innis, letter to MacGibbon, 12 June 1929, Innis Papers, University of Toronto Archives.
38 Brebner, 'Innis as Historian,' 9.
39 Innis's observation about 'schools' of thought being established on the painless extinction of the ideas of their founders applies equally well to Innis himself.
40 See Innis's letters to Mary Quayle Innis, 29 and 30 September 1922, University of Toronto Archives. Fortunately, Innis's tendency to reject intellectual 'fanaticisms' in the long term meant that Huntington's work was discarded in short order.
41 Innis, *The Fur Trade in Canada*, 393.
42 Duncan, *Marketing: Its Problems and Methods*, 5.
43 Ibid., 22.
44 Ibid., 17, 41.
45 Ibid.
46 Innis, *Essays*, 12–13.
47 Ibid., 14.
48 Ibid., 15.
49 Ibid. My emphasis.
50 Innis, *The Fur Trade in Canada*, 5–6.
51 Ibid., 19.
52 Ibid., 383–4.
53 That Innis came to establish the link between the collapse of the Amerin-

dian culture and that of our own is evident in the following passage from his *The Bias of Communication*, 141: 'The disasters which overtook North American civilization following the coming of the Europeans have been described at length. The disturbances which have characterized a shift from a culture dominated by one form of communication to another culture dominated by another form of communication ... point to the costs of cultural change.' The importance of this example of collapse is evident in Innis's encouragement of deeper investigations of this topic by a graduate student in the early 1930s. These studies eventually led to Alfred Bailey's *The Conflict of European and Eastern Algonkian Cultures 1504–1700* (see Berger, *The Writing of Canadian History*, 100).

54 See Watkins, 'The Staples Theory Revisited,' in which the same point is made in a more complex manner.

55 Innis, *The Fur Trade in Canada*, 400.

56 *Ibid.*, 386.

57 Riesman, 'The Social and Psychological Setting of Veblen's Economic Theory,' 451.

58 Ibid.

59 Ibid., 452–3.

60 See Max Lerner, 'Editor's Introduction,' *The Portable Veblen*.

61 Reisman, 'The Social and Psychological Setting of Veblen's Economic Theory,' 459.

62 As with Innis, academic debates have arisen over who Veblen 'really was.' The orthodox biography of Veblen by Joseph Dorfman, *Thorstein Veblen and His America* (New York: Viking 1934,) presents Veblen as a 'marginal man' from a poor, Norwegian farm background and has been challenged by other scholars. Stephen Edgell, author of *Veblen in Perspective* (Armonk, N.Y.: M.E. Sharpe 2001), argues that Veblen did not come from an underprivileged background or struggle with English as Dorfman suggests.

It is not my intention to delve into this heated debate. I would only contend that the image of Veblen that Innis was exposed to in the 1920s and 1930s would have more closely resembled Dorfman's and that Innis would have seen a number of close biographical parallels between Veblen and himself, far more so than occurred in the biographies of other thinkers who influenced Innis.

Dorfman presents Veblen as a 'marginal man' – the same title used for this biography of Innis. The title was chosen before I was aware of Dorfman's work. See Lauren Alexis Moses, 'The Psychology, Life and Relevance of Thorstein Veblen,' 15 April 2002; available from http://www.econ.duke.edu/Journal/DJE/dje2002/moses.pdf; accessed 10 April 2003.

63 Veblen, *The Theory of the Leisure Class*, 135.
64 Veblen, 'The Classical Economists,' *The Portable Veblen*, 270.
65 Veblen, 'Why Is Economics Not an Evolutionary Science?' *The Portable Veblen*, 233.
66 Ibid., 220–1.
67 Ibid., 240.
68 Innis, 'Industrialism and Settlement in Western Canada,' 370.
69 Ibid., 376.
70 Veblen, 'On the Merits of Borrowing,' *The Portable Veblen*, 350–5.
71 Ibid., 362.
72 Ibid., 363.
73 Innis, *The Fur Trade in Canada*, 398.
74 *Canadian Forum*, March 1929.
75 Professor Wynne Plumptre, interview by Elspeth Chisholm, 4 August 1970, University of Toronto Archives.
76 C. Wright Mills, 'Introduction' to Veblen, *The Theory of the Leisure Class*, xi.
77 Innis, 'Notes and Comments,' 5–6.
78 Innis, *Essays in Canadian Economic History*, 24–5.

5. The Great Betrayal, 1930–1940

1 Benda, *The Treason of the Intellectuals*, 185.
2 For a good overview of the debates of the period on the role of the social sciences, see Ferguson and Owram, 'Social Scientists and Public Policy from the 1920's through World War II.' They outline Innis's position as he argued for scholarly work protected by the university as opposed to the political-party engagement of the 'hot gospellers' or the technocratic engagement of their more conservative colleagues. Ferguson and Owram are particularly good at reminding us of the extent to which the professional associations of the social scientists established during this period contained significant numbers of civil servants and business people even at their governance levels.

The actual scholarly community was extraordinarily small by today's standards. As Ferguson and Owram note, 'In 1939, for example, in the English language universities in Canada, there were some 53 historians, 65 economists and political scientists, while the emerging professions of geography, psychology and sociology added only 28 people to this list ... [In 1941] only 146 social scientists were found in a population of 2,661 'professors and college principals.' Given the facts that, even in such disciplines as history, many were still oriented towards the humanist tradition and that in all areas, including the social sciences, specialization meant that large numbers had no particular ability to comment as 'experts' in public matters, the pool of academics who could claim special knowledge of understanding of contemporary policy problems was extraordinarily small' (Ferguson and Owram, 4).

3 Urwick, letter to Cody, 29 May 1936, President's Papers, University of Toronto Archives.
4 Creighton, Harold Adams Innis: Portrait of a Scholar, 95.
5 Quotation cited in Innis's obituary of Urwick, 'Edward Johns Urwick, 1867–1945.'
6 Toronto Star clipping, Innis Papers, Accession B72–003, box 28, University of Toronto Archives.
7 Urwick, 'The Role of Intelligence in the Social Process.'
8 Innis, 'The Role of Intelligence: Some Further Notes,' 282.
9 Ibid., 283.
10 Ibid., 284.
11 Ibid., 287.
12 Innis quoting Urwick in 'Edward Johns Urwick, 1867–1945,' 268.
13 Innis, 'The Role of Intelligence: Some Further Notes,' 287.
14 For an interesting summary of Innis's position on the public role of the intellectual and how these views relate to current debates on this theme, see Salter and Dahl in *Harold Innis in the New Century*, 114–34.
15 Havelock, *Harold Innis: A Memoir*.
16 Ibid., 14–15.
17 H.A. Innis, Letter to R.B. Bennett, 17 October 1932, R.B. Bennett Papers, University of New Brunswick Archives.
18 Horn, 'Free Speech Within the Law,' 33. See this article for a comprehensive summary of the incident.
19 *Evening Telegram*, 'Free Speech Urged by Varsity Faculty in Protest to Police Board,' 15 January 1931, Toronto Reference Library. The letter was also signed by thirteen classicists and eight history professors. Only six science and engineering faculty members had signed. All the rest were from the arts faculty.
20 *Globe*, Editorial, 'The Police or the Reds?' 16 January 1931.
21 Horn, *Academic Freedom for Canada*, 90.
22 Friedland, *The University of Toronto – A History*, 336. In fact, H.J. Cody, the university president, had received a decoration in 1936 from Mussolini's government for his positive remarks. He returned the decoration at the outbreak of The Second World War.

23 Gilbert Jackson, letter to President Cody, 24 January 1935, President's Papers, University of Toronto Archives.

24 One of the overlooked influences on Innis is Frank H. Knight. Knight taught Innis at Chicago and is one of the few individuals who maintained lifelong contact with him. Knight was, for instance, one of the few colleagues with whom Innis discussed his communications work.

Knight came from a similar background to Innis (eldest son of a relatively poor farm family) and was a lifelong liberal who criticized the prescriptive powers of neo-classical equilibrium theory. He argued against the social-policy pretensions of the Keynesians, socialists, and Christian social activists. Innis invited Knight to come to University of Toronto in the winter of 1933–4 to talk about the limitations of economics in the formation of social policy. The similarity of his approach with the one Innis was developing with regard to the role of the intellectual is evident in his pre-visit correspondence with Innis: 'I have been taking a stand here that in talking about current political and economic events I restrict myself to private groups, at least in the sense of excluding newspaper publicity, and preferably also any general announcement or invitation. I do not wish to prescribe anything, but this attitude is so much in harmony with what is most on my mind about the course of events and the role in connection with public affairs of "intellectuals" that it would make considerable difference in planning what I would say if I did not have assurance that the attitude of my hearers approximated the genuine student attitude. I infer from your wording that what you would want me to do would not appeal to the sensation-hunting or merely curious public anyway' (19 October 1933, letter to Harold Innis, Frank H. Knight Papers, box 60, file 16, University of Chicago Archives). The importance with which Knight's visit was viewed is underlined by the fact that at the height of the Depression, it was funded by direct out-of-pocket contributions of interested faculty members. Knight's talk triggered Urwick's article on the same subject ('The Role of the Intelligence in the Social Process') and Innis's reply. (Knight's presentation was eventually published as 'Social Science and the Political Trend,' *University of Toronto Quarterly*, 1935.)

After his talk, Knight wrote to Innis from Chicago and mentioned how much he had been impressed with two of Innis's young colleagues: Vincent Bladen and Irene Biss. In short, Innis used Frank Knight as an outside authority to support his position in the Depression era debate taking place in Canadian scholarly circles.

25 Innis, 'Introduction to Canadian Economics,' *The Canadian Economy and Its Problems*, 3.

26 Ibid., 13, 6, 7.

27 Ibid., 9.
28 Ibid., 6.
29 For instance, see Innis, 'Discussion in the Social Sciences,' 409–12:

> The difficulty of looking at Canada as a whole is almost insuperable. Emerging as a continental unit centering about the St. Lawrence during the period of the fur trade, its relation to the disunity of the Atlantic Maritimes has been loose and tentative ... A new unity was attained with the industrial revolution. The steamship, the railways, and the financial support provided in the Act of Union and the British North America Act or Confederation ...
>
> The British North American Act is essentially a credit instrument designed to install capital equipment essential to the development of Canadian unity ... The relation between the earning capacity of the capital equipment and the interest on the extent of the credit becomes of first importance. In the main, earning capacity in a country which emphasizes capital equipment rather than labour depends on exports, and these in turn fluctuate widely as a result of climate and yield in the case of wheat and as a result of world prices. Exports of lumber, pulp and paper, fish, minerals, and agricultural products such as wheat, livestock and fruit, e.g. apples, are dependent on world prices and factors influencing world prices ... Fluctuations in receipts vary in relation to the character of exports and consequently separate regions are affected differently ... Consequently it is extremely difficult to recommend a cure-all, since an advantage to one industry may not be anything like as much of an advantage to another industry, and a gain for one region may not be a gain for another ... Nor is this all. The cost of producing the article, which includes the interest paid by the government abroad and elsewhere, cannot change as rapidly as the prices of the exports or as the returns on exports, particularly if the capital equipment is involved in transportation and navigation, and if it is government-owned ... With the opening up of new industries, tariffs and railway rates are not regarded as burdens, or during a period of boom no one notices additional costs. With the trend toward exhaustion of natural resources and during a period of depression with decline in rates of growth of population and absolute decline of population, attention is drawn to the burden in a very direct way, and particularly in the regions remote from the centres.

30 Innis, 'Economics for Demos,' 393.
31 See Innis, 'The Passing of Political Economy,' 4: 'Adam Smith had the advantages of ... cultural life ... during a period in which economic activity

had witnessed marked expansion with the extension of free trade ... [he] stood at a focal point of cultural adjustment in Scotland after the union. The intellectual life of the University of Glasgow reflected the change in the introduction of lectures delivered in English rather than in Latin. The consequent stimulus to thought in philosophy and theology was in striking contrast to the intellectual impotence of Oxford.'

32 Ibid., 5.
33 Ibid., 6.
34 Ibid.
35 Innis, 'Economics for Demos,' 393.
36 S.D. Clark, interview by author, 28 June 1979.
37 Innis, 'Introduction to Canadian Economics,' 16.
38 Innis, 'Discussion in the Social Sciences,' 402.
39 Ibid., 403.
40 Ibid., 404.
41 Ibid., 405.
42 Ibid., 408, 406.
43 H.A. Innis to Irene Spry, letter, 'Tuesday,' n.d., *c.* summer 1935, Irene Biss Spry Papers, Library and Archives of Canada. For a good overall summary of Irene Biss Spry's life and intellectual contributions, see Babe, *Canadian Communications Thought*, 166.
44 Interview with Irene (Biss) Spry, Elspeth Chisholm Tapes, University of Toronto Archives.
45 Babe, *Canadian CCommunications Thought*, 166.
46 Innis to Biss, Calgary, Wednesday [1935?], Innis *Study*, *Innis College*, and Irene Biss Spry Papers, Library and Archives of Canada.
47 Ibid.
48 Ibid.
49 Ibid.
50 Irene Biss to H.A. Innis, '6.IX.35,' Department of Political Economy, University of Toronto Archives.
51 Ibid.
52 Babe, *Canadian Communications Thought*, 167.
53 Creighton, *Harold Adams Innis: Portrait of a Scholar*, 93.
54 Irene Biss Spry, letter to Creighton, 19 June 1958, Creighton Papers Library and Archives of Canada.
55 Introduction to Finding Aid, Department of Political Economy Holdings, University of Toronto Archives, vii.
56 Innis, *Political Economy in the Modern State*.
57 Irene Biss Spry, interview by David Cayley, 'The Legacy of Harold Innis,' CBC Radio, *Ideas* program, 1994.

58 For instance, during the period 1930–6, Graham Spry and Alan Plaunt lobbied vigorously for the establishment of a national public broadcaster in Canada. They co-founded the Radio League of Canada, a coalition of volunteer groups, which successfully lobbied the Conservative government of R.B. Bennett to establish public broadcasting. Spry was also active in the League for Social Reconstruction (a Canadian version of the Fabian Society) and the Co-operative Commonwealth Federation (a newly founded Canadian social-democratic political party, and forerunner of the modern-day New Democratic Party). This sort of direct party political involvement by an intellectual was an anathema to Innis. Irene Spry went on to have a distinguished career as an economist. By the end of the 1970s, she was a professor emeritus at the University of Ottawa.

59 Brebner, letter to Innis, 22 June 1950, Innis Papers, University of Toronto Archives.

60 See Babe, *Canadian Communications Thought*, for an appreciation of Graham Spry's contribution.

61 For an excellent examination of the debates in the social sciences in the 1930s, see Ferguson and Owram, 'Social Scientists and Public Policy from the 1920's through World War II.' Innis's position, which emphasized the university and scholarship, and how his stance was assailed by both the 'hot gospellers' of the left and scholarly technocrats who joined the civil service, is well presented in this essay. See also the more extended treatment in Owram's *The Government Generation: Canadian Intellectuals and the State, 1900–1945*.

62 Berger, *The Writing of Canadian History*, 151.

63 The correspondence begins in 1931 and runs through to the Second World War. At its peak – during 1937–8 – one hundred letters passed between the two men.

64 Ibid., 144–5.

65 *The Oxford Companion to Canadian History and Literature* (London: Oxford University Press 1967), 144.

66 Innis, letter to Shotwell, 15 November 1935, and 5 December 1933, Shotwell Collection, Box 286, Innis Folder, Columbia University Library.

67 As Berger points out, the map stressing the basic unity of the continent which was used as the frontispiece for each book in the series was flatly contradicted by Innis's assertion that Canada came into being because of, not in spite of, geography, and by treatments such as Creighton's. See Berger, *The Writing of Canadian History*.

68 Innis, letter to Shotwell, 11 August 1932, Shotwell Collection, Box 286, Innis Folder, Columbia University Library.

69 MacIver, letter to Shotwell, n.d., ibid.

70 Brebner, letter to Shotwell, n.d., ibid.
71 Innis, letter to Shotwell, 15 November 1933, ibid. It is interesting to note that, despite Innis's defence of scholarship free from political influence, he refers to efforts at Canadian nation-building in the past as 'our' efforts and concludes by suggesting that public-policy changes will be needed to continue these efforts in the present.
72 'W.W. McL' [McLaren?], letter to Shotwell, 'Personal Memorandum on the Innis Budget,' 23 January 1934, Shotwell Collection, Box 286. Innis Folder, Columbia University Library.
73 Shotwell, letter to Innis, 10 January 1934, ibid.
74 Innis, letter to Shotwell, 21 February 1935, ibid.
75 Shotwell, letter to Innis, 20 March 1935, ibid.
76 Innis, letter to Shotwell, 25 April 1935, ibid.
77 Innis, letter to Shotwell, Thursday, [November?] 1935, ibid.
78 Innis, letter to Shotwell, Thursday, November 1935, ibid.
79 Innis, letter to Shotwell, March 1938, ibid.
80 Shotwell, letter to Innis, 20 March 1935, ibid.
81 Innis, letter to Shotwell, received 21 January 1937, ibid.
82 Innis, letter to Shotwell, 2 April 1938, ibid.
83 For a detailed presentation of the dealings of Innis with Lorne Pierce and Ryerson Press, see Sandra Campbell, 'From Romantic History to Communications Theory: Lorne Price as Publisher of C.W. Jeffreys and Harold Innis.' However, Campbell seems unaware that, if Innis thought he could strengthen the University of Toronto Press by causing tensions between Ryerson and Shotwell/Yale, he had no hesitation in doing so. On the one hand, this presents a manipulative side to Innis that is less than honourable given his close relation with Lorne Pierce. On the other hand, it underlines the passion with which Innis pursued the building of institutions of scholarship in Canada.
84 Innis, letter to Shotwell, 13 December 1938, Shotwell Collection, box 286. Innis folder, Columbia University Library.
85 Innis, letter to Shotwell, 21 November 1938, ibid.
86 Berger, *The Writing of Canadian History*, 109.
87 Shotwell, letter to Innis, 30 August 1939, Shotwell Collection, box 286, Innis Folder, Columbia University Library.
88 Innis, letter to Shotwell, 14 November 1938, ibid.
89 For instance, in 1938 he was invited to New York to present the Canadian view on a United States/Britain trade agreement to a gathering of high-level scholars, officially called the Council on Foreign Relations. Shotwell, letter to Innis, 17 February 1938, ibid.

90 For example, Professor Trotter was the official representative of the newly created CSSRC designated to meet with the Carnegie officials. Yet Innis was encouraged by Shotwell to prepare a research plan which would be submitted to the Carnegie Endowment via the CSSRC (see Shotwell, letter to Innis, 4 September 1940, ibid.).

91 Innis, letter to Shotwell, March 1939, ibid.

92 Innis, letter to Shotwell, 13 August 1946, ibid.

93 Shotwell, letter to Innis, 28 July, 8 August , and 13 September 1938, ibid.

94 Innis, letter to Shotwell, 13 March 1939, ibid.

95 Innis, letter to Shotwell, 15 March 1939, ibid.

96 Innis, *The Cod Fisheries*, ix. My emphasis.

97 Ibid., 500.

98 Ibid., 486.

99 Ibid., 52.

100 It is worth noting here that in *The Cod Fisheries*, as in his later work, one finds implicit a theory of unequal development. In the end, the slow start of the English in the trade and their lack of inputs determines their leap ahead of the others. Ibid., 49–50.

101 Ibid., x.

102 Ibid., 499.

103 Ibid.

104 Innis, letter to Shotwell, July 1933, Shotwell Collection, box 286, Innis Folder, Columbia University Library.

105 The following are typical comments from a contemporary reviewer: 'When my friend writes something I generally make the error of trying to understand it. Here he is reviewing the whole Canadian economy (railroads, wheat, insurance, etc.) in the light of the corrective influence of fluctuating prices. But his conclusion is too profound for my comprehension.' In terms of the style of presentation and the puzzled but indulgent reaction it provoked in the reviewer, the essay was a forerunner of the communications work.

106 Innis was groping towards an elaboration of the preconditions for intellectual fluorescence later to be described much more clearly under the symbolism of Minerva's Owl. His concern with this issue as well as his obvious lack of ease in dealing with it is evident in the following passage from his *Essays in Canadian Economic History*, 272: 'Economists have reflected the confusion introduced by machine industry ... Scientific advance and the application of science to industry through inventions are characteristic of periods of prosperity. The philosophic outlook based on scientific achievement leaves its stamp on economics. Periods of prosper-

ity may be characterized by most intensive work in economics but periods of depression have been characterized by attempts at application, particularly in the field in which mathematics provides a convenient channel between science and economics, namely money ... Depressions produce deterministic systems and arguments such as have been advanced in this paper.'

107 Ibid., 271.

108 Ibid., 266–7.

6. Hunting the Snark

1 Carroll, 'The Hunting of the Snark,' *The Best of Lewis Carroll*.

2 For an excellent examination of Innis's role in this initiative, see Evenden, 'Harold Innis, the Arctic Survey and the Politics of Social Science during the Second World War.' Evenden's article is particularly interesting in its presentation of how Innis attempted to maintain an independent scholarly approach by walking a fine line between the Canadian government, which controlled access to the north and its own information on the north, and the Rockefeller Foundation, whose funding of the exercise was largely motivated by a fear of the Soviet threat and a desire to maintain control of strategic resources.

It is clear that Innis was central to this scholarly undertaking, as he was to many other initiatives during the period. 'As the chair of the grants-in-aid committee of the CSSRC, Innis had authority over the funding of the project, wrote its research scheme, and had the added trust of the Rockefeller Foundation. In the Foundation's annual report, he was listed as a recipient and sole researcher of the grant-in-aid to the CSSRC (Evenden, 54–5). Innis's experience with the Arctic Survey underlay his trip to Russia during the immediate post-war period and his plea for greater understanding of that country at a time when increasing suspicion and confrontation were the order of the day.

3 Innis had developed a reticence to fly after watching planes crash after 'dog fights' over the trenches in the First World War. See Christian, ed., *Innis on Russia: The Russian Diary and Other Writings*, for an account of the trip and its impact on Innis.

4 Innis, 'Seeing Present Day Russia,' 'Finds Moscow a City of Gaiety and Russia Very Much Like Canada,' and 'New, Stronger Contacts with Russia Seen Essential to New World Pattern,' *Financial Post*, 11, 18, 25 August 1945.

5 Friedland, *The University of Toronto*, 415–19.

6 Ibid., 350.
7 Ibid., 333.
8 Ferguson and Owram, 'Social Scientists and Public Policy from the 1920's through World War II,' 4 and 12. These authors present an overview of how the war affected the social sciences in Canada. They remind us just how small the community of social sciences was at the time. See chapter 5, n.2.
9 Horn, *Academic Freedom in Canada*, 177. Professor Walter Sage, in a letter to Innis, 14 January 1943, refers to a copy of the letter from the prime minister. This letter was probably a reply to Innis's memorandum to the prime minister concerning university policy during wartime. However, such a letter seems to have been removed from the Innis Papers before they reached the University of Toronto Archives.
10 Douglas LePan, letter to Innis, 28 January 1943, Innis Papers, University of Toronto Archives.
11 Harold Innis, 'Memorandum to Cody,' 'Personal and Confidential,' 3 December 1943, Innis Papers, Innis College Archives.
12 Ibid.
13 Horn, *Academic Freedom*, 176.
14 Innis to Cody, 3 December 1943.
15 Fisher, 'Harold Innis and the Canadian Social Science Research Council: An Experiment in Boundary Work,' *Harold Innis in the New Century*, 135–58.
16 Ibid., 150, 135.
17 Friedland, *The University of Toronto*, 418–19.
18 The Underhill controversies have been written about extensively. Innis's role in these crises is dealt with by Horn (*Academic Freedom*), Friedland (*The University of Toronto*), and Eric Havelock (*A Memoir*), as well as by Donald Creighton (Harold Adams Innis) and Carl Berger (*The Writing of Canadian History*).
19 Horn, *Academic Freedom*, 119.
20 Berger, *The Writing of Canadian History*, 82.
21 Harold Innis, 'Presentation to the President and Board of Governors,' n.d., Innis College Archives. Presumably Cody asked Innis for a copy of the presentation. The words 'statement typed for the President's personal use' are added to the top of the Archives' copy in Innis's crabbed handwriting.
22 Ibid., 1–2.
23 Ibid., 3.
24 Bartlett Brebner, letter to Harold Innis, 10 January 1941, President's Papers, University of Toronto Archives.
25 H.J. Cody, letter to Mackenzie King, 10 January 1941, and the Department

of External Affairs, letter to Cody, 15 January 1941, President's Papers, University of Toronto Archives.

26 It was not just Frank Underhill whom Innis defended during this period. He also defended Eric Havelock and F.R. Scott, two other intellectuals active in the League for Social Reconstruction, from public attack. See Havelock, letter to Innis, 23 February 1943, and Scott, letter to Innis, 24 August 1942, Innis Papers, University of Toronto Archives.

27 M.A. Trow, *The British Academics*, quoted in Horn, *Academic Freedom*, 117.

28 I am indebted to Michiel Horn for his analysis of the British and German versions of academic freedom, See *Academic Freedom*, 4–10.

29 Innis, 'The University Tradition,' *Political Economy in the Modern State*, 69.

30 Fisher, 'Harold Innis and the Canadian Social Science Research Council; An Experiment in Boundary Work,' 135.

31 Innis, 'University Economist for Russia,' *Financial Post*, October 1945.

32 Innis, letter to Mary Quayle Innis, 2 August 1941, Innis Papers, University of Toronto Archives.

33 R.B. Inch, letter to Innis, 17 February 1943, ibid.

34 George Ferguson, interview by Elspeth Chisholm, 8 November 1970, University of Toronto Archives.

35 Ibid.

36 See invitations, 9 April 1942 and 12 May 1943, Innis Papers, University of Toronto Archives.

37 For instance, McEachern, letter to Innis, 11 December 1942, ibid.: 'This week there is a Letter to the Editor on the editorial page from someone who has a lot of disagreeable things to say about commissions in the armed services for university graduates. I cut out a lot of the most virulent remarks he had to make. It is, I fear, unwise to close our column to people who want to speak their minds. But would you please see if you could get somebody out of your department who would crucify him for next week's issue.'

38 McEachern, letter to Innis, 4 February 1943, ibid.

39 Innis, letter to Bezanson, 23 November 1948, ibid.

40 Timlin, letter to Innis, 24 October 1943, ibid.

41 Innis, letter to Mary Quayle Innis, 2 August 1941, ibid.

42 Kemp, letter to Innis, 6 February 1942, ibid.

43 List compiled from Innis correspondence of the 1942–3 period, ibid.

44 Ferguson, 'Harold Innis and the Printed Word,' 4 November 1965.

45 Fisher, 'HAI and the CSSRC,' 146–7.

46 Connie McNeill (Scott), interview by author, 1979.

47 Ev Smyth, interview by Elspeth Chisholm, University of Toronto Archives.

48 John Sword, interview by author, 25 September 1979.
49 S.D. Clark, interview by author, 28 June 1979.
50 Donald Innis, interview by author, 18 July 1979.
51 Ev Smyth interview.
52 Friedland, *The University of Toronto*, 363.
53 Blodgett, 'Harold Innis as a Teacher,' 3.
54 Drummond, letter to Innis, 15 May 1944, Innis Papers, University of Toronto Archives.
55 Ernest Sirluck, interview by author, 19 March 1979.
56 Innis, letter to Cody, 29 November 1943, President's Papers, University of Toronto Archives.
57 Speech notes, n.d., Innis Papers, ibid.
58 Ibid., 1–2.
59 Ibid., 3.
60 Ibid., 5.
61 Ibid., 6.
62 Ibid., 7.
63 Ibid., 8.
64 S.D. Clark interview. Bladen, quite rightly, points out there was another element involved: Bladen was 'British' in Innis's eyes.
65 Nor did he think it inappropriate to lecture at political party 'summer schools' when called upon to do so.
66 Innis, 'Lord Kennet on Royal Commissions,' quoting from the *Journal of the Royal Statistical Society* (Part 3, 1937), Innis Papers, University of Toronto Archives.
67 Innis, *The Bias of Communication*, 213.
68 Draft letter of resignation, n.d., Innis Papers; University of Toronto Archives.
69 Innis's version of the politics of the commission presents the situation in quite a different light. Writing to George Ferguson, (Ferguson, Miscellaneous Corespondence), he says, 'I note your curiosity re the arrangement of the report. I think I told you of the difficulties between Angus and the Chairman during our meetings just before Christmas. As a result we were threatened with a minority from Angus ... and I was put in the position of being compelled to write something which would head this off. The chairman was very intent on this as a sort of annex. It had the effect of leading a substantial modification of H.F.A's position and to the appearance of a unanimous report.'
70 Creighton, *Harold Adams Innis: Portrait of a Scholar*, 139.
71 Innis's royal commission notes are similar to the communications notes:

many topics, spare style, many sentence fragments, fifty scattered subject headings arranged in alphabetical order – 'Accounting,' 'Adjusted Mileage,' 'Agreed Changes,' 'AIR,' 'Amalgamation,' and so on. See Royal Commission on Transportation, Textual records. Library and Archives Canada.

72 Hugh Innis, interview by author, 21 September 1979.
73 S.D. Clark interview.
74 Frances Halpennys interview by author, 21 June 1979.
75 Grant Robertson, interview by author, 29 March 1979.
76 For instance, Vincent Bladen reports: '[In the 1940s] Harold Innis and I were both so busy that communication broke down. It was at this time that he was moving ... to the history and philosophy of communications in the World. With his earlier (staples) work I was familiar; I had read his first drafts of articles and had discussed them. I understood him. When he moved on to the later phase, I could not follow him, we did not discuss it we were too busy' (*Bladen on Bladen*). Also, Donald Creighton commented: 'He was always ready to talk to me ... we became very close friends ... I didn't see so much of him towards the end, because I was very busy ... and he, of course, was busy with his work ... but there was never a rift, never a cloud, never anything like that ... we just were not perhaps so frequently together as we were before' (interview by Paul Kennedy, CBC Radio program, *Ideas*, University of Toronto Archives).
77 Donald Innis interview and Innis correspondence, Innis Papers, University of Toronto Archives.
78 See Donald Innis, letters to H.A. Innis, 27 January 1950, 20 March 1950, and 25 March 1950, ibid. These letters reflect Innis's discouragement at a lack of feedback from colleagues. (Donald Innis was at the University of Chicago at the time.) They also indicate that Harold Innis was very much up-to-date on the most recent developments in communications technology and their significance, in particular, magnetic tape and television.
79 Francess Halpenny interview.
80 Graham Spry, interview by author, 11 July 1979.
81 S.D. Clark interview.
82 Innis, letter to Principal Peers, 10 May 1948, correspondence of the Public Lectures Committee, Nottingham University.
83 Innis, letter to Principal Peers, 12 May 1948.
84 Even Lorne Pierce at Ryerson Press, who had worked with Innis on various publications during the 'economic history' years, declined to publish *Empire and Communications* or the *The Bias of Communications*. The extent to which Innis's methodology put people on the defensive is attested to in this note from Pierce to Innis: 'The more I read [the manuscript] the more hum-

ble I became and the deeper the blush. I hate to admit it, and I hope guard the secret well, but there is an awful lot I do not know.' (Campbell, 'From Romantic History to Communications Theory: Lorne Pierce as Publisher of C.W. Jeffreys and Harold Innis,' 111).

85 Francess Halpenny interview.

86 Note on publishing and sales statistics from R.I.K. Davidson, University of Toronto Press, 4 May 1979.

87 Note by Mary Quayle Innis on book sales, Innis Papers, University of Toronto Archives.

88 'Rambles among the Social Sciences', *The Economist*, 8 February 1947.

89 Adair, review of *The Bias of Communications*.

90 Paul Heyer, in 'Empire, History, and Communications Viewed from the Margins: The Legacies of Gordon Childe and Harold Innis,' and in his book *Harold Innis*, 52–3, presents this exchange as 'a significant meeting of kindred spirits' in which Childe praises Innis's work and Innis thanks him for his commentary. This is a fundamental misreading of the exchange. Childe's review was a full-frontal attack on *Empire and Communications*, drawing upon his specialist knowledge of archaelogy to list instance after instance of where Innis was wrong. Innis' reply was largely evasive. It avoided Childe's specific criticisms and accused him of having an archaeologist's bias of overstating the importance of artefacts that survive. In short, it was quite a vicious exchange dressed up in scholarly language.

91 Childe, review of *Empire and Communications*.

92 Ibid., 100.

93 Innis, 'Communications and Archaeology.'

94 Rudolf Coper, letter to author, 14 May 1979.

95 Family reminiscences, n.d., Innis Papers, University of Toronto Archives.

7. A Telegram to Australia: Innis's Working Methods

1 Gandz, *The Dawn of Literature*, 262.

2 Innis, 'The Problem of Space,' *The Bias of Communication*, 105.

3 Innis, 'The Church in Canada,' *Essays in Canadian Economic History*, 387.

4 Havelock maintains that the similarity of his work and Innis's during the period was a 'matter of happy coincidence' and I accept this assertion as it pertains to Innis's influence on Havelock.

5 Innis, *Empire and Communications*, xiii.

6 Comment by Mary Quayle Innis, Innis Papers, University of Toronto Archives.

7 There is some debate over the meaning of the three terms holograph,

manuscript, typescript. The proper use of the terms is as follows: holograph: a document (in this case, draft) wholly written by the person under whose name it appears. Manuscript: a document written by hand, not printed. This has come to mean, in our era in some cases, typewritten copy – the emphasis being put on the 'prepublication' sense of the meaning rather than the 'handwritten' sense. It is often thought of in the more restricted sense of the document that goes to the publishers to be transformed into printed copy. Typescript: anything typed. In short, either a holograph or a typescript can be a manuscript.

8 Even in the case of the quotations file, we have some clues to its use as a working file. At the beginning of the file there are non-alphabetic dossiers marked 'citation-check,' 'citations-sort,' '?' One of these dossiers contains regular typed pages full of small quotations from a variety of authors. As an example, one might find five quotations from a book by John Nef, typed in the order that they appear in the original book (i.e., 17, 39, 42, 44, 123), which are followed by a similar set from a book of McIver and then Sorokin, and so on. In the margin, Innis has added holograph numeration, with each source book having a distinct number. I assume that this code referenced the quotation to a bibliographic (or reading note) listing of Innis's and that the quotations were to be clipped and then sorted into the appropriate file.

Another dossier contains a more mixed group of short quotations, each with its author, title, page number, and so on identifying it. We find again a marginal numeration. This time it is entered as 'p. such and such' rather than as a straight number. Furthermore, these pages are in ascending order.

Presumably the first dossier represents lists of quotations Innis found interesting while reading source material. The second may well represent quotations he used in putting together a manuscript for publication. If this interpretation is correct, then, what we have in these dossiers is the 'deposit' and 'withdrawal' forms for his citations 'account.' It is as if in opening this file we are *in medias res* at the time of Innis's death.

This interpretation accords with the state of the quotations dossiers which follow the above. They are listed alphabetically by author. Unlike the reading notes, these files are quite limited in size. Moreover, many authors who are cited frequently in Innis's published work have little material in their dossiers (or no dossier at all). I take this to mean that Innis's throwing out the scraps of paper with citations on them (once they were incorporated in his published works) served the same function as his checking off reading notes which had been similarly employed.

9 Donald Innis, interview by author, 18 July 1979.

10 Conversation with William Christian, 27 June 1979. See also his draft introduction to *The Idea File* in the Innis College Archives, University of Toronto.
11 There are reading notes that have been preserved dating from the staples years, but these are the exceptions to the rule. They are usually notes on particularly impressive primary sources. As such, they are self-contained and have been kept through the same rationale that led Innis to retain his trip journals. His notes on the minutes of the 'Royal Company of Feltmakers' would be an example.
12 Ernest Sirluck, interview by author, 19 March 1979. See also chapter 8.
13 Innis, ironically, may be partly responsible for this loss. Oral tradition has it that, during one of the many Underhill crises, the president of the University of Toronto asked the chief librarian to show him Underhill's book check-out list so that he could get an idea of what Underhill had been reading. The chief librarian refused on the grounds that the material was of a confidential nature. Innis was one of those who supported his position. As a result, the old charge-out ledgers were apparently destroyed rather than eventually being placed in the archives. I have not been able to substantiate adequately this story. However, after a careful search I am fairly certain that the ledgers in question were in fact destroyed. This is particularly regrettable for we know that, before the general ledger was destroyed, a compilation of all the titles (with dates) that Innis had borrowed (including interlibrary loans) was prepared by the library staff. This document has since been lost.
14 A somewhat confused presentation of the relationship of these various bibliographies is to be found in Elspeth Chisholm's 'Innis' Method of Working,' Accession B72–003, ix–x, University of Toronto Archives.
15 This material has been put in order and now is contained in a separate box in Accession B72–003, the University of Toronto Archives.
16 'Dr. Stewart Wallace ... the University chief librarian ... Mrs. Innis told me, once told her that "Harold read a lot of trash as well as a lot of solid books." Mrs Innis was resentful at the time, but in recalling this as we were going over the lists of books he had read, she agreed that it was so. We both thought that the "trash" was extremely interesting, illustrating [Innis's] curiosity about the motivations of the writers. There are a great number of 'my life and times' books – sometimes by know [*sic*] authors or newspapermen, but sometimes unknown to either of us.' Elspeth Chisholm, 'Report on Trip to Toronto,' 2–7 March 1970, MG 31, E 50, vol. 12, Innis Programme File, Library and Archives of Canada.
17 See 'Basic Book List of Readings by Harold Adams Innis Listed by His Choice of Subject,' Accession B72–003, University of Toronto Archives.

18 An example of a scholarly essay depending largely on this type of biographical material would be 'Technology and Public Opinion in the U.S.A,' in *The Bias of Communication*, 156.
19 This has been deduced by the manner in which Innis's citations and reading notes from sources are grouped around certain sections of the original text.
20 Innis College Archives contains a book that has been left in this state.
21 Robertson, *The Administration of Justice in the Athenian Empire*, 7. My emphasis.
22 Innis Papers, Accession B72–003, University of Toronto Archives.
23 Foolscap is larger than normal format (8½" ×11") paper. It was blown up in the photostatting process to 10½" ×14". It has been reduced again in Figure 24.
24 Innis's first essays as a veteran were done on the back of cut-up wartime posters.
25 Hugh Innis recalls his father using primitive gel tablets for copying. However, the notes retained in the archives were made not by gelatine duplicators but by photostat technology. This involved making prints on photosensitized paper using a camera but with no negative, plate, or film intervening. The paper was then processed as a regular photographic print, producing a negative image (white on black) of the original text. Hugh Innis interview by author, 21 September 1979, and www.officemuseum.com, 'Antique Copying Machines,' 16 of 17.
26 Christian, draft introduction to *The Idea File*. Also, Joyce Wry (former secretary in the department of political economy) interview by author, 24 September 1979.
27 Author's interviews with Donald and Hugh Innis and Anne Dagg.
28 Christian, introduction to *The Idea File*.
29 Innis, *Empire and Communications*, 82.
30 See Figure 23. Copy of Innis's notes for Robertson, *The Administration of Justice in the Athenian Empire*. My emphasis.
31 These sentences are taken verbatim from Robertson, out of context, from the following pages: 72, 73, 78, 81, 88. My emphasis.
32 Innis, 'The Role of Intelligence,' 287.
33 Werner Jaeger, *Paideia: The Ideals of Greek Culture*.
34 The works of Jaeger used by Innis are: *Aristotle: Fundamentals of the History of His Development*; *Paideia: The Ideals of Greek Culture*; *Humanism and Theology*; *The Theology of the Early Greek Philosophers*; and *Demosthenes*. The citations by Innis are: *Empire*, 55 – *Theology*, 195; *Empire*, 62 – Jaeger cited by name only; *Empire*, 71 – *Paideia*, vol. 1, 229; *Empire*, 74 – ibid., vol. 1, 235ff.;

Empire, 76 – *Theology*, 155; *Empire*, 77 – ibid., 42; *Empire*, 79 – *Paideia*, vol. 1, passim.; *Empire*, 80 – ibid., vol. 1, 360, passim.; *Empire*, 84 – *Demosthenes*, n.p.; *Empire*, 127 – *Humanism*, 25–9; *Empire*, 169 – Jaeger cited by name only; *Bias*, 7 – Jaeger by name only; *Bias*, 19 – *Humanism*, 24; *Bias*, 54 – ibid., 14; *Bias*, 124 – ibid., 24.

35 Innis, *Empire and Communications*, 76–7.

36 See Accession B72–003, University of Toronto Archives.

37 Francess Halpenny, interview by author, 24 June 1979.

38 Innis, *Empire and Communications*, xiii.

39 Grant Robertson, letter to Innis, Innis Papers, University of Toronto Archives. My emphasis.

40 Brebner, letter to Innis, 22 June 1950, Innis Papers, University of Toronto Archives.

41 Buxton, 'The Bias against Communication: On the Neglect and Non-publication of the "Incomplete and Unrevised Manuscript" of Harold Adams Innis.' Available online at www.cjc-online.ca.

42 Ibid., 1–2.

43 The executors' committee for Innis's work included W.T. Easterbrook, Donald Creighton, S.D. Clark, Mary Quayle Innis, and Donald Innis.

44 Buxton, 'The Bias against Communication,' 8.

45 Chisholm Fonds, NAC, MG 31, E 50, 3 Nov 1969, Innis Manuscript – Progress Report no. 2.

46 Ibid.'Innis's Method of Working,' i.

47 Ibid., ii.

48 Chisholm Fonds, NAC, MG31, E 50, 'CITING IS FOR POSSIBLE VERIFICATION.'

49 Chisholm Fonds, George Ferguson to Elspeth Chisholm, 20 December 1969.

50 Brebner, 'Innis as Historian,' 5, University of Toronto Archives.

51 To take one example of this all-too-voluminous literature:

> The weight that Innis placed on orality, however, and the fact that although he speaks of 'culture' in the aesthetic-intellectual sense he has in mind praxis as well (including that of economic long-term planning), suggest that the argument is not quite as coherent as this account would make it seem ... Co-mingled in Innis' account are concern for the conditions for monumentalizing *poiesis* – 'continuity in form' – and a related but distinct concern for the conditions for what we might call wise praxis. The remainder is the excess of praxis over *poiesis*, with that difference masked by the moment of 'vision,' in which the eternal time of art slides into the *longe durée* of a politic addressed to civilizational matters, (Andrew Wernick, 'No Future:

Innis , Time, Sense, and Postmodernity,' in Acland and Buxton, *Harold Innis in the New Century*, 275).

This account makes Innis's argument seem more coherent? I do not think so. To me it appears to be an extreme form of the intellectual narcissism that is all too often found in Innisian commentary. Another example is found in Stamps, *Unthinking Modernity*, 91: 'The style that inspired these responses was Innis's unique adventure into theoretical non-certainty, a concrete antidote to identitarian forces that, in his view, were destroying the last vestiges of Western culture. It was also an adventure in establishing a new relationship with the reader. The key ingredients were a high degree of self-reflexivity, sharp avoidance of linear causal language, unique use of quotations, and an unusually open method of footnoting. Their product was a set of texts that must be read very actively to be read at all.' This is the academic equivalent of a torturous effort to make a 'silk purse out of a sow's ear.' To read the texts 'actively,' that is, to delve into how they were constructed, requires us to come to grips with the fact that a fundamentally questionable research methodology underlay Innis's communications works.

52 Innis, *Empire and Communications*, 56.

8. Innis and the Classicists: Imperial Balance and Social-Science Objectivity

1 Innis, *Political Economy in the Modern State*, viii.

2 Easterbrook, a humble and insightful commentator, clearly stresses the importance of classical studies and Cochrane in particular on Innis's later work, while at the same time not feeling qualified to pursue this valuable lead 'Innis and Economics,' 303.

3 'Liberal' in the sense of valourizing the role of exceptional individuals (like Adam Smith) in understanding society. However, Innis was an elitist in the sense of believing that the potential 'best minds' of any society were in limited supply and not concentrated in any particular social class. Hence the importance of an effective public-education system to identify these best minds at an early stage and hone them through education, even if this meant putting in place special measures to allow those coming from a disadvantaged background to make up for the weakness imposed by their background. Not surprisingly, this perspective made him adopt a fundamentally conservative attitude towards modern democracy.

Perhaps the best work describing Innis's political stance is Noble's essay

'Innis's Conception of Freedom,' in *Harold Innis in the New Century*, 31–45. Noble identifies Innis's affinities with 'the "old," or conservative Whig tradition of political thought – the eighteenth-century British tradition of political thinking associated with David Hume, Adam Smith, and Edmund Burke.' 'Liberal' in this sense implies 'an older kind of liberal: one who regarded liberty as a virtue of certain types of civil association rather than as a universal right derived from the very nature of human beings.'

4 Innis, *Modern State*, viii–ix.

5 Ibid., 128.

6 We should recall the quotation cited at the start of the introduction to this book. The key role of Innis in the politics of the social sciences in Canada is outlined by Bladen, Easterbrook, and Willits in 'Harold Adams Innis: 1894–1952.' 5. In a letter to the Rockefeller Foundation on grants policy, Innis writes:

> I think there are dangers in the pretensions of precision. A friend of mine has been described as more anxious to be precise than accurate, and I suspect that we are in danger of this general disease in the social sciences ... I am afraid [social scientists] deal in specific measurements and are very attracted to things which can be specifically measured. For all of this, of course, there is a very important place, but it is a limited place, and its limits ought to be very carefully studied. I keep coming back to my own particular interests in what I laughingly call 'research', namely, the persistent tendency in the field of knowledge and in the social sciences to build up monopoly or oligopoly situations. The literature builds up around the name of Keynes, or Marx, or someone else, and everything else is dropped ... What I am wondering about is whether we can reach a position in which there is a continuous discussion of vital problems ... And this is why I would like to see the drift towards the humanities – namely, to recognize the intensive work which has been done over the centuries and not so completely ignore it as is now done, or to put it in the form of first year survey courses which is probably worse. Such a drift might do something to make one alert to the possibility of the social science of totalitarianism which has become so threatening.

One is immediately struck by the intensely political tone of the letter. Innis is not talking about academic politics, the capturing of the commanding heights of a discipline by an old guard of professors with fixed ways. He is talking about the deformation of the social sciences *in general* by the demands of the modern state. He refers to 'the *social* science of totalitarianism,' not totalitarianism in the social sciences. More and more the tendency is for intellectuals to be required to work within an existing paradigm that actively serves as the ideological underpinning of a contemporary polity. In quotations such as this, Innis seems to indicate that he did not disingenuously stumble into his communications studies. It was a conscious and 'political' step.

7 Innis, 'A Plea for the University Tradition,' *Modern State*, 73, 75.
8 Innis, *The Bias of Communication*, 33, 132.
9 For instance, Marx, *Grundrisse* (London: Penguin 1973), 85, 100–2.
10 Neill, *A New Theory of Value*, 79.
11 Interestingly, a number of these classicists – for instance, Havelock himself, George Grube, and Gilbert Norwood – became closely associated with the left of the time and were heavily involved in the discussions and political activities that made a mark on Innis. See R.B. Todd, 'Oxbridge Classics Exported (1880–1920) – Some Canadian Perspectives,' a paper delivered by R.B. Todd at Trinity College, Cambridge University, 31 July 2002.
12 Cochrane's contribution to this popular-education campaign was a civics text, *This Canada of Ours*, co-authored with Innis's colleague W.S. Wallace, and a short biography of the explorer David Thompson. In this biography, Cochrane introduced Peter Pond as a 'scoundrel' to serve as a negative foil to Thompson. Innis replied to this approach with his biography of Pond, published in 1930. Innis presented the American adventurer as an unrecognized 'Father of Confederation' who had been much maligned owing to the fact that his semi-literate status left him open to attack from his historically more articulate enemies. Even at this early date, Innis was concerned with correcting the overwhelming biases of the written tradition.
13 The Cochrane and Innis families were socially quite close, with reciprocal attendance at some family affairs like weddings and birthdays.
14 Cochrane, 'The Question of Commerce.'
15 Ibid., 17.
16 Cochrane, *Christianity and Classical Culture*. It is interesting to note Innis's conception of the problems faced by the colonial intellectual surfacing at this point in the obituary. He sees them as twofold: first, a lack of recognition of the validity of the endeavour on the part of the colonial people; and secondly, a continuing pressure for the association of the intellectual with metropolitan institutions and tradition. 'It is significant that his publishers suggested that the name University College (University of Toronto) should be omitted from the title page ... as having possible depressing effects on prestige and sales.' Innis, 'Charles Norris Cochrane 1889–1945.'
17 Ibid. My emphasis.

18 Cochrane, 'The Mind of Edward Gibbon.'
19 Ibid., 17.
20 Ibid., 166.
21 Innis, 'Charles Norris Cochrane.'
22 How Innis's intensely politicized view could be in turn viewed by commentators as apolitical is obvious in a superficial reading of the following quotation from Bladen, Easterbrook, and Willits in 'Harold Adams Innis: 1894–1952,' 4: 'Culture is not concerned with (immediate) questions (of public policy). It is designed to train the individual to decide how much information he needs, to give him a sense of balance and proportion and to protect him from the fanatic.'
23 Cochrane, *Christianity and Classical Culture*, 484–5.
24 Innis, *Empire and Communications*, xiii. I would like to lay to rest the objection that Innis's relation to Cochrane is merely passive. Until to this point, one could argue that the passages cited indicate *only* a remarkable similarity of certain themes in Cochrane and Innis. Basically, I would argue that Innis's debt to the classicists goes much further; however, it is difficult to demonstrate this because Innis was extremely reluctant to write about methodology or about philosophy. The obituary is one case that clearly indicates the dependence of the communications studies on classicism. There are some other indications of the direct linkage.

Cochrane's work recurs in Innis often in an uncited fashion. Thus, Innis is in the midst of discussing the difficulties of intellectuals in Canada in 'The Church in Canada' when he suddenly resurrects the same anecdote from Herodotus cited by Cochrane (*Christianity and Classical Culture*, 468). In the passage, Innis uses this citation in exactly the same manner as Cochrane (Innis, *Essays*, 392).

In more general terms we find that Innis employs throughout his communications studies precisely those categories delineated by Cochrane as the principles of social organization in the Augustan empire. The formal discipline of *Romanitas* was presented in terms of the interplay of three principal elements. These principles in Innis became extended to apply to the concrete analysis of the history of Western empires. The three elements are *vis* or military force (which is related to changes in military technology as well as to social organization in Innis); *consilium* or policy (or what Innis called monopolies of knowledge and their relation to public opinion); and finally *authoritas* or authority, which, like *consilium*, was a multifaceted element. In Cochrane, the essence of *authoritas* was the mediation of *vis* and *consilium* through the principles of civil law, just as in Innis law is treated as a tremendous cultural innovation that arose in the consolidation of order in societies in which there existed potential balance given the presence of a number of communications systems whose biases offset one another.

One last area in which classical studies via Cochrane seems to intrude in Innis's work is found in the view of human nature developed by Hippocrates and Thucydides and presented most succinctly in Innis's comments on 'Adult Education and Universities.' This stresses the importance of natural excellence as a prerequisite for cultural training. It produces the strange amalgam of elitism and democracy that characterizes Innis's work. Access to educational training must be vigorously democratic in order to determine at an early age which individuals are in possession of that 'natural excellence' that is so rare in any community. At the higher levels, the education system then becomes dedicated to fostering the capacities of these 'best brains' rather than providing extension services to all adults. Innis, *The Bias of Communication*, 206–7.

25 Innis, *Empire and Communications*, 61: 'Generations of poets intensified the imagination of the Iliad and had a profound influence on the literature of Greece and Europe.' The footnote to this statement contains a reference to Owen's book on the *Iliad*. But when we refer to this book we find that Owen has been strongly influenced by the point of view of the 'new criticism.' In other words, he is interested in tracing the story of the poem itself without reference to its history or authorship. He is not concerned with the poem and its function in history (as Innis is) but with its craftsmanship. We do not find any clue here to his influence in suggesting 'the general problem' to which Innis addressed himself in the communications studies.

26 Owen, 'The Illusion of Thought.'

27 Ibid., 505.

28 Ibid., 506–7.

29 Ibid., 508.

30 Ibid., 510.

31 Ibid., 511.

32 Ibid. Our difficulties in reconciling the written work of Owen, aside from this philosophical treatise, and the influence attributed to him by Innis points out one of the problems involved in tracing links. Innis said many times that the tradition towards which he personally, scholarship in general, and the university as an institution were biased was the oral or living tradition. Since there is no record of discussions, we are forced to look for general thematic links in the published and preserved thought of various contemporary intellectuals. One runs into difficulties in cases such as Owen's, however, where few published documents have survived and

where these do not seem to reflect the full range of the live author's 'musings.' Anecdotes such as that cited in the next note are one of the few means of countering these difficulties.

33 Ernest Sirluck, interview by author, 19 March 1979.
34 Havelock, *Harold A. Innis – A Memoir*, 40.
35 Ibid., 41.
36 Ibid., 42.
37 Ibid., 36.
38 Ibid., 18.
39 Ibid.
40 Ibid. 17.
41 Ibid., 42.
42 Ibid., 18.
43 Ibid., 42.
44 Havelock, *Prometheus Bound*, v–vi.
45 Innis, *Changing Concepts of Time*, vi. Compare to Havelock's words, from 96 of *Prometheus Bound*: 'Intellectual man of the nineteenth century was the first to estimate with precision his total lowliness, his absolute nullity, in space and time. That is the secret shock administered to our culture. It becomes our secret despair.'
46 Havelock, *Prometheus Bound*, 17.
47 Ibid., 15.
48 Ibid., 18.
49 Ibid., 19.
50 Ibid., 20.
51 Ibid., 34.
52 Ibid., 80.
53 Ibid., 143.
54 Ibid., 145.
55 Ibid., 86.
56 Ibid., 96.
57 Ibid., 98.
58 Ibid., 99.
59 Ibid., 100. There are direct echoes of Havelock in Innis that one can trace textually. The last sentence in this quotation, for example, was one of Innis's favourite quotations from Lord Acton. However, it is not my intention here to establish the rigid causal relationship of the thought of one scholar to that of another. I am attempting only to show that the communications studies were undertaken in an intellectual environment that has not yet been fully appreciated. In particular, I am attempting to show that this

group of classical scholars was interested in themes similar to those picked up by Innis when he made the shift to communications studies.

The fact that Havelock's most important works were produced after Innis' death does not mean that they should not be examined for clues to the collective intellectual climate in which the communications studies took place. Havelock himself, in the first sentence of his preface to *The Liberal Temper of Greek Politics*, points out: 'The possibility of this book, though not yet its design, began to shape itself in my mind fifteen years ago [1941] when I was still at the University of Toronto.' (In terms of this book, Innis's position bears remarkable similarities to that of the Greek democrats.) Similarly, *Preface to Plato* (1963) and *Prologue to Greek Literacy* (1971) deal in far greater detail with themes that first appeared in print in the early 1950s. In other words, the gestation periods for Havelock's work and Innis's communications studies are contemporaneous.

The essay 'Why Was Socrates Tried?' containing many of the themes of Havelock's later works, appeared just before Innis's death. It is not unlikely that Innis saw a draft of this essay. Havelock argues that Socrates was a transitional figure of an era in which education was changing from a non-literate oral system based on learning through ecstasy to a literate but still largely oral system founded on reason and dialogue. As a result, he became a scapegoat for the anger of those who were threatened by the rapid sweeping away of the tradition of epic poetry. The article would have been of particular interest to Innis, for it dealt with the dislocations that occurred when the social space was cleared for the foundation of the first university.

In light of the above, Havelock is extraordinarily modest in his belief that he had little or no influence on Innis. He remains mildly puzzled as to why Innis contacted him without notice at Yale to give a seminar to political-economy graduate students during a visit to Canada.

In fact, we know that Innis did read in manuscript form some earlier version of the work Havelock eventually incorporated in *Preface to Plato*. On 21 May 1951 he wrote to Frank Knight at Chicago: 'Your second to last paragraph raises a further question regarding the understanding of other cultures. E.A. Havelock in a recent book *The Crucifixion of Intellectual Man*, now at Harvard and formerly a student of F.M. Cornford, has been concerned with the same problem in Greece. He thinks that the notion of Greek culture so far as the Greeks were concerned began to change between 410 and 405 B.C. when they started to educate the youth systematically. He has a manuscript on the question of the shift from the oral to the written tradition in Greek culture which he hopes to complete this summer.'

60 A.R.M. Lower, interview with Paul Kennedy for CBC Radio *Ideas* program on Innis, fall 1978, Innis Papers, University of Toronto Archives.

61 'Familiar' here is used in the sense of 'a person with whom one has constant intercourse, an intimate friend or associate,' 'a familiar spirit, a demon or evil spirit supposed to attend at a call.' *Oxford English Dictionary*.

62 Judith Stamps, in *Unthinking Modernity*, provides a good summary of the diffusion of Hegelian dialectics into the North American educational system and presents it as a key intellectual influence underlying the work of Innis. I believe she overstates the case. There is little indication in the archival material to indicate that Hegel's works directly influenced Innis to the extent, for instance, of Veblen or the classicists. There is no indication of any significant influence on Innis from Walter Benjamin, Theodor Adorno, or the Frankfurt School. The key terminology of her argument – for instance, 'negative dialectic' – was not present in Innis's writing and had to be read into it by Stamps.

I have an uneasy feeling that the work of Stamps and others who have focused on translating or interpreting Innis in relationship to other dominant academic schools of thought – Hegelianism, Marxism, American communications thought, and so on – is the realization of one of the tendencies Innis warned us about: 'The literature builds up around the name of Keynes or Marx or someone else, and everything else is dropped!' (Innis, 'The Decline in the Efficiency of Instruments Essential in Equilibrium.') Or to cite a quotation Innis was fond of: 'Perhaps it may not be for the advantage of any nation to have the arts imported from their neighbours in too great perfection. This extinguishes emulation and skirts the ardour of the generous youth' (Innis, *Bias*, 5.)

It is as if Stamps is slightly embarrassed by Innis and McLuhan and has to bring in the greater sophistication of the European thinkers to legitimate and to understand what they were saying. 'In contemporary Canadian studies,' she writes, 'the interpretations stem from an overly empiricist view of what Innis was trying to say. For both Innis and McLuhan, the phrase "technological determinism" had stuck so hard one gets the impression that "uniquely Canadian" has become a polite phrase for "uniquely crude" – a theory with a roughly hewn set of concepts that, like the fur trade, is an interesting but odd spin-off of the Pre-Cambrian Shield. (Stamps, *Unthinking Modernity*, xiv.) Or again: 'While Adorno and Benjamin had a ready, sophisticated and nuanced vocabulary with which to express their themes, Innis and McLuhan, who had little formal training in theory, invented a theoretical language that sometimes made it difficult for them to express their themes clearly. Reading Adorno and Benjamin enhances our ability to see what Innis and McLuhan were up to.' (Stamps, *Unthinking Modernity*, 22.)

If Stamps's approach had led to significant new insights into the meaning of the communications works, then she would have proved her point. See chapter 7, n.51 and chapter 8, n.70 for examples of why I do not believe this to be so.

On a minor note, Stamps also confuses the venue at which the 'Minerva's Owl' address was given (see 42.) It was not the 1948 Beit Lectures at Oxford but the 1947 presidential address to the Royal Society of Canada.

63 G.W.F. Hegel, *The Philosophy of Right*, 'Introduction' to *The Philosophy of Hegel* (New York: Modern Library, 1953), 227.

64 Innis, *The Bias of Communication*, 5.

65 Ibid., 6.

66 Ibid., citing David Hume.

67 Ibid., 30.

68 Ibid.

69 Ibid., 31.

70 A good summary of this argument has been presented by Whitaker in 'To Have Insight into Much and Power over Nothing: The Political Ideas of Harold Innis.' For a different interpretation, see Stamps, *Unthinking Modernity*, 71–2. Stamps claims that Innis viewed 'efficiency' as a 'good thing only in the negative sense': 'An empire thus would be efficient and successful in promoting civilization if it had no monopoly of communication and, by extension, no monopoly of knowledge.' This leads her to the absurd conclusion that to Innis 'the successful empire would thus be a non-empire.'

Stamps equates 'efficiency' with a particular medium and its biases. She thus concludes that 'Innis linked efficiency to the growth of monopolies and the consequent narrowing of creative thought, not to its flourishing.' I maintain that Innis used the existence of empire as an indication of efficiency in communication in the sense of a balance of media existing at that period. That is why he speaks of 'efficiency of particular media' instead of 'efficiency of a particular medium' in the quotation cited at n.75 below. That is also why Stamps believes that Innis saw empires as institutions which threatened a civilization's survival, whereas I conclude that he viewed empires incorporating a balance of media as a high point of civilization.

71 This compilation is taken from an overview of all of Innis's communications works rather than from any one book or article. Nevertheless, an excellent unpublished summary of passages encompassing this dialect of

communications can be found in the University of Toronto Archives in Liska Bridle's précis of various chapters of 'A History of Communications.' See especially the end summary, 'New Monopoly of Space.'

72 The link of Innis's concept of balance or equilibrium with the classics rather than with economics is evident in his last article. Its alternative titles were 'The Decline in the Efficiency of Instruments Essential for Equilibrium' and 'The Menace of Absolutism in Time.'

73 Innis, *Empire and Communications*, 10.

74 Ibid., 9.

75 Ibid., 9.

76 Ibid., 170. My emphasis.

77 Innis, *The Bias of Communication*, 4.

9: Time, Space, and the Oral Tradition: Towards a Theory of Consciousness

1 Einstein quote accessed at http://www.quotationspage.com/search on 5 January 2004.

2 See, for instance, Carey, 'Harold Innis and Marshall McLuhan.'

3 Ibid.

4 Consider, for instance, the significance of the title of his last book and last essay, *Changing Concepts of Time* and 'The Menace of Absolutism in Time.'

5 Innis refused to be interviewed for radio, listened to little popular radio, and did little more than glance at the headlines of newspapers.

6 Innis, 'The Role of Intelligence,' 282.

7 Innis, *The Bias of Communication*, v. The effect of Ten Broeke on Innis and others at McMaster is mentioned by S.S. Cole in a letter to Mary Quayle Innis, 20 October 1959, Innis Papers, University of Toronto Archives: '[Innis and I] majored in Philosophy at McMaster under James Ten Broeke, a very poor instructor but a scholarly man and a sterling human character. I shall never forget some of the chats Harold and I had during the closing weeks of our senior year when we were trying to pull together our deepest insights into philosophy. We came out at very much the same fundamental point of view. He defined philosophy as "a basic attitude towards life" and I ventured to affirm it was "a working hypothesis about the nature of life and reality." Ten Broeke liked our reactions and said so.'

8 These lecture notes are found in the Innis Papers at the University of Toronto Archives, in a notebook entitled 'Mental Notes.' No pagination.

9 Ibid.

10 Ibid.

11 Ibid.
12 Ibid.
13 Shotwell, 'The Discovery of Time.'
14 Ibid., 198.
15 Ibid.
16 Ibid., 200.
17 Ibid., 203.
18 Ibid., 254.
19 See Ibid., 314.
20 Ibid.
21 Ibid., 267.
22 Ibid., 310.
23 Shotwell, *An Introduction to the History of History*, 2–3.
24 Ibid., 5.
25 Ibid., 16.
26 Ibid., 10–11.
27 Innis, *The Bias of Communications*, v.
28 Shotwell, *An Introduction to the History of History*, 24.
29 Ibid., 27.
30 Ibid., 28–9.
31 Ibid., 31–2.
32 Ibid., 36.
33 Ibid., 318.
34 Ibid., 332.
35 Sorokin and Merton, 'Social Time: A Methodological and Functional Analysis.' Quotation marks from original quote.
36 Ibid., 616–17.
37 Ibid., 618–21.
38 Ibid., 627.
39 Ibid., 623. My emphasis.
40 Ibid., 628.
41 Innis, *Changing Concepts of Time*, v.
42 Lewis, *Time and Western Man*, 3–4.
43 Ibid., 24.
44 Ibid., 9–10.
45 Ibid., 12.
46 Ibid., 18.
47 Ibid., 23. Emphasis in original.
48 Ibid., 27. Emphasis in original.
49 Ibid., 437.

50 Innis, *Changing Concepts of Time*, v.
51 Lewis, *Time and Western Man*, 435–37.
52 Ibid., 406–7.
53 Ibid., 312.
54 Ibid., 313.
55 Ibid., 271, 303, 29.
56 Ibid., 408.
57 Ibid., 408–9.
58 Ibid., 412–13.
59 Ibid., 418–19.
60 Ibid., 42, 316.
61 Ibid., 138.
62 Ibid., 466.
63 Ibid., 466–7.
64 Ibid., 465.
65 Ibid., 342.
66 Ibid., 284.
67 Ibid., 285.
68 Ibid., 136, 137.
69 Ibid., 450.
70 Ibid., 41.
71 Ibid., 284–5.
72 Innis's final essay, posthumous address to the American Economics Association, 1953.
73 Lewis, *Time and Western Man*, 432, 433.
74 Woodcock, ed., *Wyndham Lewis in Canada*, 4.
75 Innis, *The Bias of Communication*, 90.
76 All page citations refer to *Time and Western Man*.
77 Ibid., 133. Emphasis in original.
78 Gandz, 'The Dawn of Literature.' 261.
79 Ibid.
80 Ibid.
81 See, for instance, ibid., 274.
82 Ibid., 281.
83 Ibid., 271.
84 Innis, *Empire and Communications*, 55.
85 Gandz, 'The Dawn of Literature,' 304.
86 Ibid., 306.
87 Ibid., 291.
88 Ibid., 309.

89 Ibid., 289.
90 Ibid., 279. Emphasis in original.
91 Ibid., 106.
92 Ibid., 463–4.
93 Ibid., 294.
94 Ibid., 294, 282.
95 Ibid., 485–6.
96 Ibid., 496.
97 Ibid., 478–80.
98 Ibid., 498–9.
99 Ibid., 413.
100 Ibid., 402.
101 Ibid., 403.
102 Ibid., 509.
103 Ibid., 277.
104 Ibid., 278.
105 Ibid., 412.
106 Ibid., 303, 311.
107 For a more extended treatment of the influence of these writers on Innis, see Watson, 'Marginal Man: Harold Innis' Communications Works in Context,' chapter 3.
108 For instance, Cornford, 'The Invention of Space,' 2/6: 'we find that space and time cannot be classified as realities of nature ... the space-time continuum ... can be crumpled and twisted and warped as much as we please without becoming one whit less true to nature – which, of course, can only mean that it is not itself part of nature. Space and time, and also their space-time product, fall into their places as mere mental frameworks of our own construction.'
109 See Nilsson, *Primitive Time-reckoning*, 105, 223, 232, 349–58.
110 Cassirer, *Language and Myth*, viii–ix.
111 Ibid., 11.
112 Innis, *The Bias of Communication*, 63 (sentences 2, 3, 4, 14, 15).
113 All pagination refers to Cassirer, *Language and Myth*.

10. At the Edge of the Precipice: The Mechanization of the Vernacular and Cultural Collapse

1 Innis's speech to Graduate Colloquium [1949?], Innis Papers, University of Toronto Archives.
2 Innis, 'Reflections on Russia,' 263.

3 Innis, *The Bias of Communication*, xvii. My emphasis.
4 Innis, *Empire and Communications*, 58.
5 Ibid., 61.
6 See, for example, ibid., 60–3.
7 It used to be believed that the true alphabet was invented by the ancient Greeks borrowing from the Phoenicians. Current scholarship now credits 'Asiatic' outsiders in Egypt who, around 2000 BC, apparently invented the true alphabet in the course of developing a way to record their Semitic language using hieroglyphic 'sound' signs employed by the Egyptians. Innis would not have been surprised that marginal outsiders to the metropolis of the day made the discovery.

   This new scholarship, it seems to me, does little to undercut the singularity of the Greeks. They still remain the only people who were able to capture an oral culture at the height of its development by introducing a true alphabet over an extended period and from the inside of that same culture. For a recent summary of the topic, see Man, *Alpha Beta*.
8 Innis, *Empire and Communications*, 59.
9 Ibid., 66.
10 Ibid., 75.
11 Innis, *The Bias of Communication*, 10.
12 Innis, *Empire and Communications*, 83.
13 Ibid., 84.
14 Raleigh Parkin to Joseph Willets, 24 June 1943, Innis Papers, University of Toronto Archives.
15 Innis to G. Ferguson, 14 August 1949, Ferguson, Miscellaneous Correspondence.
16 Innis, *Political Economy and the Modern State*, 257.
17 Ibid., 265.
18 Innis, *Empire and Communications*, 166.
19 Innis, *Political Economy and the Modern State*, 264.
20 'Russian Trip Notes,' Innis, *Innis on Russia*.
21 Innis, *Political Economy and the Modern State*, 259, 268.
22 Ibid., 267.
23 Ibid., 258.
24 Ibid.
25 Ibid., 259.
26 Memorandum to Sidney Smith, President's Papers, 1951 University of Toronto Archives.
27 Innis to Ferguson, 4 February 1951, Ferguson, Miscellaneous Correspondence.

28 W. Plumptre, interviewed by Elspeth Chisholm, University of Toronto Archives.
29 Innis, *The Bias of Communication*, 139.
30 Innis, *Innis on Russia*.
31 Innis, *Empire and Communications*, 170.
32 Innis, *Political Economy and the Modern State*, 112.
33 Ibid., 95.
34 Ibid., 102.
35 Innis, *The Bias of Communication*, 189.
36 Ibid., 189.
37 Innis, *Changing Concepts of Time*, 21, 22, 23.
38 Ibid., 38–9.
39 Ibid., 43.
40 Ibid., 50–1.
41 Ibid., 61–2.
42 Ibid., 66.
43 Ibid., 76.
44 Havelock, *Harold A. Innis*, 21.
45 Berger, *The Writing of Canadian History*, 195.
46 Innis, *Essays in Canadian Economic History*, 405.
47 Innis, *Changing Concepts of Time*, 72–4.
48 Ibid., 2–3. Innis cites the quotations as 'the words of Sir Douglas Copland, summarizing the philosophy of P.H. Roxby.'
49 Ibid., 18.
50 Ibid., 19–20.

11. Cassandra's Curse

1 Innis (quoting from Herodotus, 'The Church In Canada,' 392): 'Cassandra received the power to foretell the future from the god, Apollo. Apparently, Apollo instructed the mortal woman and taught her about the art of prophecy because he ... wished to win her affections. Cassandra accepted Apollo as a teacher, but not as a lover. Naturally, the god was insulted by this refusal. So he punished Cassandra. Apollo caused the gift that he gave to Cassandra to be twisted, making everyone who heard her true and accurate foretellings of future events believe that they were instead hearing lies. In other words, the wondrous blessing bestowed upon a mortal became instead a terrible curse.' 'The Greek Heroine Cassandra in Myth and Art,' *Mythography*, available at http://www.loggia.com/myth/cassandra.html.
2 Bonnett, in 'Communication, Complexity and Empire: The Systematic

Thought of Harold Innis,' 328, points to notes made by W.T. Easterbrook after visiting Innis on 11 July 1952. Innis's denial of the disease is made clear in this note: 'unfort letter from Willis, sorry to hear you have cancer – why didn't you tell me – but Harold, you knew this 2 years ago – still no change in attitude – my feeling is that death at this stage is so meaningless, that simply not accepted as a fact ... just simply too preposterous to believe ... Problems for Mrs. I; no thought of making will or any plans re family if death occurs.'

3 Vincent Bladen, letter to Irene Biss, 6 May 1952, Irene (Biss) Spry Papers, Library and Archives of Canada.
4 Creighton, *Harold Adams Innis: Portrait of a Scholar*, 145.
5 Mary Quayle Innis, letter to Irene Biss, 25 June 1952, Irene (Biss) Spry Papers, Library and Archives of Canada.
6 Innis, 'Autobiography,' 104, Innis Papers, University of Toronto Archives.
7 Ibid., 113.
8 Innis, *Changing Concepts of Time*, v.
9 Ibid.
10 Ibid., 128.
11 Mary Quayle Innis, 'Diary,' Innis Family Records, B1991–0029, University of Toronto Archives.
12 Mackintosh, 'Innis on Canadian Economic Development,' 19.
13 Raoy Daniells, 'For Harold Innis,' *The Checkered Shade*, Innis Family Records, B1991–0029, University of Toronto Archives.
14 Shoesmith, 'Notes from the Margin: Resurrecting the Innisian Project,' 9.
15 Lower, 'Harold Innis as I Remember Him,' 3.
16 Lower, *My First Seventy-Five Years*, 302.
17 See for instance, Katz, Peters, Liebes, and Orloff, ed., *Canonic Texts in Media Research*.
18 Donald Creighton, letter to Harold Innis, 26 April 1952, Innis Papers, University of Toronto Archives.
19 Blodgett, 'Harold Innis as a Teacher.'
20 Clark, 'The Contribution of Harold Innis to Canadian Scholarly Activity,' 1–3.
21 For a post-Innis study dealing with the oral tradition in a clear and profound manner, see Walter J. Ong, *Orality and Literacy* (London and New York: Routledge 1988).
22 Innis, 'The Decline in the Efficiency of Instruments Essential in Equilibrium,' 14.
23 See, for instance, H.M. McLuhan, letter to H.A. Innis, 14 March 1951, Innis Papers, University of Toronto Archives.

24 McLuhan, 'The Fecund Interval,' in Havelock, *Harold A. Innis: A Memoir*, 10.
25 W.T. Easterbrook, interview by author, 1 March 1979.
26 McLuhan, letter to Innis, Innis Papers, University of Toronto Archives.
27 See McLuhan's 'Introduction' to *The Bias of Communication*, xiii, and chapter 9 of this book.
28 Carey, 'Space, Time and Communications: A Tribute to Harold Innis,' 145.
29 Carey, 'The Mythos of the Electronic Revolution,' 133.
30 McLuhan's position is put precisely in the following sentence: 'The penetrative powers of any structure of technology do lie precisely ... [in] the ratio among sight and sound and touch and motion.' McLuhan, 'Effects of Improvements of Communication,' 573.
31 McLuhan, 'Introduction' to *The Bias of Communication*, xii.
32 Innis, *The Bias of Communication*, 190.
33 Innis, *Empire and Communications*, ix.
34 Ibid., vii.
35 Innis, *The Bias of Communications*, 7.
36 Ibid., xi.
37 Ibid.
38 Ibid., 100.
39 Ibid., 54.
40 Ibid., 188–9.
41 Ibid., 189.
42 Ibid., 188.
43 McLuhan, 'The Later Innis,' 390.
44 McLuhan is caught up in his own logic to such an extent that he is unable to see either the consistency or the political nature of Innis's position. The result is apparent at times in the rather shoddy scholarship behind his critique of Innis. As an example, McLuhan says:'Language itself, however, [Innis] failed to observe, was at once the greatest mass medium of communication and also the greatest time-builder of cultures and civilizations. The work of Edward Sapir ... seems to have eluded his attention.' Yet it is precisely with a reference to Sapir that Innis begins a long discussion concerning this very subject in the 'Introduction' to *Empire and Communications*. (It is in the introduction of this same volume that McLuhan made his comment.) Nowhere does Innis make clearer the fact that the oral and written traditions are not merely opposites. Innis stresses that empires must be assessed in terms of the classical dialectic of power and knowledge: '[The concept of empire] will reflect to an important extent the efficiency of particular media of communication and its possibilities in creating conditions favourable to creative thought.' Innis sees writing as opposed to print as a

development that strengthened the creative thinking already evident in the oral tradition:

> writing provided man with a transpersonal memory. Men were given an artificially extended and verifiable memory of objects and events not present to sight or recollection. Individuals applied their minds to symbols rather than things and went beyond the world of concrete experience into the world of conceptual relations created within an enlarged time and space universe. The time world was extended beyond the range of remembered things and the space world beyond the range of known places. Writing enormously enhanced a capacity for abstract thinking which had been evident in the growth of language in the oral tradition.

This is the essence of the classical vision of Innis, a balance in which the modalities of power and knowledge complemented each other. For Innis, the progressive mechanization of knowledge through modern media was seen as leading to the complete destruction of this balance in the totalitarian state which turned to force as a solution to all problems of empire. It is largely the result of the tremendous influence of McLuhan's presentation of Innis's work that this critical vision of the human condition has been lost. As always, a great deal of meaning is destroyed by translation and the more celebrated the translation the more serious the effect.

45 Innis, *Empire and Communications*, xii.

46 Ibid.

47 The intensely political critique of the present-mindedness of the American Empire is nowhere more apparent than in Innis's last article and his son's comments on how Innis planned to complete it. Not a retribalized society but Orwell's *1984* became Innis's vision of the future; he even made *1984* essential reading for his undergraduates students. Conversation with Mel Watkins, and Bladen, Easterbrook, and Willits, 'Harold Adams Innis: 1894–1952,' 25. For McLuhan, the new media abolished politics by forbidding 'the persistence of monopolies of knowledge,' a view of the modern context precisely the opposite of that of Innis. McLuhan, 'The Later Innis,' 387.

48 Innis, *Empire and Communications*, 170. My emphasis.

49 For example: 'Large-scale political organization implies a solution of problems of space in terms of administrative efficiency and of problems of time in terms of continuity. Elasticity of structure involves a persistent interest in the search for ability and persistent attacks on monopolies of knowledge. Stability involves a concern with the limitations of instruments of government as well as with their possibilities.' Ibid., 170.

50 McLuhan, 'The Later Innis,' 391–2. McLuhan is an astute commentator on Innis's work in certain fields. For instance, he is excellent on Innis's style of writing; see ibid., 389.
51 Havelock, *Harold A. Innis: A Memoir*, 106.
52 Innis, *The Bias of Communication*, 30–1.
53 Cochrane, *Christianity and Classical Culture*, 478.
54 Ibid., 468.
55 There are two pitfalls to which I am referring here. The first is to focus on one element only of Innis's theoretical triad of politics, consciousness, and technology. This usually involves focusing on technology and concluding that Innis was a media determinist. A corollary to this is a lengthy criticism of those instances where Innis's expositions on specific societies throw up examples of the media having an effect that seems contradictory to their underlying bias. I have tried to show that these instances are not contradictions at all if we take into consideration Innis's theory of politics and consciousness. The second pitfall has been for Innisian scholars to respond to the absence of an underlying articulated philosophical framework to his work by importing such a framework from intellectual circles which had little or no direct influence on him.

Epilogue

1 Innis, *The Bias of Communication*, 169.
2 Michael Ignatieff, *Empire Lite* (Toronto: Penguin 2003), 22.
3 I am not suggesting that my sons' exposure to computer games is pernicious and conversely that the higher 'story' content of my youth led to a superior consciousness. If this were the case, a parent could easily react by limiting computer (or television) access and insist on going to church and reading classic stories. This is not an option because one has to live inside of one's own times and learn to cope with them from within. This is surely how Innis developed his unique perspective. I believe that this is what he meant by the importance of 'dirt' research.
4 Of course, it is a human trait to believe that the younger generation is somehow going in the wrong direction. I am not suggesting this. I am suggesting that the degree of difference from one generation to another in their way of thinking about the world is growing. If this is the case, one would expect intergenerational conflict to increase in those societies most open to adopting new technologies of communication.
5 Innis was convinced that power was coming to dominate intelligence in modern imperial projects, or, as he sometimes put it, that the monopoly of

force at the centre of an empire no longer saw the need for allying itself with a monopoly of knowledge in order to undertake an imperial project that would endure. Force alone was now more likely to be seen as adequate for solving the problems of empire.

It is interesting to think what this means for the creative origins that are at the basis of childhood leisure pastimes. In the case of a generation that spends more time with traditional story tellers or reads more, the creative source is found directly in the oral tradition in the case of the collective composition of the former, or at one remove from the oral tradition, in the case of the individual authorship of the latter.

If it is true that these traditions so prevalent in childhood across cultures and over many generations are being destroyed, it is useful to ask where the creativity is coming from to develop new forms of childhood entertainment. In the case of the video games my sons are playing with other youngsters at a great distance, this creativity is readily traceable. The two immediate technological innovations that allow them to do so are the Internet and the software lying behind the 'first person shooter' games that they love. The former was originally developed in order to ensure that military-communications systems would not be disrupted even by the massive devastation of nuclear warfare. The latter software was originally developed to allow for training Allied military personnel for the prospect of massive tank battles with the Soviets in Europe without the high cost and significant property damage involved in real military field exercises.

In short, military necessity led to the development of the new cultural artefacts currently amusing our youth. Not surprisingly, then, the experiences of playing a computer war game and actually fighting a high-tech war are now fundamentally similar; childhood stories or even traditional childhood war games (like 'cowboys and Indians') and the actual experience of war in past generations, are not.

6 See Benjamin R. Barber, *Fear's Empire* (New York: Norton 2003).
7 Innis, *Changing Concepts of Time*, 124–5.
8 See Ron Brown, *Ghost Towns of Ontario* (Toronto: Cannonbooks 1984).
9 I believe that Innis may have been pleased, however, that the college established in his name has adopted a 'generalist,' multidisciplinary approach.

Bibliography

Archival and Unpublished Material

American Geographical Society. Secretary Wrigley's correspondence with Innis concerning articles for the *Geographical Review*. Archives of the American Geographical Society, New York City.

Baptist Church. Records of the Oxford County Churches. Canadian Baptist Archives, McMaster Divinity College, Hamilton, Ont.

Bennett, R.B. Correspondence with Innis. MG H96. University of New Brunswick Archives.

Bissell, Claude. Interview by author. Toronto, 5 December 1979.

Bladen, Vincent. Interview by author. Toronto, 1 March 1979.

Boeschstein, Herman. Interview by author. Toronto, 5 April 1979.

Bonnett, John. 'Communication, Complexity and Empire: The Systemic Thought of Harold Adams Innis.' PhD thesis. University of Ottawa 2001.

Brady, Alexander. Interview by author. Toronto, 23 October 1979.

Brebner, J.B. Papers. Columbia University Archives.

Britnell, George E. Papers. University of Saskatchewan Archives.

Brown, Margaret. Interview by author. Toronto, 20 October 1979.

Bryden, Kenneth. Interview by author. Toronto, 14 and 16 April 1979.

Carnegie Endowment Archives. Innis correspondence. Columbia University Archives.

Chisholm, Elspeth. Fonds. M31 E50. Library and Archives Canada.

– Interview by author. Toronto, 20 January 1980.

– Interviews with contemporaries of Innis students, colleagues, and university support staff, undertaken for the preparation of a 1972 CBC Radio documentary on Innis. University of Toronto Archives.

Clark, S.D. Interview by author. Toronto, 28 June 1979.

Cochrane, Charles N. Papers. B2003–00011. University of Toronto Archives.
Coper, Rudolph. Correspondence with author. 14 May 1979.
Creighton, Donald. Main holdings, including correspondence with Irene Spry. Library and Archives Canada.
Daag, Anne (Innis). Correspondence with author. 2 October 1979.
– Interview by author. Toronto, 4 April 1979.
Department of Political Economy. A96–00025. Administrative outline (Finding Aid). A76–0025. University of Toronto Archives.
– Calendars, 1921–41. Staff, Courses, Requirements. University of Toronto Archives, Reading Room.
– Departmental files, 1916–57, including Innis's departmental files and personal papers, 1916–36, and administrative files, 1937–52. University of Toronto Archives.
Duff, C. Kent. 'My Friend, Harold Innis.' Dated 15 November 1976. Harold Innis Foundation. Innis College Archives. University of Toronto.
Easterbrook, W.T. Interviews by author. Toronto, fall 1974 and 1 March 1979.
– Miscellaneous papers, including some of Innis's departmental papers left in Easterbrook's office at the time of Innis's death. University of Toronto Archives.
Ferguson, G.V. 'Harold Innis and the Printed Word.' An Address Delivered on the Occasion of the Innis College Dinner. 4 November 1965. Innis College Archives, University of Toronto.
– Miscellaneous correspondence with Innis obtained through Ferguson's son, David M. Ferguson.
Fowke, Vernon C. Papers. University of Saskatchewan Archives.
Frye, Northrop. Interview by author. Toronto, 18 April 1979.
Graham, Blanche. Interview by author. Toronto, 24 October 1979.
Grasham, W.E. Interview by author. Toronto, 11 January 1980.
Grube, George. Interview by author. Toronto, 28 March 1979.
Halpenny, Francess. Interview by author. Toronto, 21 June 1979.
Harold Innis Foundation. Sound and video recording of various conferences. B1993–0006. University of Toronto Archives.
Helleiner, Karl. Papers, including reviewer's reading notes on Innis's *History of Communications*. University of Toronto Archives.
Innis, Donald. Interview by author. Toronto, 18 July 1979.
Innis, H.A. Attestation Paper and other war service records. Library and Archives Canada.
– Birth registration and original registration book from South Norwich. Archives of Ontario. Toronto.

– COTC records. 'McMaster University – WWI.' Canadian Baptist Archives. McMaster Divinity College, Hamilton, Ont.
– Field Trip Photographic Material. B1977–0016. University of Toronto Archives.
– Harold A. Innis Family Records. B1991–0029. Includes personal files for Mary Quayle Innis and Donald Innis as well as some Innis material residual to that deposited in the above holdings. University of Toronto Archives.
– Innis Family. Family records of Harold Adams Innis and Mary Quayle Innis. B1979–0056. University of Toronto Archives.
– Innis Papers. Main holdings, including Harold A. Innis – 'Personal Records' B1972–0003 (mainly personal) and Harold A. Innis – 'Personal Records' B1972–0025 (mainly related to university activities). University of Toronto Archives. These represent the major depository of Innis's archival materials and they have been reorganized several times since research on this book began. For this reason, citations in the text have been made to 'Innis Papers' since the old box and file numbers are no longer valid. Nonetheless, tracking the original material should be facilitated by the finding aids prepared since the start of research on this book.
– Innis Photos. Mainly photos and slides used in lecturing. B1977–0016. University of Toronto Archives.
– MA thesis, 'The Returned Soldier.' McMaster University Library.
– Nova Scotia Royal Commission Materials. B1983–0001. University of Toronto Archives.
– Otterville Correspondence. B1993–0044. University of Toronto Archives.
– PhD Transcript Matric. No.71526, Registrar's Office, University of Chicago.
– War correspondence with chancellor and registrar of McMaster University. Canadian Baptist Archives. McMaster Divinity College, Hamilton, Ont.
Innis, Hugh. Interview by author. Toronto, 21 September 1979.
Innis Communications Corporation. Correspondence and other material related to the CRTC initiative to publish the 'History of Communications.' Manuscript. B1993–0043. University of Toronto Archives.
Innis Foundation. Tape recordings of proceedings of Innis conferences at Innisfree Farm. Innis Papers. B1993–0006. University of Toronto Archives
Karwin, Edwin. 'History of the Otterville Baptist Church.' Canadian Baptist Archives. Baptist Divinity College. McMaster University, Hamilton, Ont.
Knight, Frank H. Papers. Series VII and VIII correspondence. University of Chicago Archives.
Leitch, Elizabeth. Interview by author. Toronto, 24 September 1979.
Lewis, Gail. Interview by author. Otterville, 7 July 2003.

London University. Stamp Lecture information. Palaeography Room, University of London Library.
Lower, A.R.M. Papers. Queen's University Archives.
Mackintosh, W.A. Papers. Queen's University Archives.
McMaster University. Student correspondence from the First World War, Chancellor's and Registrar's Correspondence Files. Canadian Baptist Archives. McMaster Divinity College, Hamilton, Ont.
McNeill, Connie (Scott). Interview by author. Toronto, 24 September 1979.
Mitchell, W.C. Innis Correspondence. Columbia University Archives.
Norwich Township Archives. Various records of the Innis Family. Norwich, Oxford County, Ont.
Nottingham University. Cust Lecture Material, including Innis correspondence. Manuscripts Department, University of Nottingham Library.
Oxford University. Beit Lecture Information. Bodleian University Archives.
President's Papers. University of Toronto. Correspondence of presidents Falconer, Cody, and Smith with Innis and subjects related to Innis. University of Toronto Archives.
Riendeau, Roger. 'On the Critical Edge: A History of Innis College, 1964–2004,' chapters 1–2. Draft manuscript.
Robertson, Grant. Interview by author. Toronto, 29 March 1979.
Royal Commission on Transportation. Textual records. Library and Archives Canada.
Shotwell, J.T. Innis Correspondence. Columbia University Archives.
Sirluck, Ernest. Interview by author. Toronto, 19 March 1979.
Sisler, William James. Papers. Archives of Manitoba.
Spry, Graham. Interview by author. 11 July 1979.
Spry, Irene (Biss). Correspondence from Innis. University of Toronto Archives.
– Interview by author. Ottawa, 23 April and 11 July 1979.
– Letters to Harold Innis (1935). Department of Political Economy. A76–0025. University of Toronto Archives.
– Main holdings, including correspondence, research, publications, photographs, etc. MG30 C249. Library and Archives Canada.
– 'The Technological Trap.' Convocation Address. University of Toronto. 3 June 1971. Innis College Archives, University of Toronto.
S.S. #1. South Norwich. School Trustees' minutebook. Otterville Museum. Otterville, Ont.
Sword, John. Interview by author. Toronto, 25 September 1979.
University of Chicago – Department of Economics Papers. C.W. Wright correspondence. University of Chicago Archives.
Veterans of the First World War. Interviews with Canadians who fought in the

war, undertaken in preparation from CBC Radio documentary 'Flanders Fields' in 1964. CBC Radio Archives, Toronto.
Waters, F.W. Original Manuscript: 'A Century of Philosophy at McMaster.' Especially chapter 2, 'The Ten Broeke Years.' In the author's possession, Hamilton, Ont.
– Interview by author. Hamilton, Ont., 8 November 1979.
Watson, A. John. 'Marginal Man: Harold Innis' Communication Works in Context.' PhD thesis, University of Toronto 1981.
Woodstock Collegiate Institute. Local History Files. Woodstock Public Library. Woodstock, Ont.
Wrye, Joyce. Interview by author. Toronto, 24 September 1979.
Zerker, Sally. Interview by author. 6 March 1979.

Published Material

This section of the bibliography includes those published works cited or mentioned in the text as well as a number of secondary sources that may be of interest to Innis scholars. It is not meant to provide a complete listing of Innis's published works. Such a bibliography has been produced by Jane Ward in 'The Published Works of H.A. Innis,' *CJEPS* 19 (1953): 233–44. This bibliography, with corrections and additions, has been reproduced in Robin Neill, *A New Theory of Value*. A further improved and comprehensive bibliography of Innis's works was carried out by the Harold Innis Foundation in April 1973 and is available from that source.

I have attempted to give as the source of my citations the most readily available version in existence at the time of my research. As a rule, those of Innis's essays that were republished in *Political Economy in the Modern State*, *The Bias of Communication*, and *Essays in Canadian Economic History* have been cited usually in reference to these books rather than in reference to their original publication in scholarly journals. Accordingly, these essays may not be listed separately below. The dates given below, therefore, do not necessarily indicate the date of first publication of the work but the date of publication of the version cited.

The listing incorporates the following abbreviations of scholarly journals: *American Economic Review* (AER); *Contributions to Canadian Economics: University of Toronto Studies in History and Economics* (CCE); *Canadian Historical Review* (CHR); *Canadian Journal of Economics and Political Science* (CJEPS); *Economic Journal* (EJ); *Journal of Canadian Studies* (JCS); *University of Toronto Quarterly* (UTQ).

Acland, Charles R., and William J. Buxton. *Harold Innis in the New Century*. Montreal and Kingston: McGill-Queen's University Press 1999.

Adair, E.R. Review of *The Bias of Communication*. *CHR* 33 (December 1953): 393–4.

Angus, Ian. *A Border Within: National Identity, Cultural Plurality, and Wilderness*. Montreal and Kingston: McGill-Queen's University Press 1997.

– 'Orality in the Twilight of Humanism: A Critique of the Communications Theory of Harold Innis.' *Continuum: The Australian Journal of Media and Culture* 7 (1993): 1.

Babe, Robert E. *Canadian Communication Thought*. Toronto: University of Toronto Press 2000.

Beale, Alison. 'Harold Innis and Canadian Cultural Policy in the 1940's.' *Continuum: The Australian Journal of Media and Culture* 7 (1993): 1.

Benda, Julien. *The Treason of the Intellectuals*. New York: William Morrow 1928.

Berger, Carl. *The Writing of Canadian History*. Toronto: Oxford University Press 1976.

Berton, Pierre. *Vimy*. Toronto: McClelland and Stewart 1986.

Bickerton, James. 'Too Long in Exile: Innis and Maritime Political Economy.' In Charles R. Acland and William J. Buxton, eds., *Harold Innis in the New Century*. 225–39.

Black, J. David. '"Both of Us Can Move Mountains": Mary Quayle Innis and Her Relationship to Harold Innis' Legacy.' *Canadian Journal of Communication 28* (2003): 433–47.

Bladen, V.W. *Bladen on Bladen*. Toronto: privately published, 1979.

– with W.T. Easterbrook and J.H. Willits. 'Harold Adams Innis: 1894–1952.' *AER* 43 (1953): 1–15.

Blodgett, George S. 'Harold Innis as a Teacher.' *Harold Innis Foundation Newsletter* (1987): 2–3.

Bonn, M.J. Review of *Empire and Communications*. *Time and Tide* (July 1950): 29.

Brady, A. 'Harold Adams Innis: 1894–1952.' *CJEPS* 19 (1953): 87–96.

Brebner, J.B. 'Harold Adams Innis.' *EJ* 63 (1953): 728–33.

– 'Innis as Historian.' Speech notes for the Canadian Historical Association. Innis Papers, University of Toronto Archives.

Butlin, N.G. Review of *The Bias of Communication*. In *Economic Record* (May 1953): 140–3.

Buxton, William J. 'The Bias against Communication: On the Neglect and Non-Publication of the "Incomplete and Unrevised Manuscript" of Harold Adams Innis.' *Canadian Journal of Communication* 26 (2001): 2–3.

– 'Harold A. Innis's "History of Communications" Manuscript.' In Paul Heyer, *Harold Innis*, 2003. 103–11.

– 'Harold Innis' Excavation of Modernity: The Newspaper Industry, Communications, and the Decline of Public Life.'*Canadian Journal of Communication* 23 (1998): 2.

Campbell, Sandra. 'From Romantic History to Communications Theory: Lorne Pierce as Publisher of C.W. Jefferys and Harold Innis.' *JCS* 30, no.3 (fall 1995): 91–116.

Canadian Journal of Economics and Political Science. Review of *The Bias of Communications*. August 1952. 388–9.

Carey, James W. *Communication as Culture*. New York: Routledge 1992.

– 'Culture, Geography and Communications: The Work of Harold Innis in an American Context.' In William H. Melody, Liora Salter, and Paul Heyer, eds., *Culture, Communication and Dependency*. 73–91.

– 'Harold Innis and Marshall McLuhan.' *Antioch Review* 27, no.1 (1967): 5–39.

– 'Innis "in" Chicago: Hope as the Sire of Discovery.' In Charles R. Acland and William J. Buxton, eds., *Harold Innis in the New Century*. 81–104.

– 'Innis and the Chicago School.' Presentation at Innisfree Farm conference. Innis Papers, University of Toronto.

– 'Introduction' to *Changing Concepts of Time*. Lanham, Md., and Boulder, Colo.: Rowman and Littlefield 2004.

Carroll, Lewis. *The Best of Lewis Carroll*. Secaucus, N.J.: Castle, n.d.

Cassirer, Ernst. *Language and Myth*. New York: Dover 1946.

Cayley, David. 'The Legacy of Harold Innis.' Transcript of CBC Radio program. Toronto: CBC RadioWorks 1994.

Childe, V. Gordon. Review of *Empire and Communications*. *CJEPS* 17 (1951): 98–100.

Christian, William, ed. *The Idea File of Harold Innis*. Toronto: University of Toronto Press 1980.

– *Innis on Russia: The Russian Diary and Other Writings*. Toronto: Harold Innis Foundation 1981.

Clark, J.M. *Studies in the Economics of Overhead Costs*. Chicago: University of Chicago Press 1923.

Clark, S.D. *Church and Sect in Canada*. Toronto: Ryerson 1948.

– 'The Contribution of H.A. Innis to Canadian Scholarship.' In William H. Melody, Liora Salter, and Paul Heyer, eds., *Culture, Communication and Dependency*. 27–35.

Cochrane, C.N.. *Christianity and Classical Culture*. London: Oxford University Press 1944.

– *David Thompson, the Explorer*. Toronto: Macmillan 1924.

– 'The Mind of Edward Gibbon.' *UTQ* 11 (1942): 1–17, 146–66.

– 'The Question of Commerce.' *University of Toronto Monthly* (1932).

– *This Canada of Ours*. Toronto: J.M. Dent 1931.

Cooper, Tom. 'The Unknown Innis.' *JCS* 12, no.5 (winter 1977):111–18.

Cornford, F.M. 'The Invention of Space.' In *Essays in Honour of Gilbert Murray*. London: George Allen 1936.

Cotton's Weekly/The Observer. Starting September 1908. Toronto Reference Library, Microfilm Periodicals Holdings.

Creighton, Donald. 'Harold Adams Innis: 1894–1952.' *CHR* 33 (1952): 405–6.

– *Harold Adams Innis: Portrait of a Scholar*. Toronto: University of Toronto Press 1957.

– 'Harold Innis – An Appraisal.' In William H. Melody, Liora Salter, and Paul Heyer, eds., *Culture, Communication and Dependency*. 13–25.

Department of Political Economy. 'Research in Economic and Social Studies Relating to the Dominion of Canada.' *Bulletin*. Toronto: University of Toronto 1925.

Dudley, Leonard M. 'Space, Time, Number: Harold A. Innis as Evolutionary Theorist.' *Canadian Journal of Economics* 28, no.4a (November 1995): 754–69.

Duncan, C.S. *Marketing: Its Problems and its Materials*. New York: Macmillan 1920.

Eacott, John, and the Norwich and District Historical Society. *Of Other Times*. Norwich, Ont: Huddleson and Barney 1980.

Easterbrook, W.T. 'Innis and Economics.' *CJEPS* 19 (1953): 291–303.

Eccles, W.J. 'A Belated Review of Harold Adams Innis, *The Fur Trade in Canada*.' *CHR* 60 (1979): 419–41.

– 'A Response to Hugh M. Grant on Innis.' *CHR* 62 (1981): 323–9.

The Economist. 'Rambles among the Social Sciences.' Review of *Political Economy in the Modern State*. 8 February 1947.

– Review of *Empire and Communications*. 4 March 1950.

Evenden, Matthew D. 'Harold Innis, the Arctic Survey, and the Politics of Social Science during the Second World War.' *CHR* 70 (March 1998): 36–67.

Evening Telegram [Toronto]. 15–26 Jan. 1931.

Fay, C.R. 'The Toronto School of Economic History.' *Economic History* (January 1934): 170–1.

Ferguson, Barry, and Doug Owram. 'Social Scientists and Public Policy from the 1920's through World War II.' *JCS* 15, no.4 (winter 1980–1): 3–17.

Fisher, Donald. 'Harold Innis and the Canadian Social Science Research Council: An Experiment in Boundary Work.' In Charles R. Acland and William J. Buxton, eds., *Harold Innis in the New Century*. 135–58.

Friedland, Martin L. *The University of Toronto – A History*. Toronto: University of Toronto Press 2002.

Fussell, Paul. *The Great War and Modern Memory*. London: Oxford University Press 1977.

Gagnon, Alain-G., and Sarah Fortin. 'Innis in Québec: Conjectures and Conjuntures.' In Charles R. Acland and William J. Buxton, eds., *Harold Innis in the New Century*. 209–24.

Gandz, Solomon. 'The Dawn of Literature: Prolegomena to a History of Unwritten Literature.' *Osiris* 7 (1939): 261–522. Originally published by Uitgeverij De Tempel, Bruges.

Gauvreau, Michael. 'Baptist Religion and the Social Science of Harold Innis.' *CHR* 76 (1995): 161–204.

Globe [Toronto]. 15–26 Jan. 1931.

Grant, George. *George Grant in Process: Essays and Conversations*. Larry Schmidt, ed. Toronto: Anansi 1978.

– 'Ideology in Modern Empires.' In *Perspectives on Empires*. London: Longman's 1973.

– *Technology and Empire*. Toronto: Anansi 1969.

Grant, Hugh M. 'One Step Forward, Two Steps Back: Innis, Eccles, and the Canadian Fur Trade.' *CHR* 62 (1981): 304–22.

Gwyn, Sandra. *Tapestry of War*. Toronto: Harper Collins 1992.

Harold Innis Research Foundation. *Innis Research Bulletin*. Editors: Mel Watkins and Roger Riendeau. Toronto: Innis College 1994–5.

Havelock, E.A. *Harold A. Innis: A Memoir*. Toronto: Harold Innis Foundation 1982.

– *Preface to Plato*. Oxford, U.K.: Basil Blackwood 1963.

– *Prologue to Greek Literacy*. Cincinnati: University of Cincinnati Press 1971.

– *Prometheus Bound: The Crucifixion of Intellectual Man*. Boston: Beacon 1950.

– 'Why Was Socrates Tried?' In *Studies in Honour of Gilbert Norwood*. Toronto: University of Toronto Press 1953.

Heyer, Paul. 'Empire, History and Communications Viewed from the Margins: The Legacies of Gordon Childe and Harold Innis.' *Continuum: The Australian Journal of Media and Culture* 7 (1993): 1.

– *Harold Innis*. Lanham, Md., and Boulder, Colo.: Rowman and Littlefield 2003.

– 'Innis and the History of Communication: Antecedents, Parallels, and Unsuspected Biases.' In William H. Melody, Liora Salter, and Paul Heyer, eds., *Culture, Communication and Dependency*. 247–59.

– and David Crowley. 'Introduction' to *The Bias of Communication* (1991 ed.): ix–xxvi.

Horn, Michiel. *Academic Freedom in Canada: A History*. Toronto: University of Toronto Press 1999.

– '"Free Speech Within the Law": The Letter of the Sixty-Eight Toronto Professors, 1931.' *Ontario History* 72 (March 1980): 27–48.

– *The League for Social Reconstruction: Intellectual Origins of the Democratic Left in Canada 1930–42*. Toronto: University of Toronto Press 1980.
Hotson, Zella M. *Pioneer Baptist Work for Oxford County*. Woodstock: Commercial Printing 1939.
Innis, H.A. 'Between the Gold and Iron Curtain.' *Commerce Journal* (1949): 11–13.
– *The Bias of Communication*. Toronto: University of Toronto Press 1951.
– 'A Bibliography of Publications on Canadian Economics, 1927–1929.' *CCE* 1 (1928): 86–100.
– 'A Bibliography of Recent Publications on Canadian Economics, 1925–27, 1928–29.' *CCE* 2 (1929): 98–102.
– 'Bibliography of Research Work.' *CCE* 1 (1928): 69–85.
– *The Canadian Economy and Its Problems*. Edited with A.R.W. Plumtre. Toronto: Canadian Institute of International Affairs 1934. 'Introduction,' 3–24.
– *Changing Concepts of Time*. Toronto: University of Toronto Press 1952.
– 'Charles Norris Cochrane, 1889–1945.' *CJEPS* 12 (1946): 95–7.
– *The Cod Fisheries: The History of an International Economy*. In *The Relations of Canada and the United States* series. Toronto: Ryerson Press 1940.
– 'The Commerce Course.' *Commerce Journal* (1941): 1–2.
– 'The Decline in the Efficiency of Instruments Essential in Equilibrium.' *AER* 43 (1953): 16–22.
– 'Discussion in the Social Sciences.' *Dalhousie Review* 15 (1936): 401–13.
– 'Economic Conditions in Canada in 1931-2.' *EJ* 42 (1932): 1–16.
– 'Economic Nationalism.' *Papers and Proceedings of the Canadian Political Science Association for 1934* 6 (1934): 17–31.
– 'The Economic Problems of Army Life.' *McMaster University Monthly* (Christmas 1918): 106–9.
– 'Economics for Demos.' *UTQ* 3 (1934): 389–95.
– 'Editor's Introduction.' *Labor in Canadian-American Relations*. Toronto: Ryerson Press 1937. v–xxxi.
– 'Edward Johns Urwick, 1867–1945. *CJEPS* 11 (1945): 265–8.
– *Empire and Communications*. Oxford, U.K.: Clarendon Press 1950. Revised ed. Toronto: University of Toronto Press 1972.
– *Essays in Canadian Economic History*. Edited by Mary Quayle Innis. Toronto: University of Toronto Press 1956.
– 'Foreword.' *Commerce Journal* (1945): xi.
– 'Foreword' to Percy H. Wright, 'Smoothing the Bumps in Business.' Ryerson Essay no.55. Toronto: Ryerson Press 1932. 5–8. – 'The Fur Trade of Canada.' *University of Toronto Studies in Economics* 5 (1927): 1.

– *The Fur Trade in Canada: An Introduction to Canadian Economic History*. New Haven, Conn.: Yale University Press 1930.
– Graduating class entry for Innis. *McMaster University Monthly* (1916).
– *The History of the Canadian Pacific Railway*. Revised ed. Toronto: University of Toronto Press 1971.
– 'Industrialism and Settlement in Western Canada.' *Report of Proceedings of the International Geographical Congress, Cambridge, July, 1928* (1930): 369–76.
– *Innis on Russia – The Russian Diary and Other Writings*. Edited by William Christian. Toronto: Harold Innis Foundation 1981.
– 'The Necessity of Research in Marketing.' *Commerce Journal* (1940): 12–14.
– Note: 'Communications and Archaeology.' *CJEPS* 17 (1951): 237–40.
– Note: 'Social Sciences in the Post-war World.' *CHR* 22 (1941): 118–20.
– 'Notes and Comments." *CCE* 2 (1929): 5–6.
– 'The Passing of Political Economy.' *Commerce Journal* (1938): 3–6.
– *Peter Pond: Fur Trader and Adventurer*. Toronto: Irwin and Gordon 1930.
– *Political Economy in the Modern State*. Toronto: Ryerson Press 1946.
– 'Preface' to Mary Quayle Innis, *An Economic History of Canada*. Toronto: Ryerson Press 1935.
– *Problems of Staple Production in Canada*. Toronto: Ryerson Press 1933.
– 'Public Utilities.' *Encyclopedia of Canada* 5 (1937): 176–80.
– 'Pulp and Paper Industry.' *Encyclopedia of Canada* 5 (1937): 180–6.
– 'Recent Books on Arctic Exploration and the Canadian Northland.' *CHR* 21 (1940): 197–205.
– *Report of the Royal Commission, Provincial Economic Inquiry, Province of Nova Scotia*. Halifax: Province of Nova Scotia 1934.
– Review of A. Guthrie, 'The Newspaper Industry: An Economic Analysis'; L.T. Stevenson, 'The Background and Economics of American Papermaking'; C. McNaught, 'Canada Gets the News'; O. Gramling, 'AP: The Story of News'; H.M. Hughes, 'News and the Human Interest Story.' *CJEPS* 7 (1941): 578–63.
– Review of H.E. Stephenson and C. McNaught, *The Story of Advertising in Canada: A Chronicle of Fifty Years*; R.M. Hower, *The History of an Advertising Agency: N.W. Ayer and Son at Work, 1869–1939*. *CJEPS* 7 (1941): 109–12.
– Review of M.I. Newbigin, *Canada, The Great River, the Lands and the Men*. *AER* 17 (1927): 497–8.
– Review of 'Recent Books on the North American Arctic.' *CHR* 23 (1942): 401–7.
– Review of 'Recent Books on the North American Arctic.' *CHR* 25 (1944): 54–60.
– Review of 'Sub Specie Temporis.' *CJEPS* 17 (1951): 553–7.

– 'The Role of Intelligence: Some Further Notes.' *CJEPS* 1 (1935): 280–7.
– 'Some English-Canadian University Problems.' *Queen's Quarterly* 50 (1943): 30–6.
– *Staples, Markets, and Cultural Change: Harold A. Innis. Edited by* Daniel Drache. Montreal and Kingston: McGill-Queen's University Press 1995.
– 'A Tragedy of the Bluffs.' *McMaster University Monthly* (April 1916): 266–70.
Innis, Mary Quayle. *The Clear Spirit*. Toronto: University of Toronto Press 1966.
– *An Economic History of Canada*. Toronto: Ryerson Press 1935.
– *Stand on the Rainbow*. New York: Duell, Sloan, and Pearce 1944.
Jaeger, Werner. *Paideia: The Ideals of Greek Culture*. London: Oxford University Press 1965.
Jhally, Sut. 'Communications and the Materialist Conception of History: Marx, Innis and Technology.' *Continuum: The Australian Journal of Media and Culture* 7 (1993):1.
Johnson, Jane. 'From Silence to Communications? What Innisians Might Learn by Analysing Relations.' In Charles R. Acland and William J. Buxton, eds., *Harold Innis in the New Century*. 177–95.
Johnston, Charles M. *McMaster University*. Toronto: University of Toronto Press 1976.
Katz, Elihu, John Durham Peters, Tamar Liebes, and Avril Orloff. *Canonic Texts in Media Research*. Cambridge, U.K.: Polity Press 2003.
Keast, Ronald. '"It Is Written – But I Say Unto You": Innis on Religion.' *JCS* 20, no.4 (1986): 12–25.
Knight, M.M. Review of *The Bias of Communication*. *AER* 43 (March 1953): 179–80.
Lee, Dennis. 'The Death of Harold Ladoo.' In *Nightwatch: New and Selected Poems, 1968–1996*. Toronto: McClelland and Stewart 1996.
Leed, Eric J. *No Man's Land*. London: Cambridge University Press 1979.
Lewis, Wyndham. *Time and Western Man*. London: Chatto and Windus 1927.
Lower, A.R.M. 'Harold Innis As I Remember Him.' *JCS* 20, no.4 (winter 1985–6): 3–11.
– *My First Seventy-Five Years*. Toronto: Macmillan 1967.
Mackintosh, W.A. 'Innis on Canadian Economic Development.' *Journal of Political Economy* 61, no.3 (1953): 185–94.
Man, John. *Alpha Beta*. London: Headline 2000
McLuhan, H. Marshall. 'The Fecund Interval.' 'Preface' to Eric A. Havelock. *Harold A. Innis: A Memoir*. Toronto: Harold Innis Foundation 1982.
– 'Foreword' to *Empire and Communication*. Toronto: University of Toronto Press 1972.

– 'Introduction' to *The Bias of Communication*. Toronto: University of Toronto Press 1964.
– 'The Later Innis.' *Queen's Quarterly* 60 (1953): 385–94.
Melody, William, Liora Salter, and Paul Heyer, eds. *Culture, Communication and Dependency*. Norwood, N.J.: Ablex Publishing 1981.
Moses, Lauren Alexis. 'The Psychology, Life, and Relevance of Thorstein Veblen.' Available at: http://www.econ.duke.edu/JOURNALS/DJE/dje2002/moses.pdf. Accessed 10 April 2003.
Neill, Robin. *A New Theory of Value*. Toronto: University of Toronto Press 1972.
Nilsson, Martin P. *Primitive Time-reckoning*. London: Humphrey Milford 1920.
Noble, Richard. 'Innis's Conception of Freedom.' In Charles R. Acland and William J. Buxton, ed., *Harold Innis in the New Century*. 31–45.
Office Museum. 'Antique Copying Machines.' Available at: at www.officemuseum.com. Accessed 20 July 2004.
Parker, Ian. 'Innis, Marx and the Economics of Communication: A Theoretical Aspect of Canadian Political Economy.' In William H. Melody, Liora Salter, and Paul Heyer, eds., *Culture, Communication and Dependency*. 27–143.
Patterson, Graeme. *History and Communications*. Toronto: University of Toronto Press 1990.
Ong, Walter J. *Orality and Literacy*. London and New York: Routledge 1988.
Owen, E.T. 'The Illusion of Thought.' *Journal of Philosophy* 45, no.19 (1948): 505–11.
Owram, Doug. *The Government Generation: Canadian Intellectuals and the State 1900–1945*. Toronto: University of Toronto Press 1986.
Ray, A.J. 'Introduction' to *The Fur Trade In Canada*. Toronto: University of Toronto Press 1999.
Reid, W. Stanford. *The Scottish Tradition in Canada*. Toronto: McClelland and Stewart 1974.
Riesman, David. 'The Social and Psychological Setting of Veblen's Economic Theory.' *Journal of Economic History* 13, no.4 (1953): 449–61.
Robertson, Grant. *The Administration of Justice in the Athenian Empire*. Toronto: University of Toronto Library 1924.
Robertson, Heather. *A Terrible Beauty*. Toronto: James Lorimer 1974.
Salée, Daniel. 'Innis and Québec: The Paradigm That Would Not Be.' In Charles R. Acland and William J. Buxton, eds., *Harold Innis in the New Century*. 196–208.
Salter, Liora, and Cheryl Dahl. 'The Public Role of the Intellectual.' In Charles R. Acland and William J. Buxton, eds., *Harold Innis in the New Century*. 114–34.

Shoesmith, Brian. 'Notes from the Margin: Resurrecting the Innisian Project.' *Innis Research Bulletin* (May 1994): 1.
– and Ian Angus, ed. 'Dependency/Space/Policy.' *Continuum: The Australian Journal of Media and Culture* 7 (1993): 1.
Shotwell, James T. 'The Discovery of Time.' *Journal of Philosophy* (1915): 198–206, 254–316. – *An Introduction to the History of History*. New York: Columbia University Press 1922.
Sorokin, P.A., and R.K. Merton. 'Social Time: A Methodological and Functional Analysis.' *American Journal of Sociology* 42 (1936–7).
Stamps, Judith. 'Innis in the Canadian Dialectic Tradition.' In Charles R. Acland and William J. Buxton, eds., *Harold Innis in the New Century*. 46–66.
– *Unthinking Modernity: Innis, McLuhan, and the Frankfurt School*. Montreal and Kingston: McGill-Queen's University Press 1995.
Stark, F.M. 'Harold Innis and the Chicago School.' *JCS* 29, no.3 (fall 1994): 131–45.
Ten Broeke, James. 'Educational Sermon: Education a Means to Personal Development.' Toronto: Baptist Convention of Ontario and Quebec 1898.
Tuchman, Barbara. *The Guns of August*. New York: Bantam 1976.
Urwick, E. J. 'The Role of Intelligence in the Social Process.' *CJEPS* 1 (1935): 64–76.
Usher, Peter J. 'Staple Production and Ideology in Northern Canada.' In William H. Melody, Liora Salter, and Paul Heyer, eds., *Culture, Communication and Dependency*. 177–86.
Veblen, Thorstein. *The Portable Veblen*. New York: Viking 1950.
– *The Theory of the Leisure Class*. New York: New American Library (Mentor ed.) 1954.
Watkins, M.H. 'A Staple Theory of Economic Growth.' *CJEPS* 29 (1963): 141–58.
– 'The Staples Theory Revisited.' *JCS* 12, no.5 (winter 1977): 83–95, and in William H. Melody, Liora Salter, and Paul Heyer, eds., *Culture, Communication and Dependency*. 53–71.
Watson, A. John. 'Harold Innis and Classical Scholarship.' *JCS* 12, no.5 (winter 1977): 45–61.
Wernick, Andrew. 'No Future: Innis, Time, Sense and Postmodernity.' In Charles R. Acland and William J. Buston, eds., *Harold Innis in the New Century*. 261–80.
Whitaker, Reginald. '"To Have Insight into Much and Power over Nothing": The Political Ideas of Harold Innis.' *Queen's Quarterly* 90, no.3 (1983): 818–31.
Woodcock, George. *Wyndham Lewis in Canada*. Victoria: University of British Colombia Press 1971.

Index